HEATING, VENTILATING AND AIR CONDITIONING LIBRARY VOLUME I

Heating Fundamentals

Furnaces • Boilers

Boiler Conversions

by James E. Brumbaugh

Macmillan Publishing Company
New York

Collier Macmillan Publishers
London

SECOND EDITION

Macmillan Publishing Company
866 Third Avenue, New York, N.Y. 10022
Collier Macmillan Canada, Inc.

Library of Congress Cataloging-in-Publication Data

Brumbaugh, James E.
    Heating, ventilating, and air conditioning library.

    Includes indexes.
    Contents: v. 1. Heating fundamentals, furnaces, boilers, boiler conversions—v. 2. Oil, gas, and coal burners, controls, ducts, piping, valves.
    1. Heating. 2. Ventilation. 3. Air conditioning.
I. Title.
[TH7222.B78 1986]   697   86-8735
ISBN 0-672-23389-4 (v. 1)

Macmillan books are available at special discounts for bulk purchases for sales promotions, premiums, fund-raising, or educational use. For details, contact:

> Special Sales Director
> Macmillan Publishing Company
> 866 Third Avenue
> New York, N.Y. 10022

10 9 8 7 6

Printed in the United States of America

Foreword

The purpose of this series is to provide the layman with an introduction to the fundamentals of installing, servicing, and repairing the various types of equipment used in residential heating, ventilating, and air conditioning systems. Consequently, it was written, not only for the engineer or skilled serviceman, but also for the average homeowner. A special effort was made to remain consistent with the terminology, definitions, and practices of the various professional and trade associations involved in the heating, ventilating, and air conditioning fields.

Volume 1 begins with a description of the principles of thermal dynamics and ventilation, and proceeds from there with a general description of the various heating systems used in residences and small commercial buildings. Volume 2 contains descriptions of the working principles of various types of equipment and other components used in these systems. Volume 3 includes detailed instructions for installing, servicing, and repairing these different types of equipment and components. Those sections in Volume 3 that deal with air conditioning follow a similar format.

The author wishes to acknowledge the cooperation of the many organizations and manufacturers for their assistance in supplying valuable data in the preparation of this book. Every effort was made to give credit and courtesy lines for materials and illustrations used in this edition.

JAMES E. BRUMBAUGH

Contents

CHAPTER 5

CHAPTER 6

CHAPTER 7

CHAPTER 8

natural vacuum systems—mechanical vacuum system—combined atmospheric pressure and vacuum systems—exhaust steam heating—proprietary systems—high-pressure, steam heating systems—steam boilers—control components—Hartford return connection—pipes and piping details—steam traps—pumps—heat-emitting units—air supply and venting—unit heaters—air conditioning

CHAPTER 9

CHAPTER 10

CHAPTER 11

CHAPTER 12

CHAPTER 13

CHAPTER 14

CHAPTER 15

water pressure-reducing valves—combination valves (dual controls)—
air separators—purging air from the system—air supply and venting—
induced draft fans—control of excessive draft on gas-fired boilers—tankless water heaters—leaking coils—blowing down a boiler—boiler
operation and maintenance—steam boilers—hot-water space heating
boilers—boiler water—cleaning boilers—troubleshooting boilers—
boiler repair—boiler insulation—water chillers

CHAPTER 16

CHAPTER 1

Introduction

This series is an introduction to the basic principles of heating, ventilating, and air conditioning. Each represents a systematic attempt to control the various aspects of the environment within an enclosure, whether it be a room, a group of rooms, or a building.

Among those aspects of the immediate environment that people first sought to control were heat and ventilation. Attempts at controlling heat date from prehistoric times and probably first developed in colder climates where it was necessary to produce temperatures sufficient for both comfort and health. Over the years the technology of heating advanced from simple attempts to keep the body warm to very sophisticated systems of maintaining stabilized environments in order to reduce heat loss from the body or the structural surfaces of the room.

Ventilation (covered in Volume 3) also dates back to very early periods in history. Certainly the use of slaves to wave large fans or fanlike devices over the heads of rulers was a crude early attempt to solve a ventilating problem. Situating a room or a building so that it took advantage of prevailing breezes and winds was a more sophisticated attempt. Nevertheless, it was not until the nineteenth century that any really significant advances were made in ventilating. During that period, particularly in the early stages of the Industrial Revolution, ventilating acquired increased importance. Work efficiency and the health of the workers necessitated the creation of ventilation systems to remove contaminants from the air. Eventually, the interrelationship of heating and ventilating became such that it is now regarded as a single subject.

Air conditioning (Volume 3) is a very recent development and encompasses all aspects of environmental control. In addition to the control of temperature, both humidity (i.e., the moisture content of the air) and air cleanliness are also regulated by air conditioning. The earliest attempts at air conditioning involved the placing of wet cloths over air passages (window openings, entrances, etc.) to cool the air. Developments in air conditioning technology did not progress much further than this until the nineteenth century. From about 1840 on, several systems were devised for both cooling and humidifying rooms. These were first developed by textile manufacturers in order to reduce the static electricity in the air. Later, adaptations were made by other industries.

Developments in air conditioning technology increased rapidly in the first four decades of the nineteenth century, but widespread use of air conditioning in buildings is a phenomenon of the post-World War II period (i.e., 1945 to the present). Today, air conditioning is found not only in commercial and industrial buildings but in residential dwellings as well. Unlike the early forms of air conditioning, which were designed to cool the air or add moisture to it, modern air conditioning systems can control temperature, air moisture content, air cleanliness, and air movement. The modern systems *condition* the air rather than simply cool it.

HEATING AND VENTILATING SYSTEMS

Many different methods have been devised for heating buildings. Each has its own characteristics, and most methods have at least one objectionable aspect (e.g., high cost of fuel, expensive equipment, or inefficient heating characteristics). Most of these heating methods can be classified according to one of the following four criteria:

1. The heat conveying medium.
2. The fuel used.
3. The nature of the heat.
4. The efficiency and desirability of the method.

The term "heat conveying medium" means the substance or combination of substances that carries the heat from its point of origin to the area being heated. There are basically four mediums for conveying heat. These four mediums are:

1. Air.
2. Water.
3. Steam.
4. Electricity.

Different types of wood, coal, oil, and gas have been used as fuels for producing heat. Electricity may be considered both as a fuel and a heat conveying medium. Each heating fuel has its own characteristics; the advantage of one type over another depends upon such variables as availability, efficiency of the heating equipment (which, in turn, is dependent upon design, maintenance, and other factors), and cost. A detailed analysis of the use and effectiveness of the various heating fuels is found in Chapter 5 (Heating Fuels).

Heating methods can also be classified with respect to the nature of the heat applied. For example, the heat may be of the exhaust steam variety or it may consist of exhaust gases from internal combustion engines. The nature of the heat applied is inherent to the heat system and can be determined by reading the various chapters that deal with each type of heating system (Chapters 6 through 9) or with heat-producing equipment (e.g., Chapter 11, Gas-Fired Furnaces).

11

The various heating methods differ considerably in efficiency and desirability. This is due to a number of different but often interrelated factors, such as heat cost, conveying medium employed, and type of heating unit. The integration of these interrelated components into a single operating unit is referred to as a *heating system*.

Because of the different conditions met with a practice, there is a great variety in heating systems, but most fall into one of the following broad classifications:

1. Warm-air heating system (Chapter 6).
2. Hot-water heating systems (Chapter 7).
3. Steam heating systems (Chapter 8).
4. Electric heating systems (Chapter 9).

You will note that these classifications of heating systems are based on the heat conveying method used. This is a convenient method of classification because it includes the vast majority of heating systems used today.

As mentioned, ventilating is so closely related to heating in its various applications that the two are very frequently approached as a single subject. In this series, specific aspects of ventilating are considered in Chapter 6 (Ventilation Principles) and Chapter 7 (Fan Selection and Operation) of Volume 3.

The type and design of ventilating system employed depends on a number of different factors, including:

1. Building use or ventilating purpose.
2. Size of building.
3. Geographical location.
4. Heating system used.

A residence will have a different ventilating system from a building used for commercial or industrial purposes. Moreover, the requirements of a ventilating system used to provide fresh air, as opposed to one that must remove noxious gases or other dangerous contaminants from the enclosure, result in fundamental design differences.

The size of a building is a factor that also must be considered. A large building with internal areas far from the points where outside air would initially gain access presents certain ventilating problems. These can usually be solved by giving special attention to the overall design of the ventilating system.

Buildings located in the tropics or semitropics present different ventilating problems from those found in temperature zones. The differences are so great that they often result in different architectural forms. At least this was the case *before* the advent of widespread use of air conditioning. The typical southern house of the nineteenth century was constructed with high ceilings (heat tends to rise); large porches that sheltered sections of the house from the hot, direct rays of the sun; and large window areas to admit the maximum amount of air. They were also usually situated so that halls, major doors, and sleeping areas faced the direction of the prevailing winds. Today, with air conditioning so widely used, these considerations are not as important—at least not until the power fails or the equipment breaks down.

AIR CONDITIONING

Although the major emphasis in this series has been placed on the various aspects of heating and ventilating, some attention has also been given to air conditioning. The reason for this, of course, is the increasing use of year-round air conditioning systems that provide heating, ventilating, and cooling. These systems *condition* the air by controlling its temperature (warming or cooling it), cleanliness, moisture content, and movement. This is the true meaning of the term "air conditioning." Unfortunately, it has become almost synonymous with the idea of cooling, and this is becoming increasingly misrepresentative of the true function of an air conditioning system. Air conditioning, particularly the year-round air conditioning systems, are examined in detail in Chapters 8, 9, and 10 of Volume 3.

SELECTING A SUITABLE HEATING, VENTI- LATING, OR AIR CONDITIONING SYSTEM

There are a number of different types of heating, ventilating, and air conditioning equipment and systems available for installation in the home. The problem is choosing the most efficient one in terms of the installation and operating costs. These factors, in turn, are directly related to one's particular heating and cooling requirements. The system must be the correct size for the home. Any reputable building contractor or heating and air conditioning firm should be able to advise you in this matter.

If you are having a heating and ventilating or air conditioning system installed in an older house, be sure to check the construction. Weather stripping is the easiest place to start. All doors and windows should be weather-stripped to prevent heat loss. Adequate weather stripping can cut heating costs by as much as 15 to 20 percent. If the windows provide suitable protection (they should be double- or triple-glazed) from the winter cold, check the calking around the edge of the glass. If it is cracking or crumbling, replace it with fresh calking. You may even want to go to the expense of insulating the ceilings and outside walls. This is where a great deal of heat loss and air leakage occur.

You have several advantages when you are building your own house. For example, you may be able to determine the location of your house on the lot. This should enable you to establish the direction in which the main rooms and largest windows face. If you position your house so that these rooms and windows face south, you will gain maximum sunlight and heat from the sun during the cold winter months. This will reduce the heat requirement and heating costs. The quality of construction depends on how much you wish to spend and the reliability of the contractor. It is advisable to purchase the best insulation you can afford. Your reduced heating costs will eventually pay for the added cost of the insulation. If you suspect that your building contractor cannot be trusted, you can reduce opportunities for cheating and careless work by making frequent and unexpected visits to the construction site.

CAREER OPPORTUNITIES

Many career opportunities are available in heating, ventilating, air conditioning, and heating fields, and they extend over several levels of education and training. Accordingly, the career opportunities open to an individual seeking employment in these fields can be divided roughly into four categories, each dependent upon a different type or degree of education and/or training. This relationship is shown in Table 1-1.

Table 1-1. Relationship between Career Category and Type of Work or Education and/or Training Required

Career Category	Type of Work	Education/Training
Engineer	Design and development	4 yrs. or more of college
Technician	Practical application	Technical training school and/or college
Skilled worker	Installation, maintenance and repair	Apprentice program or on-the-job (OJT) training
Apprentice or OJT worker	Training for skilled-worker position	High school degree or equivalency

Engineers are usually employed by laboratories, universities and colleges, or frequently by the manufacturers of materials and equipment used in heating, ventilating, air conditioning, and related industries. Of workers in these fields, engineers receive the highest pay, but they also undergo the longest periods of education and training. Their primary responsibility is designing, developing, and testing the equipment and materials used in these fields. In some cases, particularly when large buildings or district heating to several buildings is employed, they also supervise the installation of the entire system. Moreover, industry codes and standards are usually the results of research conducted by engineers.

Technicians obtain their skills through technical training schools, some college, or both. Many assist engineers in the practical application of what the latter have designed. Technicians are particularly necessary during the developmental stages. Other technicians are found in the field working for contractors in the larger companies. Their pay often approximates that of engineers, depending on the size of the company for which they work.

Skilled workers are involved in the installation, maintenance, and repair of heating, ventilating, and air conditioning equipment. Apprentices and OJT (on-the-job training) workers are in training for the skilled positions and are generally expected to complete at least a 2- to 5-year training program. Most skilled workers and trainees are employed by local firms that install or repair equipment in residential, commercial, and industrial buildings. Some also work on the assembly lines of factories that manufacture such equipment. Their pay varies, depending on the area, their seniority, and the nature of the work. Most employers require that both skilled workers and trainees have at least a high school diploma or its equivalent (e.g., the GED). The requirement for a high school diploma may be waived if the individual has already acquired the necessary skills on a previous job. The pay for skilled workers and trainees is lower than that earned by engineers and technicians but compares favorably to salaries received by skilled workers or equivalent trainees in other occupations.

Pipe fitters, plumbers, steam fitters, and sheet-metal workers may occasionally do some work with heating, ventilating, and air conditioning equipment. Both pipe fitters and plumbers (especially the former) are frequently called upon to assemble and install pipes and pipe systems that carry the heating or cooling conveying medium from the source. Both are also involved in repair work, and some pipe fitters can install heating and air conditioning units.

Steam fitters can assemble and install hot-water or steam heating systems. The installation of boilers, stokers, oil and gas burners, radiators, radiant heating systems, and air conditioning systems can also be done by many steam fitters.

Sheet-metal workers can also assemble and install heating, ventilating, and air conditioning systems. Their skills are particularly necessary when assembling sheet-metal ducts and duct systems.

Some special occupations, such as those performed by air conditioning and refrigeration mechanics or stationary engineers, are limited to certain functions in the heating, ventilating, and air conditioning fields. Mechanics are primarily involved with assembling, installing, and maintaining both air conditioning and refrigeration equipment. Stationary engineers maintain and operate heating, ventilating, and air conditioning equipment in large buildings and factories. Workers in both occupations require greater skills and longer training periods than most skilled workers.

It should be readily apparent by now that the heating, ventilating, and air conditioning fields offer a variety of career opportunities. The pay is generally good, and the nature of the work provides considerable job security. Both the type of work and the level at which it is done depend solely on the amount and type of education and training acquired by the individual.

PROFESSIONAL ORGANIZATIONS

A number of professional organizations have been established for those who work in the heating, ventilating, and air conditioning industries or who handle their products. These organizations (frequently referred to as *associations, societies,* or *institutes*) provide a number of different services to members and nonmembers alike, including:

1. The compilation of safety codes and other standards.
2. Educational programs and training schools.
3. Market research and various technical services.
4. The compilation of industry-wide statistics.
5. The dissemination of information in the form of journals, books, pamphlets, movies, slides, and tapes.

Some professional organizations have established permanent libraries as resource centers for those seeking to improve their skills or wishing to keep abreast of current developments in their fields. In many instances, research programs are conducted in cooperation with laboratories, colleges, and universities.

Anyone, member or not, can write to these professional organizations for information or assistance. Most seem very willing to comply with any reasonable request. The only difficulty that may be encountered is determining the current name of the particular organization and obtaining its address. Unfortunately, these professional organizations have shown a strong proclivity toward mergers over the years, with resulting changes of names and addresses.

17

The best and most current guide to the names and addresses of professional organizations is *The Encyclopedia of Associations,* which can be found in the reference department of most public libraries. It is published in three volumes, but everything you will need can be found in the first one. At the back of this volume is a section called the "Alphabetical & Key Word Index." By looking up the key word (e.g., "heating" or "ventilating") of the subject that interests you, you can find the page number and full name of the professional organization (or organizations) concerned with the particular area.

There are over twenty different professional organizations in the United States concerned with heating, ventilating, air conditioning or related fields (e.g., fuels or piping). Many of them address themselves to the problems and interest of specific groups. For example, there are professional organizations for manufacturers, wholesalers, jobbers, distributors, and journeymen. Some organizations represent an entire industry, while others restrict their scope to only a segment of it. Every aspect of heating, ventilating, and air conditioning is covered by one or more of these professional organizations.

Among the professional organizations currently active are the following:

1. Air-Conditioning and Refrigeration Institute (ARI).
2. Air-Conditioning and Refrigeration Wholesalers (ARW).
3. Air Diffusion Council (ADC).
4. Air Moving and Conditioning Association (AMCA).
5. American Gas Association (AGA).
6. American Society of Heating, Refrigeration, and Air Conditioning Engineers (ASHRAE).
7. American Society of Mechanical Engineers (ASME).
8. Association of Industry Manufacturers (Plumbing—Heating—Cooling—Piping).
9. Electric Energy Association (EEA).
10. Hydronics Institute.
11. International Solar Energy Society.
12. Mechanical Contractors Association of America (MCAA).
13. National Association of Oil Heating Service Managers (NAOHSM).

14. National Association of Plumbing-Heating-Cooling Contractors.
15. National Environmental Systems Contractors Association (NESCA).
16. National LP-Gas Association (NLPGA).
17. National Oil Fuel Institute (NOFI).
18. National Warm Air Heating and Air Conditioning Association (NWAH & ACA).
19. Plumbing-Heating-Cooling Information Bureau (PHCIB).
20. Refrigeration Service Engineers Society (RSES).
21. Sheet Metal and Air Conditioning Contractors National Association (SMACNA).

Some professional organizations of long standing have been merged with others or have been disbanded. For example, the Steel Boiler Institute (formerly the Steel Heating Boiler Institute), which maintained standards in the heating industry with its SBI Rating Code, is now defunct. The Institute of Boiler and Radiator Manufacturers (source of the old IBR Code) merged with the Better Heating-Cooling Council to form the Hydronics Institute. A recent attempt to contact the Steam Heating Equipment Manufacturers Association has resulted in the return of a letter marked "no forwarding address." It seems very likely that it, too, has joined the list of defunct professional organizations.

Appendix A (Professional and Trade Associations) at the end of this book gives a listing of professional organizations. It also contains their present addresses, the names of some of their publications, and a brief synopsis of their backgrounds and who they represent.

CHAPTER 2

Heating Fundamentals

There is still considerable disagreement about the exact nature of heat, but most authorities agree that it is a particular form of energy. Specifically, heat is a form of energy not associated with matter and *in transit* between its source and destination point. Furthermore, heat energy exists as such only between these two points. In other words, it exists as heat energy only while flowing between the source and destination.

So far this description of heat energy has been practically identical to that of work energy, the other form of energy in transit not associated with matter. The distinguishing difference between the two is that heat energy is energy in transit as a result of temperature differences between its source and destination point, whereas work energy in transit is due to other, nontemperature factors.

BRITISH THERMAL UNIT

Heat energy is measured by the British thermal unit (Btu). Each thermal unit is regarded as equivalent to one unit of heat (heat energy).

Since 1929, British thermal units have been defined on the basis of one Btu being equal to 251.996 IT (International Steam Table) calories, or 778.26 ft.-lb. of mechanical energy units (work). Taking into consideration that one IT calorie equals ⅟₈₆₀ of a watt-hour, one Btu is then equivalent to about ⅓ watt-hour.

Prior to its 1929 redefinition, 1 Btu was defined as the amount of heat necessary to raise 1 lb. of water 1°F. Because of the difficulty in determining the exact value of a Btu, it was later redefined in terms of the more fundamental electrical unit.

RELATIONSHIP BETWEEN HEAT AND WORK

Energy is the ability to do work or move against a resistance. Conversely, work is the overcoming of resistance through a certain distance by the expenditure of energy.

Work is measured by a standard unit called the *foot-pound*, which may be defined as the amount of work done in raising one pound the distance of one foot, or in overcoming a pressure of one pound through a distance of one foot (Fig. 2-1).

The relationship between work and heat is referred to as the *mechanical equivalent of heat;* one unit of heat is equal to 778.26 ft.-lb. This relationship (i.e., the mechanical equivalent of heat) was first established by experiments conducted in the nineteenth century. In 1843 Dr. James Prescott Joule (1818–1889) of Manchester, England, determined by numerous experiments that when 772 ft.-lb. of energy had been expended on 1 lb. of water, the temperature of water had risen 1°F and the relationship between heat and mechanical work was found (Fig. 2-2). The value 772 ft.-lb. is known as *Joule's equivalent.* More recent experiments give higher figures and the value 778 (1 Btu = 778.26 ft.-lb.). (See the preceding section).

ONE FOOT POUND
(OF WORK)

ONE POUND

ONE FOOT

Fig. 2-1. Man raising 1 pound 1 foot to illustrate the foot-pound standard unit.

HEAT TRANSFER

When bodies of unequal temperatures are placed near each other, heat leaves the hotter body and is absorbed by the colder one until the temperature of each is equal. The rate by which the heat is absorbed by the colder body is proportional to the difference of temperature between the two bodies—the greater the difference in temperature, the greater the rate of flow of the heat.

23

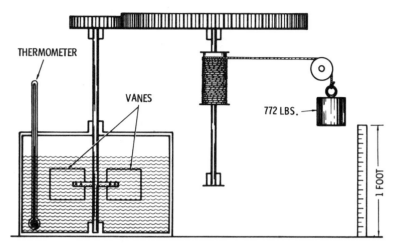

Fig. 2-2. The mechanical equivalent of heat.

Heat is transferred from one body to another at lower temperature by any one of the following means (Fig. 2-3):

1. Radiation.
2. Conduction.
3. Convection.

Radiation, insofar as heat loss is concerned, refers to the throwing out of heat in rays. The heat rays proceed in straight lines, and the intensity of the heat radiated from any one source becomes less as the distance from the source increases.

The amount of heat loss through radiation depends upon the temperature of the floor, ceiling, and walls in the room or building. The colder these surfaces are, the faster and greater will be the heat loss from a human body standing within the enclosure. If the wall, ceiling, and floor surfaces are *warmer* than the human body within the enclosure they form, heat will be radiated from these surfaces to the body. In these situations a person may complain that the room is too hot.

Knowledge of the (mean radiant) temperature of the surfaces of an enclosure is important when dealing with heat loss by radiation. The *mean radiant temperature* (MRT) is the weighted aver-

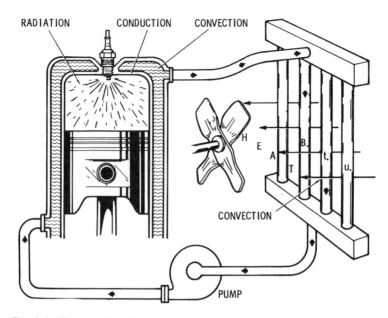

RADIATION   CONDUCTION   CONVECTION

CONVECTION

PUMP

Fig. 2-3. The transfer of heat by radiation, conduction, and convection.

age temperature of the floor, ceiling, and walls. The significance of the mean radiant temperature is determined when compared with the clothed body of an adult (80°F, or 26.7°C). If the MRT is below 80°F, the human body will lose heat by radiation to the surfaces of the enclosure. If the MRT is higher than 80°F, the opposite effect will occur.

Conduction is the transfer of heat through substances, for instance, from the boiler plate to another substance in contact with it (Fig. 2-4). Conductivity may be defined as the relative value of a material, compared with a standard, in affording a passage through itself or over its surface for heat. A poor conductor is usually referred to as a *nonconductor* or *insulator*. Copper is an example of a good conductor. Fig. 2-5 illustrates the comparative heat conductivity rates of three frequently used metals. The various materials used to insulate buildings are poor conductors. It should be pointed out that any substance that is a good conductor of electricity is also a good conductor of heat.

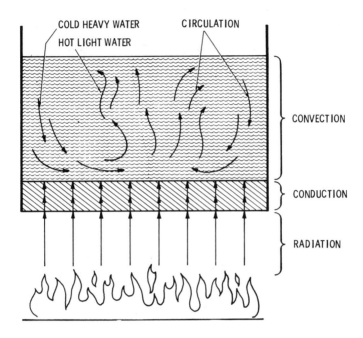

Fig. 2-4. Radiation, conduction, and convection in boiler operation.

Convection is the transfer of heat by the motion of the heated matter itself. Because motion is a required aspect of the definition of convection, it can take place only in liquids and gases.

Fig. 2-4 illustrates how radiation, conduction, and convection are often interrelated. Heat from the burning fuel passes to the metal of the heating surface by radiation, through the metal by conduction, and is transferred to the water by convection (i.e., circulation). Circulation is caused by a variation in the weight of the water due to temperature differences. That is, the water next to the heating surface receives heat and expands (becomes lighter) and immediately rises as a result of displacement by the colder and heavier water above.

Proper circulation is very important because its absence will cause a liquid, such as water, to reach the spheroidal state. This, in turn, causes the metal of the boiler to become dangerously overheated. A liquid that has reached the spheroidal state is easy

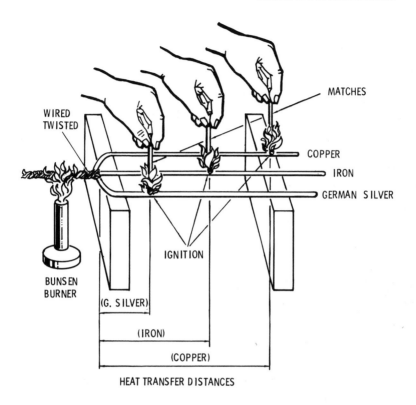

Fig. 2-5. Conductivity of various metals.

to recognize by its appearance. When liquid is dropped upon the surface of a highly heated metal, it rolls about in spheroidal drops (Fig. 2-6) or masses without actual contact with the heated metal. This phenomenon is caused by the repulsive force of heat and the intervention of a cushion of steam.

SPECIFIC, SENSIBLE, AND LATENT HEAT

The *specific heat* of a substance is the ratio of the quantity of heat required to raise its temperature one degree Fahrenheit to

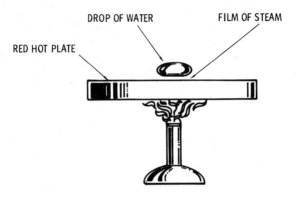

Fig. 2-6. Drop of water on a hot plate illustrating the spheroidal state.

the amount required to raise the temperature of the same weight of water one degree Fahrenheit (Fig. 2-7). This may be expressed in the following formula:

$$\text{Specific heat} = \frac{\text{Btu to raise temp. of substance } 1°F}{\text{Btu to raise temp. of same weight water } 1°F}$$

The standard used in water at approximately 62 to 63°F, which receives a rating of 1.00 on the specific heat scale. Simply stated, specific heat represents the Btu required to raise the temperature of one pound of a substance one degree Fahrenheit.

Sensible heat is the part of heat that provides temperature change and that can be measured by a thermometer. It is referred to as such because it can be "sensed" by instruments or the touch.

Latent heat is the quantity of heat that disappears or becomes concealed in a body while producing some change in it other than a rise of temperature. Changing a liquid to a gas and a gas to a liquid are both activities involving latent heat. The two types of latent heat are:

1. Internal latent heat.
2. External latent heat.

These are explained in detail in the next section under "Steam".

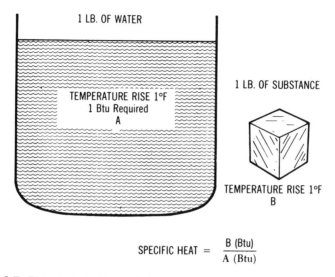

$$\text{SPECIFIC HEAT} = \frac{B \ (Btu)}{A \ (Btu)}$$

Fig. 2-7. The principle of specific heat.

HEAT CONVEYING MEDIUMS

As mentioned in Chapter 1, several methods are used to classify heating systems. One method is based on the medium that conveys the heat from its source to the point being heated. When the majority of heating systems in use today are examined closely, it can be seen that there are only four basic heat conveying mediums involved:

1. Air.
2. Steam.
3. Water.
4. Electricity.

Air

Air is a gas consisting of a mechanical mixture of 23.2% oxygen (by weight), 75.5% nitrogen, and 1.3% argon with small amounts of other gases. It functions as the heat conveying medium for hot-air heating systems.

29

Atmospheric pressure may be defined as the force exerted by the weight of the atmosphere in every point with which it is in contact (Fig. 2-8), and is measured in inches of mercury or the corresponding pressure in pounds per square inch (psi).

The pressure of the atmosphere is approximately 14.7 psi at sea level. The standard atmosphere is 29.921 inches of mercury (in. Hg) at 14.696 psi. "Inches of mercury" refers to the height to which the column of mercury in the barometer will remain suspended to balance the pressure caused by the weight of the atmosphere.

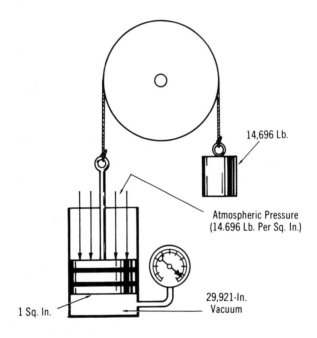

Fig. 2-8. Atmospheric pressure.

Atmospheric pressure will vary due to elevation by decreasing approximately ½ lb. for every 1000 ft. ascent above sea level. Atmospheric pressure in pounds per square inch is obtained from the barometer reading by multiplying the barometer reading in inches by 0.49116. Examples are given in Table 2-1.

Table 2-1. Atmospheric Pressure per Square Inch for Various Barometer Readings

Barometer, in. Hg	Pressure, psi	Barometer, in. Hg	Pressure, psi
28.00	13.75	29.921	14.696
28.25	13.88	30.00	17.74
28.50	14.00	30.25	14.86
28.75	14.12	30.50	17.98
29.00	14.24	30.75	15.10
29.25	14.37	31.00	15.23
29.50	14.49		
29.75	14.61		

Gauge pressure is pressure whose scale starts at atmospheric pressure. *Absolute pressure*, on the other hand, is pressure measured from true zero or the point of no pressure. When the hand of a steam gauge is at zero, the absolute pressure existing in the boiler is approximately 14.7 psi. Thus, by way of example, 5 lb. pressure measured by a steam gauge (i.e., gauge pressure) is equal to 5 lb. plus 14.7 lb., or 19.7 psi of absolute pressure.

When air is compressed, both its pressure and temperature are changed in accordance with Boyle's and Charles' laws. According to Robert Boyle (1627–1691), the English philosopher and founder of modern chemistry, the absolute pressure of a gas at constant temperature varies inversely as its volume. Jacques Charles (1746–1823) established that the volume of a gas is proportional to its absolute temperature when the volume is kept at constant pressure.

If the cylinder in Fig. 2-9 is filled with air at atmospheric pressure (14.7 psi absolute), represented by volume A, and the piston B moved to reduce the volume to, say, ⅓ A, as represented by B, then according to Boyle's law, the pressure will be trebled (14.7 × 3 = 44.1 lb. absolute, or 44.1 − 14.7 = 29.4 gauge pressure). According to Charles' law, a pressure gauge on the cylinder would at this point indicate a *higher* pressure than 29.4 gauge pressure because of the increase in temperature produced by compressing the air. This is called *adiabatic compression* if no heat is lost or is received externally.

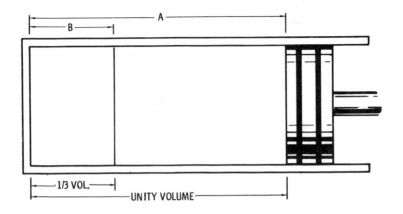

Fig. 2-9. Elementary air compressor illustrating the phenomenon of compression as stated in Boyle's and Charles' law.

Steam

Those who design, install, or have charge of steam heating plants certainly should have some knowledge of steam and its formation and behavior under various conditions.

Steam is a colorless expansive and invisible gas resulting from the vaporization of water. The white cloud associated with steam is a fog of minute liquid particles formed by condensation, that is to say, finely divided condensation. This white cloud is caused by the exposure of the steam to a temperature lower than that corresponding to its pressure.

If the inside of a steam heating main were visible, it would be filled partway with a white cloud; in traversing the main, the little particles combine, forming drops of condensation too heavy to remain in suspension, which accordingly drop to the bottom of the main and drain off as condensation. This condensation flows into a drop leg of the system and finally back into the boiler, together with additional condensation draining from the radiators.

Although the word "steam" should be applied only to saturated gas, the five following classes of steam are recognized:

1. Saturated steam.
2. Dry steam.
3. Wet steam.
4. Superheated steam.
5. Highly superheated or gaseous steam.

Three of these classes of steam (wet, saturated, and super-heated) are shown in the illustration of a safety valve blowing in Fig. 2-10. It should be pointed out that neither saturated steam nor superheated steam can be seen by the naked eye.

Saturated steam may be defined as steam of a temperature due to its pressure. Steam containing intermingled moisture, mist, or spray is referred to as *wet steam*. *Dry steam* is steam containing no moisture. It may be either saturated or superheated. Finally, *superheated steam* is steam having a temperature higher than that corresponding to its pressure.

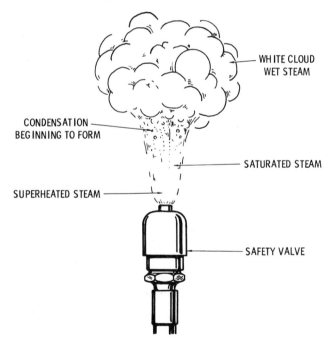

Fig. 2-10. Three types of steam.

The various changes that take place in the making of steam are known as *vaporization* and are shown in Fig. 2-11. For the sake of illustration, only one bubble is shown in each receptacle. In actuality there is a continuous procession upward of a great multiplicity of bubbles.

The amount of heat necessary to cause the generation of steam is the sum of the sensible heat, the internal latent heat, and the external latent heat. As mentioned elsewhere in this chapter, sensible heat is the part of the heat that produces a rise in temperature as indicated by the thermometer. The *internal latent heat* is the amount of heat that water will absorb at the boiling point without a change in temperature—that is, before vaporization begins. *External latent heat* is the amount of heat required when vaporization begins to push back the atmosphere and make room for the steam.

Another important factor to consider when dealing with steam is the boiling point of liquids. By definition, the *boiling point* is the temperature at which a liquid begins to boil (Fig. 2-12), and it

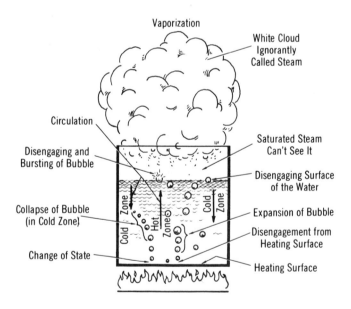

Fig. 2-11. The phenomenon of vaporization.

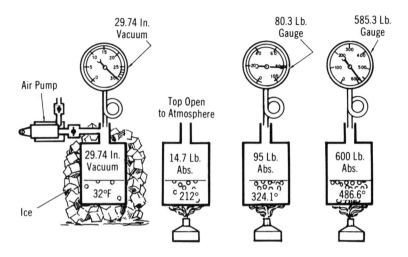

Fig. 2-12. Variation of the boiling point when pressure changes.

depends upon both the pressure and nature of the liquid. For instance, water boils at 212°F and ether at 9°F under atmospheric pressure of 14.7 psi.

The relationship between boiling point and pressure is such that there is a definite temperature or boiling point corresponding to each value of pressure. When vaporization occurs in a closed vessel and there is a temperature rise, the pressure will rise until the equilibrium between temperature and pressure is reestablished.

One's knowledge of the fundamentals of steam heating should also include an understanding of the role that condensation plays. By definition, *condensation* is the change of a substance from the gaseous to the liquid (or condensate) form. This change is caused by a reduction in temperature of the steam below that corresponding to its pressure.

The condensation of steam can cause certain problems for steam heating systems unless they are designed to allow for it. The water from which the steam was originally formed contained, mechanically mixed with it, $\frac{1}{20}$, or 5 percent, of air by volume (at atmospheric pressure). This air is liberated during

35

vaporization and does not recombine with the condensation. As a result, trouble is experienced in heating systems when one attempts to get the air out and keep it out. Suitable air valves are necessary to correct the problem.

Water

Water is a chemical compound of two gases, oxygen and hydrogen, in the proportion of two parts by weight of hydrogen to sixteen parts by weight of oxygen having mixed with it about 5 percent of air by volume at 14.7 lb. absolute pressure. It may exist as ice, water, or steam due to changes in temperature (water freezes at 32°F and boils at 212°F when the barometer reads 29.921 in.).

One cubic foot of water weighs 62.41 lb. at 32°F and 59.82 lb. at 212°F. One U.S. gallon of water (231 cu. in.) weighs 8.33111 lb. (ordinarily expressed as 8⅓ lb.) at a temperature of 62°F. At any other temperature, of course, the weight will be different (Table 2-2).

Water changes in weight with changes of temperature. That is, the higher the temperature of the water, the less it weighs. It is this property of water that causes circulation in boilers and in hot-water heating systems. The change in weight is due to expansion and a reduction in water volume. As the temperature rises, the water expands, resulting in a unit volume of water containing less water at higher temperature than lower temperature.

Fill a vessel with cold water and heat it to the boiling point. Note that boiling causes it to overflow due to expansion. Now let the water cool. You will note that when the water is cold, the vessel will not be as full because the water will have contracted.

The point of maximum density of water is 39.1°F. The most remarkable characteristic of water is its expansion below *and* above its point of maximum density. Imagine 1 lb. of water at 39.1°F placed in a cylinder having a cross-sectional area of 1 sq. in. (Fig. 2-13). The water having a volume of 27.68 cu. in. will fill the cylinder to a height of 27.68 in. If the water is cooled, it will expand, and at, say, 32°F (the freezing point) will rise in the tube to a height of 27.7 in. before freezing. If the water is heated, it

Table 2-2. Weight of Water per Cubic Foot at Different Temperatures

Temp., deg. F	lb. per cu. ft.	Temp., deg. F	lb.per cu. ft.	Temp., deg. F	lb. per cu. ft.	Temp., deg. F	lb. per cu. ft.
32	62.41	62	62.35	91	62.10	121	61.69
33	62.41	63	62.34	92	62.08	122	61.68
34	62.42	64	62.34	93	62.07	123	61.66
35	62.42	65	62.33	94	62.06	124	61.64
36	62.42	66	62.32	95	62.05	125	61.63
37	62.42	67	62.32	96	62.04	126	61.61
38	62.42	68	62.31	97	62.02	127	61.60
39	62.42	69	62.30	98	62.01	128	61.58
40	62.42	70	62.30	99	62.00	129	61.56
41	62.42	71	62.29	100	61.99	130	61.55
42	62.42	72	62.28	101	61.98	131	61.53
43	62.42	73	62.27	102	61.96	132	61.51
44	62.42	74	62.26	103	61.95	133	61.50
45	62.42	75	62.25	104	61.94	134	61.48
46	62.41	76	62.25	105	61.93	135	61.46
47	62.41	77	62.24	106	61.91	136	61.44
48	62.41	78	62.23	107	61.90	137	61.43
49	62.41	79	62.22	108	61.89	138	61.41
50	62.40	80	62.21	109	61.87	139	61.39
51	62.40	81	62.20	110	61.86	140	61.37
52	62.40	82	62.19	111	61.84	141	61.36
53	62.39	83	62.18	112	61.86	142	61.34
54	62.39	84	62.17	113	61.81	143	61.32
55	62.38	85	62.16	114	61.80	144	61.30
56	62.38	86	62.15	115	61.78	145	61.28
57	62.38	87	62.14	116	61.77	146	61.26
58	62.37	88	61.13	117	61.75	147	61.25
59	62.37	89	62.12	118	61.74	148	61.23
60	32.36	90	62.11	119	61.72	149	61.21
61	62.35			120	61.71		

will also expand and rise in the tube; and at the boiling point (for atmospheric pressure 212°F) it will occupy the tube to a height of 28.88 cu. in.

The elementary hot-water heating system in Fig. 2-14 illustrates the principle of thermal circulation. The weight of the hot and expanded water in the upflow column C, being less than that of the cold and contracted water in the downflow column C', upsets the equilibrium of the system and results in a continuous circulation of water as indicated by the arrows. In other words,

Table 2-2. Weight of Water per Cubic Foot at Different Temperatures (Cont'd)

Temp., deg. F	lb. per cu. ft.	Temp., deg. F	lb. per cu. ft.	Temp., deg. F	lb. per cu. ft.	Temp., deg. F	lb. per cu. ft.
150	61.19	180	60.57	208	59.92	440	51.87
151	61.17	181	60.55	209	59.90	450	51.28
152	61.15	182	60.53	210	59.87	460	51.02
153	61.13	183	60.51	211	59.85	470	50.51
154	61.11	184	60.49	212	59.82	480	50.00
155	61.09	185	60.46	214	59.81	490	49.50
156	61.07	186	60.44	216	59.77	500	48.78
157	61.05	187	60.42	218	59.70	510	48.31
158	61.03	188	60.40	220	59.67	520	47.62
159	61.01	189	60.37	230	59.42	530	46.95
160	60.99	190	60.35	240	59.17	540	46.30
161	60.97	191	60.33	250	58.89	550	45.66
162	60.95	192	60.30	260	58.62	560	44.84
163	60.93	193	60.28	270	58.34	570	44.05
164	60.91	194	60.26	280	58.04	580	43.29
165	60.89	195	60.23	290	57.74	590	42.37
166	60.87	196	60.21	300	57.41	600	41.49
167	60.85	197	60.19	310	57.08	610	40.49
168	60.83	198	60.16	320	56.75	620	39.37
169	60.81	199	60.14	330	56.40	630	38.31
170	60.79	200	60.11	340	56.02	640	37.17
171	60.77	201	60.09	350	55.65	650	35.97
172	60.75	202	60.07	360	55.25	660	34.48
173	60.73	203	60.04	370	54.85	670	32.89
174	60.71	204	60.02	380	54.47	680	31.06
175	60.68	205	59.99	390	54.05	690	28.82
176	60.66	206	59.97	400	53.62	700	25.38
177	60.64	207	59.95	410	53.19	706.1	19.16
178	60.62			420	52.74		
179	60.00			430	52.33		

the heavy low-temperature water sinks to the lowest point in the boiler (or system) and displaces the light high-temperature water, thus causing continuous circulation as long as there is a temperature difference in different parts of the boiler (or system). This is referred to as *thermal circulation*.

Electricity

Electricity differs from air, steam, or water in that it does not actually convey heat from one point to another; therefore, includ-

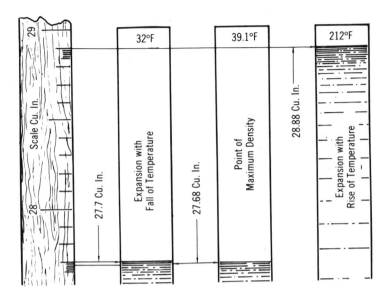

Fig. 2-13. The point of maximum density.

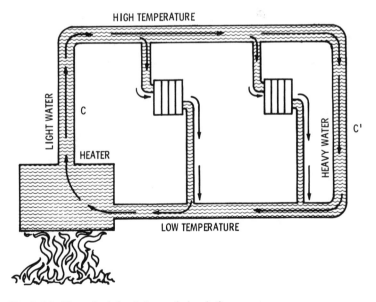

Fig. 2-14. The principle of thermal circulation.

ing it in a list of heat conveying mediums can be misleading at first glance.

Electricity can best be defined as a quantity of electrons either in motion or in a state of rest. When these electrons are at rest, they are referred to as being *static* (hence the term "static electricity"). Electrons in *motion* move from one atom to another creating an electrical current and thereby a medium for conveying *energy* from one point to another. Many different devices have been created to change the energy conveyed by an electric current into heat, light, and other forms of energy. Electric-fired furnaces and boilers are examples of devices used to produce heat.

CHAPTER 3

Insulation Principles

This chapter is not concerned with the theoretical aspects of heat loss or gain, or with heating and cooling calculations per se. These problems are discussed in considerable detail in other chapters of this book, particularly in Chapter 2 (Heating Fundamentals), Chapter 4 (Heating Calculations), and Chapter 9 (Electrical Heating Systems), as well as in Volume 3. The purpose of this chapter is to describe the various *types* of insulation used in construction, the methods of applying insulation, and the thermal properties of building materials.

THE NEED FOR INSULATION

An uninsulated structure or one that is poorly insulated will experience a heat gain in the summer (or a heat loss in the winter) as high as 50 percent. The exact percentage will depend upon the materials used in its construction because all construction mate rial will have some insulating effect. In any event, it has been observed that approximately 30 percent of heat gain (or loss) is experienced through ceilings and about 70 percent occurs through walls, glass, window and door cracks, or as a result of air ventilation.

Any attempt to heat or cool an *uninsulated* structure will be terribly inefficient *and expensive*. Obviously, the possibility of anyone being so shortsighted as to attempt to install a heating or cooling system in such a structure is fairly remote, but inadequate insulation in existing structures or the failure to provide for proper insulation in proposed construction is a common occurrence. The provision for some form of barrier to reduce the rate of heat flow to an acceptable level is therefore of prime importance. Insulation materials serve this purpose with varying degrees of effectiveness. As much as 80 to 90 percent of heat loss (or gain) through ceilings and 60 percent of heat loss (or gain) through walls can be prevented by properly insulating these areas.

The inefficiency of ordinary building materials in resisting the passage of heat brought about the need for the development of materials specifically designed for insulation. A *good* insulating material should be lightweight, contain numerous air pockets, and exhibit a high degree of resistance to heat transmission. It should also be specifically treated to resist fire and the attacks of rodents and insects. Finally, a good insulating material should react well to excess moisture by drying out and retaining its resistance to heat flow, rather than disintegrating.

Many inexperienced workers seem to feel that the more insulation one uses, the better the insulating effect. Unfortunately, this does not hold true on a one-to-one basis. For example, twice as much insulation does *not* insulate twice as well. Apparently a law of diminishing returns operates here. The first layer of insulation is the most effective, with successive layers decreasing in the effectiveness to impede the rate of heat flow.

The capacity of a heating or cooling system is determined by the amount of heat that must be supplied (heating applications) or removed (cooling applications) to maintain the desired temperature within the structure or space. In heating applications, the amount of heat supplied to a space at constant temperature should roughly equal the amount of heat lost. In cooling applications, the heat removed from a space should be roughly equal to the amount of heat gained. Both heat loss and heat gain will be controlled by the way in which the structure is insulated. Good insulation will reduce the rate of heat flow through construction materials to an acceptable level, and this will mean that a heating or cooling system will be able to operate much less expensively and with greater efficiency. It will also mean that the capacity of the heating or cooling unit you wish to install will be less than the one required for a poorly insulated structure. This, in turn, will reduce the initial equipment and installation costs. As you can see, this type and quality of insulation in a structure is an important factor to be considered when designing a heating or a cooling system.

The roof or attic floor of any structure can be insulated, as can the walls of any frame building, whether of stucco, clapboard, shingles, brick, or stone veneer. The form in which the insulation is applied depends on whether the structure is already built or is being constructed.

PRINCIPLES OF HEAT TRANSMISSION

Heat flows through a barrier of building materials (e.g., walls, ceilings, and floors) by means of conduction and radiation. When it has passed to the other side of this barrier, it continues its movement away from the source by means of conduction, radiation, and convection. The *direction* of heat flow depends on the relationship of outdoor and indoor temperatures because warmed air will always move toward a colder space. Thus, in the summer months, the warmer outdoor air will move toward the cooler indoor spaces. This is referred to as *heat leakage* and is described in Chapter 9 of Volume 3 (Air Conditioning Calculations). In the winter months, the reverse is true. The warmer indoor air moves to the cooler outdoor spaces, and this phenomenon is referred to as *heat loss* (see Chapter 4, Heating Calculations). The purpose of insulation is to reduce the rate of heat flow to an acceptable level.

Heat Transfer Values

The American Society of Heating, Refrigerating, and Air Conditioning Engineers (ASHRAE) has conducted considerable research into the thermal properties of building materials, and these data are used in the calculation of heat loss and heat gain. Tables of computed values for many construction materials and combinations of construction materials are made available through ASHRAE publications. One table gives the design values (i.e., conductivity values, conductance values, and resistance values) of various building and insulating materials. These values are used to calculate the coefficients of heat transmission (U-values) for different types of construction (e.g., masonry walls and frame partitions). Examples of these two tables have been taken from the *1960 ASHRAE Guide* (see Table 3-1 and 3-2).

The four heat transfer values contained in these tables are: (1) the k-value (thermal conductivity); (2) the C-value (thermal conductance); (3) the R-value (thermal resistance); and (4) the U-value (overall coefficient of heat transmission). Each of these heat transfer values will be examined before their use in heating and cooling calculations is explained.

Thermal Conductivity

The *k-value* (thermal conductivity) represents the amount of heat that flows through one square foot of a homogeneous material one inch thick in one hour for each degree of temperature difference between the inside and the outside temperatures.

To illustrate, we shall use an example from the ASHRAE data shown in Tables 3-1 and 3-2. You will note that the k-value for face brick listed in Table 3-1 is 9.0. In other words, 9.0 Btu of heat pass through 1 sq. ft. of 1-in.-thick face brick each hour. The thermal resistance (R-value) of this same 1-in.-thick face brick is the reciprocal of its k-value:

$$R = \frac{1}{k} = \frac{1}{9} = 0.11$$

Remember, this is the thermal resistance for a 1-in.-thick face brick. Because the masonry wall in Table 3-2 contains a 4-in. thickness of face brick, you will have to multiply the R-value

Table 3-1. Conductivities (*k*), Conductances (*C*), and Resistances (*R*) of Various Building and Insulating Materials

Material	Position	Heat Flow	Thickness	Density (lb. per cu. ft.)	(*k*)	(*C*)	Per inch thickness $\frac{1}{k}$	For thickness listed $\frac{1}{C}$
AIR SPACES								
	Horizontal	Up (winter)	3/4–4 in.	—	—	1.18	—	0.85
	Horizontal	Up (summer)	3/4–4 in.	—	—	1.28	—	0.78
	Horizontal	Down (winter)	3/4 in.	—	—	0.98	—	1.02
	Horizontal	Down (winter)	1 1/2 in.	—	—	0.87	—	1.15
	Horizontal	Down (winter)	4 in.	—	—	0.81	—	1.23
	Horizontal	Down (winter)	8 in.	—	—	0.80	—	1.25
	Horizontal	Down (summer)	3/4 in.	—	—	1.18	—	0.85
	Horizontal	Down (summer)	1 1/2 in.	—	—	1.07	—	0.93
	Horizontal	Down (summer)	4 in.	—	—	1.01	—	0.99
	Sloping, 45°	Up (winter)	3/4–4 in.	—	—	1.11	—	0.90
	Sloping, 45°	Down (summer)	3/4–4 in.	—	—	1.12	—	0.89
	Vertical	Horizontal (winter)	3/4–4 in.	—	—	1.03	—	0.97
	Vertical	Horizontal (summer)	3/4–4 in.	—	—	1.16	—	0.86

Table 3-1. Conductivities (*k*), Conductances (*C*), and Resistances (*R*) of Various Building and Insulating Materials (Cont'd)

	Position	Heat Flow		k	C		R
AIR SURFACES							
Still Air	Horizontal	Up	—	—	1.63	—	0.61
	Up Sloping (45°)	Up	—	—	1.60	—	0.62
	Vertical	Horizontal	—	—	1.46	—	0.68
	Sloping (45°)	Down	—	—	1.32	—	0.76
	Horizontal	Down	—	—	1.08	—	0.92
15 mph Wind	Any position—any direction (for winter)		—	—	6.00	—	0.17
7½ mph Wind	Any position—any direction (for summer)		—	—	4.00	—	0.25
BUILDING BOARD							
Boards, Panels, Sheathing, etc.	Asbestos-cement board		120	4.0	—	0.25	—
	Asbestos-cement board	⅛ in.	—	—	33.00	—	0.03
	Gypsum or plaster board	⅜ in.	50	—	3.10	—	0.32
	Gypsum or plaster board	½ in.	50	—	2.25	—	0.45
	Plywood		34	0.80	—	1.25	—
	Plywood	¼ in.	34	—	3.20	—	0.31
	Plywood	⅜ in.	34	—	2.12	—	0.47
	Plywood	½ in.	—	—	1.60	—	0.63
	Plywood or wood panels	¾ in.	—	—	1.07	—	0.94
	Wood fiber board, laminated or homogenous		26	0.42	—	2.38	—
	Wood fiber—hardboard type		31	0.50	—	2.00	—
	Wood fiber—hardboard type	¼ in.	65	1.40	5.60	0.72	0.18
	Wood—fir or pine sheathing	25/32 in.	65	—	1.02	—	0.98
	Wood—fir or pine	1⅝ in.	—	—	0.49	—	2.03

Table 3-1. Conductivities (*k*), Conductances (*C*), and Resistances (*R*) of Various Building and Insulating Materials (Cont'd)

Category	Material	Thickness	Density	*k*	*C*	*R* (1/*k*)	*R* (1/*C*)
BUILDING PAPER	Vapor—permeable felt		—	—	16.70	—	0.06
	Vapor—seal, 2 layers of mopped 15-lb. felt		—	—	8.35	—	0.12
	Vapor—seal, plastic film		—	—	—	—	Negl
FLOORING MATERIALS	Asphalt tile	⅛ in.	120	—	24.80	—	0.04
	Carpet and fibrous pad		—	—	0.48	—	2.08
	Carpet and rubber pad		—	—	0.81	—	1.23
	Ceramic tile	1 in.	—	—	12.50	—	0.08
	Cork tile		25	0.45	—	2.22	—
	Cork tile	⅛ in.	—	—	3.60	—	0.28
	Felt, flooring		—	—	16.70	—	0.06
	Floor tile or linoleum—av. value	⅛ in.	—	—	20.00	—	0.05
	Linoleum	⅛ in.	80	—	12.00	—	0.08
	Plywood subfloor	⅝ in.	—	—	1.28	—	0.78
	Rubber or plastic tile	⅛ in.	110	—	42.40	—	0.02
	Terrazzo	1 in.	—	—	12.50	—	0.98
	Wood subfloor	25⁄32 in.	—	—	1.02	—	0.98
	Wood, hardwood finish	¾ in.	—	—	1.47	—	0.68
INSULATING MATERIALS **Blankets and Batts**	Cotton fiber		0.8–2.0	0.26	—	3.85	—
	Mineral wool, fibrous form, processed from rock, slag, or glass		1.5–4.0	0.27	—	3.70	—
	Wood fiber		3.2–3.6	0.25	—	4.00	—
	Wood fiber, multilayer, stitched expanding		1.5–2.0	0.27	—	3.70	—

Table 3-1. Conductivities (k), Conductances (C), and Resistances (R) of Various Building and Insulating Materials (Cont'd)

INSULATING MATERIALS		9.5	0.25	—	4.00	—
(Cont'd)						
Board						
Glass fiber		—	—	—	—	—
Wood or cane fiber						
Acoustical tile	½ in.	—	—	0.84	—	1.19
Acoustical tile	¾ in.	—	—	0.56	—	1.78
Interior finish (plank, tile, lath)		15.0	0.35	—	2.86	—
Interior finish (plank, tile, lath)	½ in.	15.0	—	0.70	—	1.43
Roof deck slab						
Approx.	1½ in.	—	—	0.24	—	4.17
Approx.	2 in.	—	—	0.18	—	5.56
Approx.	3 in.	—	—	0.12	—	8.33
Sheathing (impreg. or coated)		20.0	0.38	—	2.63	—
Sheathing (impreg. or coated)	½ in.	20.0	—	0.76	—	1.32
Sheathing (impreg. or coated)	²⁵/₃₂ in.	20.0	—	0.49	—	2.06
Board and Slabs						
Cellular glass		9.0	0.40	—	2.50	—
Corkboard (without added binder)		6.5–8.0	0.27	—	3.70	—
Hog hair (with asphalt binder)		8.5	0.33	—	3.00	—
Plastic (foamed)		1.62	0.29	—	3.45	—
Wood shredded						
(cemented in preformed slabs)		22.0	0.55	—	1.82	—
Loose Fill						
Macerated paper or pulp products		2.5–3.5	0.28	—	3.57	—
Mineral wood (glass, slag, or rock)		2.0–5.0	0.30	—	3.33	—
Sawdust or shavings		8.0–15.0	0.45	—	2.22	—
Vermiculite (expanded)		7.0	0.48	—	2.08	—
Wood fiber: redwood, hemlock, or fir		2.0–3.5	0.30	—	3.33	—

Table 3-1. Conductivities (*k*), Conductances (*C*), and Resistances (*R*) of Various Building and Insulating Materials (Cont'd)

	Density	*k*	*C*	*R*	*R*
ROOF INSULATION					
All types					
Preformed, for use above deck					
Approx. ½ in.	—	—	0.72	—	1.39
Approx. 1 in.	—	—	0.36	—	2.78
Approx. 1½ in.	—	—	0.24	—	4.17
Approx. 2 in.	—	—	0.19	—	5.26
Approx. 2½ in.	—	—	0.15	—	6.67
Approx. 3 in.	—	—	0.12	—	8.33
MASONRY MATERIALS					
Concretes					
Cement mortar	116	5.0	—	0.20	—
Gypsum-fiber concrete 87½ % gypsum, 12½ % wood chips	51	1.66	—	0.60	—
Lightweight aggregates, incuding expanded shale, clay or slate; expanded slags; cinders; pumice; perlite; vermiculite; also cellular concretes	120	5.2	—	0.19	—
	100	3.6	—	0.28	—
	80	2.5	—	0.40	—
	60	1.7	—	0.59	—
	40	1.15	—	0.86	—
	30	0.90	—	1.11	—
	20	0.70	—	1.43	—
Sand and gravel or stone aggregate (oven dried)	140	9.0	—	0.11	—
Sand and gravel or stone aggregate (not dried)	140	12.0	—	0.08	—
Stucco	116	5.0	—	0.20	—

Table 3-1. Conductivities (k), Conductances (C), and Resistances (R) of Various Building and Insulating Materials (Cont'd)

MASONRY UNITS		120 / 130	5.0 / 9.0	—	0.20 / 0.11	—
Brick, common		—	—	—	0.20	—
Brick, face		—	—	—	0.11	—
Clay tile, hollow:						
1 cell deep	3 in.	—	—	1.25	—	0.80
1 cell deep	4 in.	—	—	0.90	—	1.11
2 cells deep	6 in.	—	—	0.66	—	1.52
2 cells deep	8 in.	—	—	0.54	—	1.85
2 cells deep	10 in.	—	—	0.45	—	2.22
3 cells deep	12 in.	—	—	0.40	—	2.50
Concrete blocks, three oval core:						
Sand and gravel aggregate	4 in.	—	—	1.40	—	0.71
	8 in.	—	—	0.90	—	1.11
	12 in.	—	—	0.78	—	1.28
Cinder aggregate	3 in.	—	—	1.16	—	0.86
	4 in.	—	—	0.90	—	1.11
	8 in.	—	—	0.58	—	1.72
	12 in.	—	—	0.53	—	1.89
Gypsum partition tile:						
3 × 12 × 30 in. solid		—	—	0.79	—	1.26
3 × 12 × 30 in. 4-cell		—	—	0.74	—	1.35
3 × 12 × 30 in. 3-cell		—	—	0.60	—	1.67
Lightweight aggregate (expanded shale,	3 in.	—	—	0.79	—	1.27
clay, slate, or slag; pumice)	4 in.	—	—	0.67	—	1.50
	8 in.	—	—	0.50	—	2.00
	12 in.	—	—	0.44	—	2.27

Table 3-1. Conductivities (*k*), Conductances (*C*), and Resistances (*R*) of Various Building and Insulating Materials (Cont'd)

		Density	k	C	R	R
MASONRY UNITS (Cont'd)		—	—	—	0.08	—
Stone, lime, or sand		116	12.50	—	0.08	—
PLASTERING MATERIALS						
Cement plaster, sand aggregate		—	5.0	—	0.20	—
Sand aggregate	½ in.	—	—	10.00	—	0.10
Sand aggregate	¾ in.	—	—	6.66	—	0.15
Gypsum plaster						
Lightweight aggregate	½ in.	45	—	3.12	—	0.32
Lightweight aggregate	⅝ in.	45	—	2.67	—	0.39
Lightweight aggregate on metal lath	¾ in.	—	—	2.13	—	0.47
Perlite aggregate		45	1.5	—	0.67	—
Sand aggregate		105	5.6	—	0.18	—
Sand aggregate	½ in.	105	—	11.10	—	0.09
Sand aggregate	⅝ in.	105	—	9.10	—	0.11
Sand aggregate on metal lath	¾ in.	—	—	7.70	—	0.13
Sand aggregate on wood lath		—	—	2.50	—	0.40
Vermiculate aggregate		45	1.7	—	0.59	—
ROOFING						
Asbestos-cement shingles		120	—	4.76	—	0.21
Asphalt roll roofing		70	—	6.50	—	0.15
Asphalt shingles		70	—	2.27	—	0.44
Builtup roofing	⅜ in.	70	—	3.00	—	0.33
Slate	½ in.	—	—	20.00	—	0.05
Sheet metal		—	400+	—	Negl	—
Wood shingles		—	—	1.06	—	0.94

Table 3-1. Conductivities (k), Conductances (C), and Resistances (R) of Various Building and Insulating Materials (Cont'd)

SIDING MATERIALS (On Flat Surface)					
Shingles					
Wood, 16-in. 7½-in. exposure	—	—	1.15	—	0.87
Wood, double, 16-in., 12-in. exposure	—	—	0.84	—	1.19
Wood, plus insul. backer board ⁵/₁₆ in.	—	—	0.71	—	1.40
Siding					
Asbestos-cement, ¼ in., lapped	—	—	4.76	—	0.21
Asphalt roll siding	—	—	6.50	—	0.15
Asphalt insulating siding (½ in. bd.)	—	—	0.69	—	1.45
Wood, drop, 1 × 8 in.	—	—	1.27	—	0.79
Wood, bevel, ½ × 8 in., lapped	—	—	1.23	—	0.81
Wood, bevel, ¾ × 10 in., lapped	—	—	0.95	—	1.05
Wood, plywood, ³/₈ in., lapped	—	—	1.59	—	0.59
Structural glass	—	—	10.00	—	0.10
WOODS					
Maple, oak, and similar hardwoods	45	1.10	—	0.91	—
Fir, pine, and similar softwoods	32	0.80	—	1.25	—

Courtesy ASHRAE 1960 Guide

Table 3-2. Coefficients of Heat Transmission for Solid Masonry Walls

Resistances used are given below in this table.		
Construction	Resistance	(R)
1. Outside surface (15 mph wind)		0.17
2. Face brick (4 in.)		0.44
3. Common brick (4 in.)		0.80
4. Air space		0.97
5. Gypsum lath (⅜ in.)		0.32
6. Plas. (sand agg.) (½ in.)		0.09
7. Inside surface (still air)		0.68
Total resistance		3.47
$U = 1/R = 1/3.47 =$		0.29
See value 0.29 in boldface type in table below		
Assume plain wall—no furring or plaster		
Total resistance		3.47
Deduct 4. Air space	0.97	
5. Gypsum lath (⅜ in.)	0.32	
6. Plas. (sand agg.) (½ in.)	0.09	1.38
Total resistance		2.09
$U = 1/R = 1/2.09 =$		0.48

Table 3-2. Coefficients of Heat Transmission for Solid Masonry Walls (Cont'd)

Exterior Construction		None	Plas. ⅝ in. on Wall		Metal Lath and ¾ in. Plas. on Furring		Interior Finish					Wood Lath and ½ in. Plas.	No.
							Gypsum Lath (⅜ in.) and ½ in. Plas. on Furring			Insul. Bd. Lath (½ in.) and ½ in. Plas. on Furring				
Resistance		U	(Sand agg.) 0.11 U	(Lt. wt. agg.) 0.39 U	(Sand agg.) 0.13 U	(Lt. wt. agg.) 0.47 U	No plas. 0.32 U	(Sand agg.) 0.41 U	(Lt. wt. agg.) 0.64 U	No plas. 1.43 U	(Sand agg.) 1.52 U	(Sand agg.) 0.40 U	
Material	R	A	B	C	D	E	F	G	H	I	J	K	
Brick (face and common)													
(6 in.)	0.61	0.68	0.64	0.54	0.39	0.34	0.36	0.35	0.33	0.26	0.25	0.35	1
(8 in.)	1.24	0.48	0.45	0.41	0.31	0.28	0.30	0.29	0.27	0.22	0.22	0.29	2
(12 in.)	2.04	0.35	0.33	0.30	0.25	0.23	0.24	0.23	0.22	0.19	0.19	0.23	3
(16 in.)	2.84	0.27	0.26	0.25	0.21	0.19	0.20	0.20	0.19	0.16	0.16	0.20	4
Brick (common only)													
(8 in.)	1.60	0.41	0.39	0.35	0.28	0.26	0.27	0.26	0.25	0.21	0.20	0.26	5
(12 in.)	2.40	0.31	0.30	0.27	0.23	0.21	0.22	0.22	0.21	0.18	0.17	0.22	6
(16 in.)	3.20	0.25	0.24	0.23	0.19	0.18	0.19	0.18	0.18	0.16	0.15	0.18	7
Stone (lime and sand)													
(8 in.)	0.64	0.67	0.63	0.53	0.39	0.34	0.36	0.35	0.32	0.26	0.25	0.35	8
(12 in.)	0.96	0.55	0.52	0.45	0.34	0.31	0.32	0.31	0.29	0.24	0.23	0.31	9
(16 in.)	1.28	0.47	0.45	0.40	0.31	0.28	0.29	0.28	0.27	0.22	0.22	0.29	10
(24 in.)	1.92	0.36	0.35	0.32	0.26	0.24	0.25	0.24	0.23	0.19	0.19	0.24	11

Table 3-2. Coefficients of Heat Transmission for Solid Masonry Walls (Cont'd)

Hollow clay tile													
(8 in.)	1.85	0.36	0.36	0.32	0.26	0.24	0.25	0.25	0.23	0.20	0.19	0.25	12
(10 in.)	2.22	0.33	0.31	0.29	0.24	0.22	0.23	0.22	0.21	0.18	0.18	0.23	13
(12 in.)	2.50	0.30	0.29	0.27	0.22	0.21	0.22	0.21	0.20	0.17	0.17	0.21	14
Poured concrete													
30 lb. per cu. ft.													
(4 in.)	4.44	0.19	0.19	0.18	0.16	0.15	0.15	0.15	0.14	0.13	0.13	0.15	15
(6 in.)	6.66	0.13	0.13	0.13	0.12	0.11	0.11	0.11	0.11	0.10	0.10	0.11	16
(8 in.)	8.88	0.10	0.10	0.10	0.09	0.09	0.09	0.09	0.09	0.08	0.08	0.09	17
(10 in.)	11.10	0.08	0.08	0.08	0.08	0.07	0.08	0.07	0.07	0.07	0.07	0.08	18
80 lb. per cu. ft.													
(6 in.)	2.40	0.31	0.30	0.27	0.23	0.21	0.22	0.22	0.21	0.18	0.17	0.22	19
(8 in.)	3.20	0.25	0.24	0.23	0.19	0.18	0.19	0.18	0.18	0.16	0.15	0.18	20
(10 in.)	4.00	0.21	0.20	0.19	0.17	0.16	0.16	0.16	0.15	0.14	0.14	0.16	21
(12 in.)	4.80	0.18	0.17	0.17	0.15	0.14	0.14	0.14	0.14	0.12	0.12	0.14	22
140 lb. per cu. ft.													
(6 in.)	0.48	0.75	0.69	0.58	0.41	0.36	0.38	0.37	0.34	0.27	0.26	0.37	23
(8 in.)	0.64	0.67	0.68	0.53	0.39	0.34	0.36	0.35	0.32	0.26	0.25	0.35	24
(10 in.)	0.80	0.61	0.57	0.49	0.36	0.32	0.34	0.33	0.31	0.25	0.24	0.33	25
(12 in.)	0.96	0.55	0.52	0.45	0.34	0.31	0.32	0.31	0.29	0.24	0.23	0.31	26

Courtesy ASHRAE 1960 Guide

(0.11) of the 1-in.-thick face brick by four (4 in.) to obtain the R-value (0.44) for the larger (4-in.) section.

Thermal Conductance

The C-value (thermal conductance) represents the amount of the heat flow through a square foot of material per hour per degree of temperature difference between the inside and outside temperatures. The C-value is based on the stated construction or thickness of the material. This distinguishes it from the k-value, which is based on a 1-in. thickness of a homogeneous material having no fixed thickness. For example, under "loose fill" in Table 3-1, you will see vermiculite listed and given a k-value of 0.48. Why a k-value? It is given a k-value (rather than a C-value) because vermiculite is a loose-fill insulating material that can be poured into a space to form whatever thickness is desired. It is convenient, then, to give numerical values (k-values) based on 1-in. thicknesses of these materials. If, say, vermiculite was used to insulate the top-floor ceiling (that is to say, the attic floor) and poured to a depth of 4 in., the R-value (thermal resistance) would be calculated as follows:

$$k\text{-value} = 0.48$$

$$R\text{-value} = \frac{1}{0.48} = 2.08 \quad (1 \text{ in.})$$

$$R\text{-value} = 2.08 \times 4 = 8.32 \quad (4 \text{ in.})$$

Under "masonry units" in Table 3-1, you will find hollow clay tile listed with six different *stated* thicknesses (e.g., "1 cell deep, 3 in.; 1 cell deep, 4 in."). Each is given a C-value rather than a k-value because the thickness of the material listed in the table is the same thickness that will be used in the construction of a structural section.

The masonry wall in Table 3-2 also contains gypsum plaster (sand aggregate) ½ in. thick. The C-value is given as 11.10, and its R-value as 0.09 (Table 3-1). In other words, the R-value (thermal resistance) is calculated as follows:

$$R = \frac{1}{C} = \frac{1}{11.10} = 0.09$$

Examine the data in Table 3-1 once more. You will probably note that each building material has either a k- or a C-value, but not both. Each material is offered on the basis of either a stated thickness or an arbitrary thickness. This was explained above. Note also that each type of building material has an R-value determined by either of the following two equations:

$$R = \frac{1}{k}$$

$$R = \frac{1}{C}$$

The R-value is simply the reciprocal of either the k-value or the C-value.

Thermal Resistance

The *R-value* (thermal resistance) of a building material is its *resistance* to the flow of heat. As explained above, the R-value is the reciprocal of the k-value or the C-value, and represents heat flow resistance expressed as a numerical value. The higher the R-value of a material (or a combination of materials), the greater its resistance to the flow of heat. Ceilings should have an R-value of R-19 or better. Floors and walls should be insulated to at least R-11 (Fig. 3-1).

Returning to the masonry wall illustrated in Table 3-2, you will note that all seven R-values have been added to produce the total resistance of a solid masonry wall (3.47). In other words, the *total* resistance to heat flow through a solid masonry wall of the type of construction illustrated in Table 3-2 is equal to the sum of the resistances of the wall components:

$$R_t = R_1 + R_2 + R_3 \ldots R_n$$

The total resistance (R_t) of a particular type of construction is used to determine its U-value, which represents the total *insulating* effect of a structural section. The U-value is the reciprocal of the total resistance (R_t). Using the data supplied for the masonry wall illustrated in Table 3-2, its U-value is determined as follows:

$$U = \frac{1}{R} = \frac{1}{3.47} = 0.29$$

Overall Coefficient of Heat Transmission

The U-value, or U-factor (overall coefficient of heat transmission), represents the amount of heat (in Btu) that will pass through one square foot of a structural section per hour per degree Fahrenheit. This is the difference between the air on the inside and the air on the outside of the section. The U-value is the reciprocal of the sum of the thermal resistance values (R-values) of each element of the structural section. The total resistance (R_t) for the solid masonry wall illustrated in Table 3-2 is 3.47. Since the U-value is the reciprocal of R_t, the following is obtained:

$$U = \frac{1}{R_t} = \frac{1}{3.47} = 0.29$$

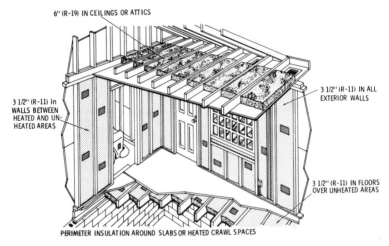

6" (R-19) IN CEILINGS OR ATTICS

3 1/2" (R-11) in WALLS BETWEEN HEATED AND UN-HEATED AREAS

3 1/2" (R-11) IN ALL EXTERIOR WALLS

3 1/2" (R-11) IN FLOORS OVER UNHEATED AREAS

PERIMETER INSULATION AROUND SLABS OR HEATED CRAWL SPACES

Courtesy Owens-Corning Fiberglass Corp.

Fig. 3-1. Minimum levels of effective insulation.

The overall coefficient of heat transmission (U-value, or U-factor) is used in the formula for calculating heat loss or heat gain. The use of this formula is described in Chapter 4 of this book (Heating Calculations) and in Chapter 9 of Volume 3 (Air Conditioning Calculations).

CONDENSATION

Air will always contain a certain amount of moisture (water vapor), and the amount of moisture it contains will depend upon its temperature. Generally speaking, warm air is capable of containing more moisture than cold air.

Air will lose moisture in the form of condensation when its temperature falls below dew point. The dew point is the temperature at which moisture begins to condense, in the form of tiny droplets or dew. Any reduction in temperature below the dew point will result in condensation of some of the water vapor present in the air.

The attic is one place where condensation often occurs. In the winter, the warm air from the occupied spaces leaks past the insulation on the top-floor ceiling and comes into contact with the cooler air in the attic. This causes the water vapor to condense. The moisture resulting from condensation works its way down through the ceiling and walls, eventually causing damage to the insulation, wood, and other building materials.

The solution to condensation in the attic is twofold. Adequate ventilation should be provided in the attic to remove the humid air. This can be accomplished by installing louvers at each end of the attic or in roof overhangs to provide cross ventilation (Fig. 3-2). The louvers should have a total area of 1 sq. ft. for each 300 sq. ft. of attic floor space. For example, if the attic has an area of 900 sq. ft., the total louver area should be at least 3 sq. ft. In addition to suitable attic ventilation, you should install a vapor barrier between the rafters. This is particularly true if the attic is occupied or used extensively.

Foundation crawl spaces are also subject to condensation. A vapor barrier between the floor joists prevents excess moisture

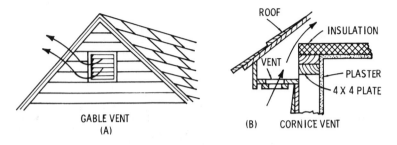

Fig. 3-2. Two methods of attic ventilation.

from accumulating and causing dampness and eventually damage to the structure. Ventilating the crawl space so that the humid air can be removed is also very effective. Finally, placing an insulating paper (e.g., roofing paper) on the ground will reduce the amount of moisture released by the soil (Fig. 3-3).

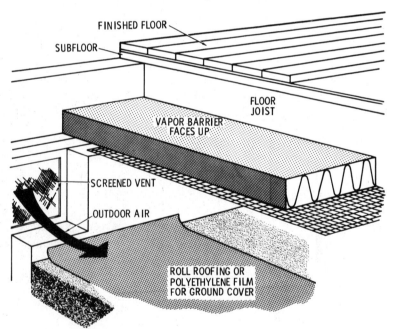

Fig. 3-3. Crawl-space construction showing vent location and vapor barrier installation.

Vapor Barriers

Insulation must be protected against moisture or it will lose its ability to reduce thermal transmission. An unprotected insulation material will absorb moisture and lose its insulating value. This can be very expensive because the insulation must be replaced completely. In most cases, there is no way to restore it to its original condition.

A *vapor barrier* is a nonabsorbent material (e.g., plastic, asphalt, metal foil, and roofing paper) designed to protect insulation from moisture or prevent moisture from entering a space (Fig. 3-4). It can completely enclose the insulation or be applied to only one side. If the latter situation is the case, the vapor barrier must be installed so that it faces inward toward the heated rooms and spaces.

In addition to protecting insulation from absorbing moisture, a vapor barrier is an effective means of retaining moisture in the rooms and spaces in the structure or preventing it from entering these areas. As a result, it functions as an important factor in maintaining interior humidity conditions.

If you add insulation over an existing vapor barrier, you should make a number of cuts in the barrier to avoid moisture entrapment. Remember, the "name of the game" is *dry insulation*. Take whatever measure necessary to protect it from moisture.

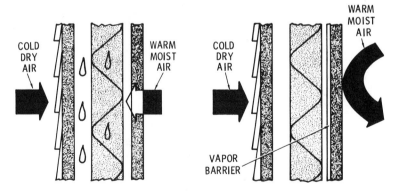

Fig. 3-4. The location of a vapor barrier in respect to insulation. Note that vapor barrier faces the warm, moist air.

INSULATING MATERIALS

Insulating materials are specifically designed to reduce the rate of heat transmission through ordinary construction materials to an acceptable level. A dry material of low density is considered a good insulator; however, in addition to this characteristic, it must also have a conductivity value of less than 0.5.

The conductivity value of a material is a purely arbitrary one determined by the amount of heat that flows through a one-inch thickness of a material one square foot in area with the temperature exactly one degree Fahrenheit higher on one side of the material than on the other.

Using the arbitrary conductivity value as a guide, the following materials are regarded as providing the best insulation:

1. Expanded polyurethane (0.17).
2. Batts and blankets (0.27).
3. Mineral wood (0.27).
4. Corkboard (0.30).
5. Insulating board (0.33).
6. Expanded perlite fill (0.35).
7. Wood fiber sheathing (0.380).
8. Vermiculite (0.48).

Air spaces or air spaces bounded by either ordinary building materials or aluminum foil also provide some insulation, but not to the degree formerly thought possible. Dead air spaces in building walls were once considered capable of preventing heat transmission in a manner similar to the space in the walls of a thermos bottle. Later research proved this to be a somewhat false analogy because the air in such spaces often circulates and transmits heat by convection.

Air circulation can be checked by filling the hollow space with an insulating material that contains a great number of small *confined* air spaces per unit volume. This stoppage of air circulation is what produces the insulating effect, and not necessarily the existence of the air space. Under these circumstances, it is obvious that the most practical method of insulation is to fill the area in the walls with a material containing these minute air spaces.

Several manufacturers produce insulating materials in a variety of shapes and forms for installation in houses and other buildings. Frequently, instructions for the installation of the products will also be provided by the manufacturer. Local building supply outlets and lumber yards will often be very helpful, too, and will usually recommend the best way to install the insulation material. Some of these materials and their applications are described in the following paragraphs.

Insulating Board

Insulating board (Fig. 3-5) is a rigid or semirigid synthetic product commonly referred to as *structural insulating board, rigid board, or sheathing board,* and produced under a variety of trade names. It is available in ½ - to 1-in. thicknesses and in board sizes generally 4 ft. by 8 ft.

Some insulating board is produced specifically for use on interior walls. This type is generally made from wood fibers, and one side is prefinished. Insulating board designed for use as structural insulation is produced from compressed fibrous materials other than wood and does not have a prefinished surface.

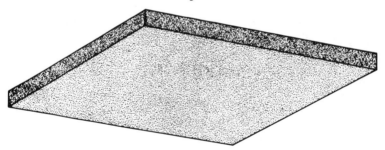

Fig. 3-5. Rigid insulation board.

Slab Insulation

Rigid *slab insulation* is designed especially for application to flat roofs. The minimum thickness of slab insulation is 1 in., with widths ranging up to 2 ft. and lengths up to 4 ft. Foamed plastic or mineral wood impregnated with asphalt is used in the production of slab insulation; each provides excellent protection against moisture.

Reflective Insulation

Reflective insulation (Fig. 3-6) is designed to reflect heat rather than resist its rate of flow. This characteristic distinguishes it from other insulating materials. Aluminum foil is commonly used as the reflective surface, although coated papers have also been employed.

Air spaces must be provided on either side of a layer of reflective insulation. This type of insulating material has proven to be a good moisture barrier and exhibits a high degree of resistance to heat transmission.

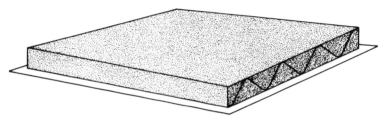

Fig. 3-6. Reflective insulation board.

Batt-Type Insulation

Batts (Fig. 3-7) are essentially smaller examples of blanket insulation, ranging in thicknesses from 1 to 6 in. and available in 4- and 8-ft. lengths.

The batts are manufactured with a vapor seal paper adhered to one face, covering the entire surface and extending 1½ in. on the long sides of the batt. These 1½-in. laps, neatly folded against the membrane backing in manufacture, are turned out and tacked or stapled to the studs, rafters, or joists in application. The paper backing helps to prevent the passage of vapor and resists the penetration of moisture from excess water in fresh plaster or other sources.

Blanket-Type Insulation

Blanket insulation (Fig. 3-8) is produced in the form of a continuous flexible strip and is generally available in rolls or packs in the following three thicknesses: thick (3 in.), medium (2 in.), and thin (1 in.). Widths are commonly 15 in., but wider strips up to 33

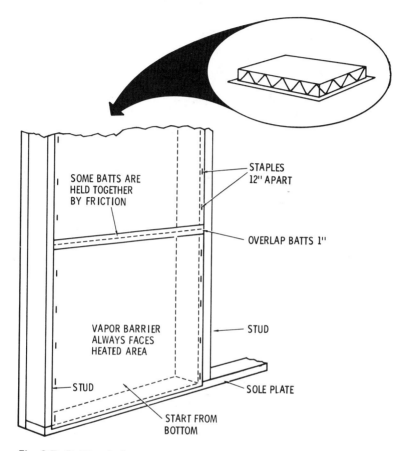

SOME BATTS ARE
HELD TOGETHER
BY FRICTION

STAPLES
12" APART

OVERLAP BATTS 1"

VAPOR BARRIER
ALWAYS FACES
HEATED AREA

STUD

STUD

SOLE PLATE

START FROM
BOTTOM

Fig. 3-7. Batt insulation.

in. can be obtained on special order. Blanket insulation lengths range from 36 to 48 ft.

Sometimes both sides of the blanket insulation strip are covered with aluminum foil or plain paper, which serves as a vapor seal. Application is made by tacking or stapling through 1½-in. laps extending from both sides of the strip.

Loose-Fill Insulation

Loose-fill insulation (Fig. 3-9) is available in granular, fibrous, or powdered form and is sold in bulk (usually in bags). It is

65

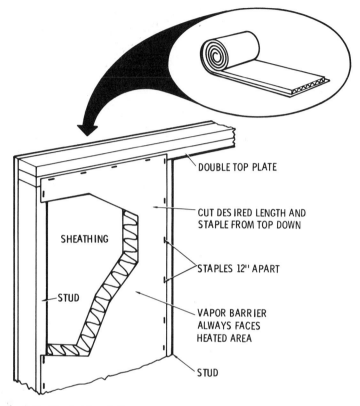

DOUBLE TOP PLATE

CUT DESIRED LENGTH AND
STAPLE FROM TOP DOWN

SHEATHING

STAPLES 12" APART

STUD

VAPOR BARRIER
ALWAYS FACES
HEATED AREA

STUD

Fig. 3-8. Blanket insulation.

commonly produced from vermiculite or other granular materials, vegetable fibers, or mineral woods and is especially suitable for pouring between floor joists.

Blown-in Insulation

Blown-in insulation (Fig. 3-10) is a loose-fill insulating material that is blown under pneumatic pressure into the spaces between joists and wall studs in an existing structure. Water vapor barriers are applied first and are available in the form of a moisture-resistant, spray-on paint. Blown-in insulation is most commonly used to insulate interior walls. Because special equipment must be used for its application, blown-in insulation should be applied by individuals trained in its use.

BUILDING CONSTRUCTION AND LOCATION

A new building (particularly a residence) should be located so that the large windows in the main rooms face south to receive the maximum sunlight during the winter months (Fig. 3-11). If possible, the building should be built in a location that offers some natural protection from the prevailing winter winds. Tight, well-insulated construction should be incorporated in the design of the building from the beginning. Although the initial costs will be somewhat higher, they will be effectively offset by the reduction in heating and cooling costs.

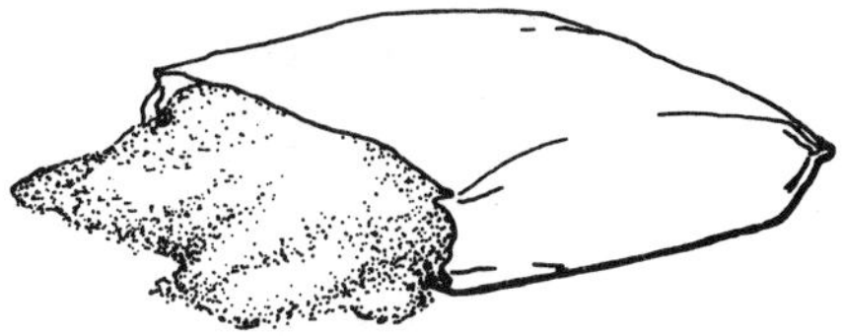

Fig. 3-9. Loose-fill insulation.

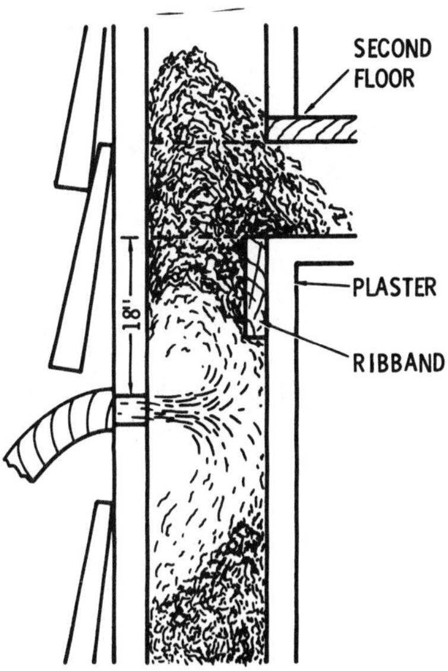

Fig. 3-10. Blown-in insulation

SECOND FLOOR

PLASTER

RIBBAND

18"

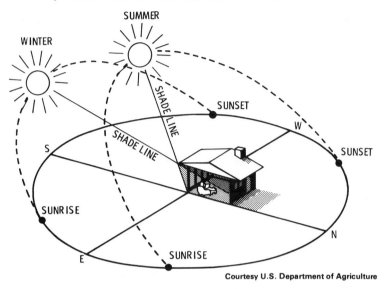

Fig. 3-11. Large glass window area should face south to take advantage of the winter sun.

If a new heating or cooling system is planned for an *older* structure, the existing insulation should be checked and, if necessary, repaired or replaced before the new system is installed.

Installing Windows and Doors

Windows and doors account for at least 30 percent of the total heat loss (or heat gain) in a structure. Much of this results from air infiltration through cracks around the windows and doors or in the walls of the structure. Weather stripping is used for window and door cracks, and calking is used for wall cracks (Fig. 3-12).

Cold draft (i.e., cold-air infiltration) and warm-air leakage can be greatly reduced if the windows and doors of a house are properly weather-stripped and caulked. This method of reducing air leakage or infiltration is much cheaper than installing storm windows and will reduce heating and cooling costs up to as high as 30 percent.

Calk the frame around all windows and doors and install weather strip along movable joints (e.g., the edges of a door). If this is done properly, a mist or fog will form on windows located on the downwind side of the structure. Not only will calk and weather stripping reduce heating and cooling costs, they will also provide high and healthy humidity conditions inside the structure. Typical methods of applying weather stripping to windows are illustrated in Fig. 3-13.

Using storm windows or windows made from insulating glass will cut heat loss by as much as 40 to 50 percent. Although the initial cost for installing storm windows may seem high, it will be more than offset by the lowered heating costs. The savings will be highest in those areas in which snow lies on the ground all winter.

Fig. 3-12. Calking around windows and weather-stripping doors.

Storm windows reduce not only heat loss in the winter but also heat *gain* in the summer. That is, they are equally effective in reducing the amount of summer heat that infiltrates the interior of the structure.

Various methods for double glazing (i.e., installing an additional layer of glass) and adding storm sash are illustrated in Figs. 3-14 and 3-15.

Overhead and Sidewall Insulation

Overhead insulation (Fig. 3-16) refers to the layer of insulation covering the top-floor ceiling. This layer should be a *minimum* of

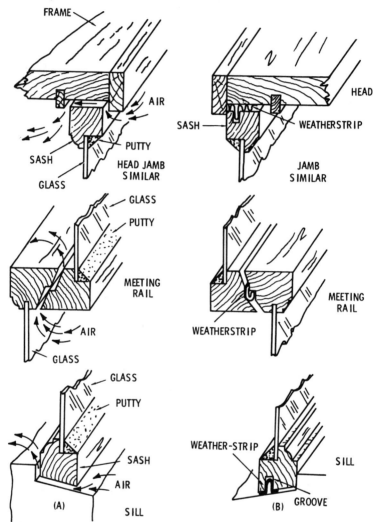

Fig. 3-13. Typical method of applying weather-stripping.

6 in. thick over the entire area. By carefully insulating this section of the structure, heating costs can be reduced by 10 to 15 percent. Air conditioning costs in the summer will also be reduced. The overhead insulation should be checked in an older structure

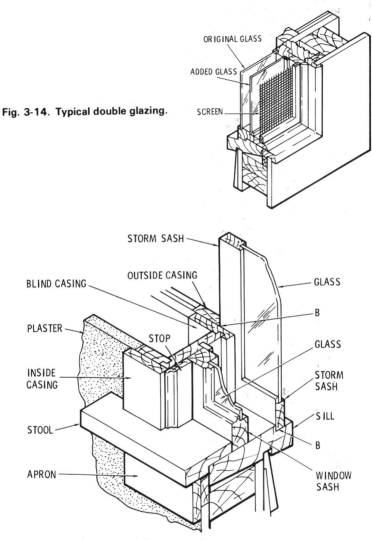

ORIGINAL GLASS

ADDED GLASS

Fig. 3-14. Typical double glazing.   SCREEN

STORM SASH

BLIND CASING

OUTSIDE CASING

GLASS

B

PLASTER

STOP

GLASS

INSIDE
CASING

STORM
SASH

STOOL

SILL

B

APRON

WINDOW
SASH

Fig. 3-15. Typical storm sash construction.

because some building contractors will only install 3 to 4 in. of the material. Adding an additional 2 to 3 in. is not only easy to do but well worth the effort.

Tacks or staples can be used to attach batt or blanket insulation

71

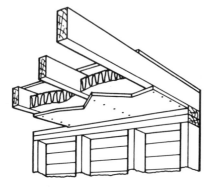

Fig. 3-16. Overhead insulation.

flanges to the wood joists supporting the ceiling. Blanket, batt, *and loose* insulation require a water vapor barrier placed between the insulating material and the ceiling lath (Fig. 3-17). Reflective insulation is attached to the *middle* of the joists so that there is an air space both above and below it. No water vapor barrier is necessary. When using reflective insulation, additional protection is required for the insulating material, which usually takes the form of lath (below) and an attic floor (above).

Sidewall insulation (Fig. 3-18) refers to insulation installed in the exterior sidewalls of a house or building. In new construction, this should be incorporated as a part of the design before ground is broken. An existing house or building presents a different problem because there is always the possibility of moisture condensation forming within the walls. If you feel you lack the expertise, then expert technical advice should be obtained before you attempt to install insulation in a sidewall.

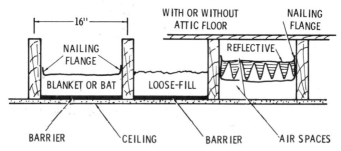

Fig. 3-17. Three methods of insulating an attic.

INSULATING ATTICS, ATTIC CRAWL SPACES, AND FLAT ROOFS

Most houses and many other structures are constructed with an attic or attic crawl space between the roof and the ceiling of the top floor. Unless this space is properly insulated and ventilated, it will produce the following two problems:

1. Condensed moisture formation.
2. Excessive heat loss or gain.

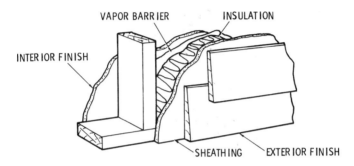

VAPOR BARRIER   INSULATION

INTERIOR FINISH

SHEATHING   EXTERIOR FINISH

Fig. 3-18. Sidewall insulation.

During the winter months, the heat from the occupied lower spaces moves upward (because warm air is lighter) into the cooler attic or attic crawl space. As it moves, the heat follows a number of different pathways, including:

1. Through poorly insulated ceilings.
2. Around cracks in attic stairway doors or attic pulldown stairways.
3. Through ceiling fixture holes.
4. Along plumbing vents and pipes.
5. Along air ducts.
6. Through air spaces within interior partitions.

Each of these pathways for heat transmission can be blocked by insulation. Typical attic insulation methods are illustrated in Fig. 3-19. Attic door cracks can be blocked with weather stripping. Pipe, duct, vent, and fixture holes can be sealed off by stuffing them with loose insulation torn from batts or blankets.

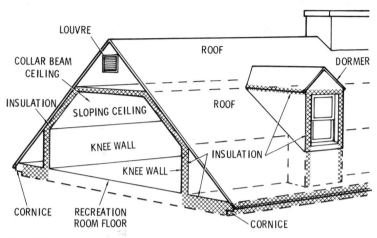

Fig. 3-19. Attic insulation methods.

The ceiling of the top floor should be covered with a minimum of 6 in. of suitable insulation material to be properly effective in reducing the heat loss from the occupied spaces during the winter months or the heat gain during the summer. This can be installed during the initial construction stages of a new house or building, or it can be added to the attic floor surface of an existing structure. It is not a particularly difficult thing to do (see "Overhead and Sidewall Insulation" in this chapter). Sealing or filling the air spaces within interior partitions is relatively easy to do if it is a question of new construction. An existing structure presents a more difficult problem, one that generally requires considerable experience in installing insulation materials.

Sometimes a half wall (i.e., a knee wall) is installed between the attic floor and the roof. This is most commonly insulated with blanket insulation, as illustrated in Fig. 3-20.

The condensation of moisture in the attic or attic crawl space is caused by the moisture in the warm air rising from the lower occupied spaces. This moisture comes in contact with the colder air of the attic, condenses, and causes the insulation and other building materials to become damp. In time, this dampness will have a damaging effect. Ventilation is an effective method of reducing or eliminating condensation in attics and attic crawl spaces. The air enters and leaves through vents placed in the gables and cornices. The size of the vents will be determined by the area being ventilated.

A flat roof can be effectively insulated by installing 2-in.-thick rigid insulation board before the roofing material is applied. Roofing felt should be placed between the bare wood boards of the roof deck and the insulation. Additional insulation is placed *under* the deck boards (i.e., between the deck boards and the ceiling of the room below). (See Fig. 3-21.)

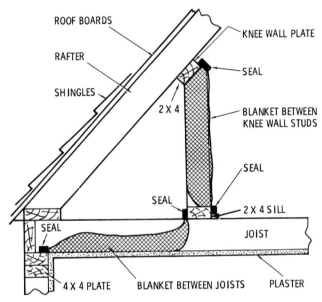

ROOF BOARDS
KNEE WALL PLATE
RAFTER
SEAL
SHINGLES
2 X 4
BLANKET BETWEEN KNEE WALL STUDS
SEAL
SEAL
2 X 4 SILL
SEAL
JOIST
SEAL
4 X 4 PLATE
BLANKET BETWEEN JOISTS
PLASTER

Fig. 3-20. Insulating a knee wall with blanket insulation.

FOUNDATION CRAWL SPACES

Some houses and buildings are constructed so that the first floor is raised slightly above the ground, leaving a crawl space between the two. These foundation crawl spaces can be effectively insulated by attaching blanket or batt insulation between the joists supporting the floor or by nailing rigid insulation to the bottom of the joists (Fig. 3-22). Placing waterproof paper on the surface of the ground in the crawl space and extending it up

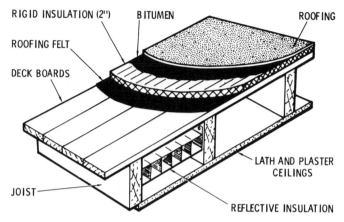

RIGID INSULATION (2")   BITUMEN   ROOFING

ROOFING FELT

DECK BOARDS

LATH AND PLASTER CEILINGS

JOIST

REFLECTIVE INSULATION

Fig. 3-21. Insulating a flat roof.

along the inside of the foundation walls will increase the effectiveness of the insulation.

Condensation can also occur in foundation crawl spaces. The accumulation of moisture resulting from condensation can be prevented or effectively reduced by providing adequate ventilation. Foundation crawl-space vents should be placed in the sidewalls.

SLAB CONSTRUCTION INSULATION PROBLEMS

Houses and buildings constructed on concrete slab are in direct contact with the ground. Unless they are properly insulated, the slab can become extremely cold and damp. Some form of insulation should be placed between the slab and the ground before the concrete is poured (Fig. 3-23). Insulation can be added to an existing structure by excavating around the edges of the foundation to a level below the frost line and adding 2 in. of waterproof insulating material to the sides of the slab (Fig. 3-24).

INSTALLING BATTS AND BLANKETS

Batt or blanket insulation is usually available faced with paper on both the front and back surfaces. Paper flanges extending along both sides of the batt or blanket are used for nailing or stapling the insulation to studs or joists.

Some batts or blankets are manufactured with a vapor barrier on one side, usually in the form of asphalt-impregnated paper. When this is the case, the surface with the vapor barrier should be

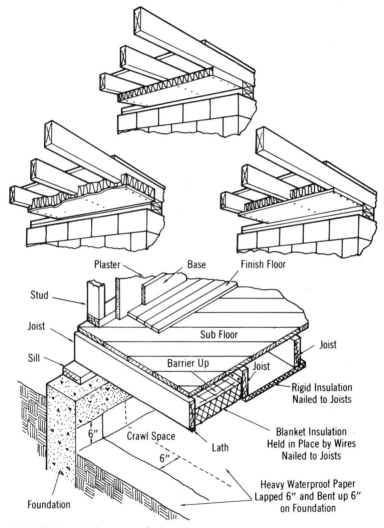

Fig. 3-22. Insulating a crawl space.

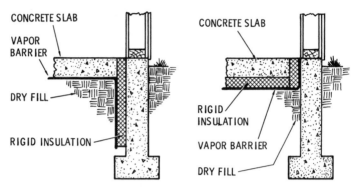

Fig. 3-23. Insulating a concrete slab.

installed so that it faces inward toward the heated rooms and spaces. For example, when installing insulation between top-floor ceiling joists, place the insulation so that the surface with the vapor barrier faces downward toward the heated occupied rooms and spaces below.

When the batt or blanket insulation is installed directly over an insulating board panel ceiling, ordinary wood lath spaced approximately 6 in. on centers should be laid across the furring strips to support the insulation.

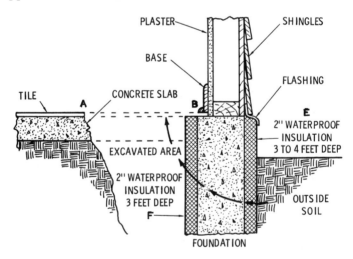

Fig. 3-24. Insulating the foundation on concrete slab construction.

Insulation should be secured in place by tacking or stapling through the paper flanges extending along the sides of the batt or blanket insulation to the frame members. Space fasteners on 6-in. centers. Where lath is to be applied after insulation has been installed, fasteners sufficient only to hold the insulation in place temporarily need to be used. Overlap the flanges on the face of each stud or joist to be certain there will be no break in the vapor barrier.

Insulation should be placed *under* the bridging between joists on ceilings if there is sufficient clearance. If there is insufficient clearance, cut the insulation and fit it around the bridging, but extend the vapor barrier under it. Stuff handfuls of insulation in the spaces in the bridging. Remember to place the vapor barrier so that it faces down toward the occupied spaces below.

Start tacking or stapling insulation at the plate at the top of walls and work down toward the floor. Compress the insulation so that it extends around the *back* of pipes, ducts, outlet boxes, electrical receptacles, and other structural components that must remain exposed to the room (i.e., the interior) side of the walls. If it is not possible to extend the batt or blanket insulation behind them, then cut (or tear) the insulation and work it around the back, making certain the fit is as tight as possible. Be particularly careful to refit the vapor barrier so that there is no break.

Floors can be protected with a vapor barrier over the subfloor or between the floor joists and the subfloor. Waterproof roofing paper should be placed over the ground in foundation crawl spaces and extended up the insides of the exterior foundation walls. The exterior walls of crawl spaces should also be insulated with 2 to 3 in. of insulation. This will reduce the amount of cold air seepage entering the foundation crawl-space area.

APPLYING BLOWN INSULATION

Blown insulation (not batt or blanket insulation) is recommended for insulating the exterior walls of *existing* structures. It requires the use of pneumatic equipment and the knowledge of where to drill access holes in the exterior walls and how much insulation to use to completely fill each stud space.

The equipment and tools required for blown insulation are:

1. A blowing machine with hoses.
2. Electric drills.
3. Circular saws.
4. Screwdrivers.

The circular saws should have teeth set for wood cutting. The screwdrivers are used for removing plugs for the saws (if not self-ejecting).

The insulation, when blown into the walls and attic spaces with compressed air, fills every space with heat-resistant material. This method is the only practical way of insulating structures already constructed, whether of wood, brick or stone veneer, or stucco.

The insulation is usually blown into the walls at 2 psi pressure or less. This pressure gives a firm, even pack, which ensures

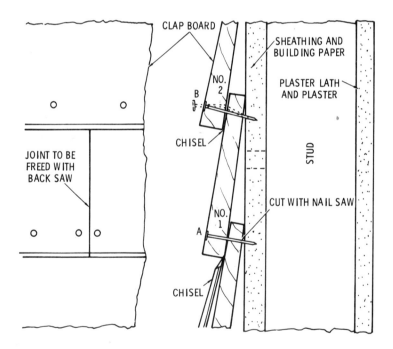

Fig. 3-25. Removing clapboard siding.

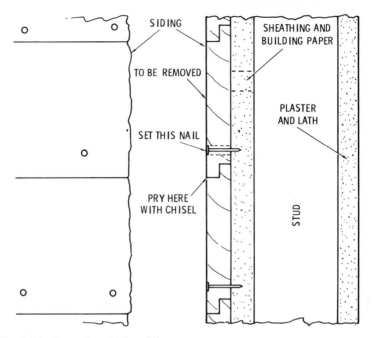

Fig. 3-26. Removing shiplap siding.

maximum thermal efficiency. Furthermore, it puts the material under an initial compression so that any subsequent vibration or building movement will not cause the insulation to settle; rather, it will expand and thus retain its full insulating value.

The insulation work, for the most part, is done from the outside of the building with a minimum of litter, dust, or disturbance inside the building.

The openings for blowing are made near the top of wall panels. The insulation is installed by inserting the hose nozzle of the blowing machine into each hole and blowing the material into the panel.

Before starting to blow the insulation into the interior spaces of the walls, soundings should be made with a plumb bob to locate cross framing. Where such framing is found, additional openings are made to ensure that all spaces will be filled.

For the third story of houses having one or more rooms above the ceiling joists of the second floor, it is necessary to get between

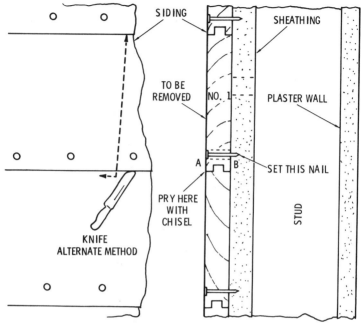

Fig. 3-27. Removing tongue-and-groove siding.

the roof and the walls of the upper-story room and insulate the sidewalls (kneewalls) of the room from behind. This is done by boarding up the back of the studs with insulating board or by nailing diamond crosscord paper or a similar material across the back of the studs and blowing behind it. It may be necessary to make an opening in the roof large enough to allow workmen to enter. The entire ceiling, including the portion from the wall line of the eaves to the walls of the third-story room, is insulated by blowing.

All portions around windows should be insulated thoroughly, as considerable air infiltration takes place around window framing.

In the case of open joists on unflooring attics, the operator of the nozzle stands on the joists and sprays the material to the desired thickness between the joists. Because these areas are not blown under pressure, there is no settlement problem. Therefore, the one objective is to lay the material as evenly as possible in order to eliminate the extra labor of leveling or screening by hand.

If the attic is floored, it is necessary to remove a few boards and insert the hose between each pair of joists, pushing the hose in as far as necessary to fill the space to the eave line. The material is then blown, and as the space fills up to the required depth, the hose is slowly withdrawn.

It is recommended, as a general practice, that the size of holes made for blowing the insulation material be no greater than 3 in. and no less than 2 in. in diameter. A $2\frac{1}{2}$-in.-diameter hole is sufficient to take the entire output from a $2\frac{1}{2}$-in.-diameter hose.

Removing building materials from the exterior walls in order to drill holes for blowing the insulation requires the skill of an experienced worker. Each type of construction (e.g., clapboard, wood siding shingles, brick veneer) presents a different problem. For example, Figs. 3-25 to 3-28 illustrate the problems involved in

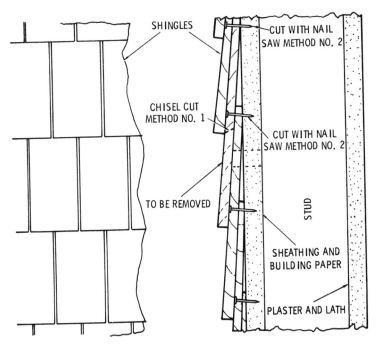

Fig. 3-28. Removing wood shingle siding.

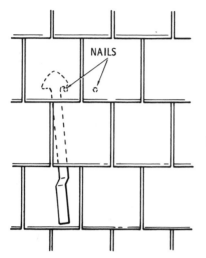

Fig. 3-29. Removing a shingle with a ripper.

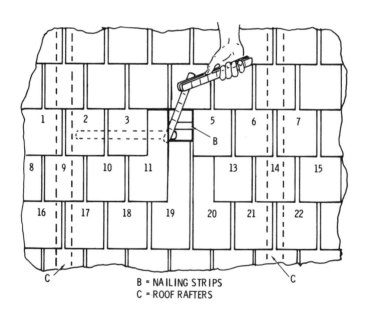

B = NAILING STRIPS
C = ROOF RAFTERS

Fig. 3-30. Sequence suggested for removing shingles.

removing clapboard, siding joined with a shiplap, siding joined with a tongue-and-groove, and wood siding shingles.

A shingle should be removed by inserting a shingle ripper under the course *above* the shingles to be removed (Fig. 3-29). The upper and lower course of nails is cut by hammering the ripper downward and upward. Then insert the ripper under the shingle to be removed and cut two sets of nails, upper and lower, by pulling or hammering downward on the ripper.

Removing shingles should be done with a definite sequence in mind. This is illustrated in Fig. 3-30. For example, if five shingles are to be removed from the top row, the number should start at the left. Shingle 1 remains in the roof; shingles 2 to 6 are removed; and 7 remains on the roof. Shingles 1, 7, 8, 15, 16, and 22 remain in place. Remove 2 to 6, 9 to 14, and 17 to 21 for opening (more if necessary).

Two methods for removing brick veneer are illustrated in Fig. 3-31. Care must be taken to match the color of the old mortar when refilling the hole.

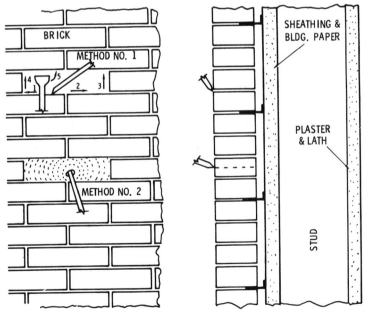

Fig. 3-31. Two methods used for removing brick veneer.

The method of blowing exterior stud walls, after the removal of that part of the outside surface essential for the preparation of holes for blowing, is usually a standard procedure. It is recommended that no greater vertical height than 7 ft. be blown from any one wall opening. Blowing upward from any opening should not exceed a height of 2 ft.

CHAPTER 4

Heating Calculations

The size of the heating system is directly related to the amount of heat lost from the hose or building. *All* structures lose heat to the outdoors or to adjacent unheated or partially heated spaces when the temperatures of the outdoor air or adjacent spaces are colder than those inside the structure. The heat within the building is normally lost by transmission through the building materials and by infiltration around doors and windows.

The loss of heat from a structure must be replaced at the same rate that it is lost. Consequently, determining the correct size of the heat system and the rated capacity of the heating plant required by the steam are very important. It should be obvious that an oversized heating plant will cost more to install than a smaller one and will provide more heat than the structure requires. On the other hand, an undersized heating plant lacks the capacity to provide sufficient heat, particularly during cold spells. The heating system and the selection of the heating plant must be carefully planned in order to adequately and efficiently replace the lost heat.

This chapter describes several methods for calculating heat loss, ranging from rule-of-thumb methods to the more precise method of using overall coefficients of heat transmission (*U*-values) computed for the various construction materials and combinations of construction materials through which heat is commonly transmitted.

USING COEFFICIENTS OF HEAT TRANSMISSION

The more precise methods of calculating heat loss from a structure require a thorough knowledge of the thermal properties of many materials and combinations of materials used in its construction. The term "thermal property" is used here to mean the overall coefficient of heat transmission (i.e., the rate of heat flow through a material). Each construction material (or combination of materials) will have its own coefficient of heat transmission. Before continuing any further, it would be advisable to review the sections of Chapter 3 (Insulation Principles) that specifically pertain to the problem of heat loss (e.g., see "Principles of Heat Transmission," "Thermal Conductance," and "Thermal Resistance").

ASHRAE and other authorities suggest the following basic steps (in the sequence given) for calculating heat loss:

1. Decide upon the desired inside air temperature for the structure.
2. Obtain the winter outside design temperature for the location of the structure from published lists or a local weather bureau.
3. Determine the design temperature difference (the difference between the temperatures found in Steps 1 and 2).
4. Identify on the heat loss worksheet each room or space in the structure.
5. List every structural section in each identified room or space

that has an outer surface exposed to the outdoors or to an unheated or partially heated space.

6. Determine the coefficient of heat transmission (U-value) for each structural section (e.g., walls, glass, ceiling).
7. Calculate the infiltration heat loss for each identified room or space.
8. Calculate the *total* area for each exposed surface (i.e., total outside wall area, total floor area).
9. Calculate the area for each door and window in the walls and add these figures together to obtain the total area for these openings.
10. Subtract the total area for doors and windows from the *gross* outside wall area (obtained in Step 8) to determine the total *net* outside wall area. Enter the amount on the heat loss worksheet.
11. Multiply the total net wall area determined in Step 10 by the U-value (Step 2) by the design temperature difference (Step 3) to obtain the total heat loss for walls (expressed in Btuh).
12. Make the same calculations for the other surface areas (floors, ceilings, etc.) determined in Step 8.
13. Add the heat loss figures calculated for each surface category and the infiltration heat loss to obtain the total heat loss for the identified room or space.
14. Repeat the procedure outlined in Steps 1 to 11 for each identified room or space in the structure.
15. Add the various totals to obtain the total heat loss from the structure (expressed in Btuh).

These fifteen basic steps for calculating heat loss are provided as a useful means of reference for the more detailed description of the procedure contained in the paragraphs that follow.

OUTSIDE DESIGN TEMPERATURE

In heating calculations, the *outside design temperature* is the coldest outside temperature expected for a *normal* heating season. It is *not* the coldest temperature on record, but the lowest one recorded for a particular locale over a three- to five-year period.

Lists of outside design temperatures are published for selected localities throughout the United States (Table 4-1). If a locality is

Table 4-1. Winter Outside Design Temperatures for Major Cities in the United States

State	City	Outside Design Temperature Commonly Used	State	City	Outside Design Temperature Commonly Used
Alabama	Birmingham	10	New Hampshire	Concord	−15
Arizona	Tucson	25	New Jersey	Atlantic City	5
Arkansas	Little Rock	5		Trenton	0
California	San Francisco	35	New Mexico	Albuquerque	0
	Los Angeles	35	New York	Albany	−10
Colorado	Denver	−10		Buffalo	−5
Connecticut	New Haven	0		New York City	0
Dist. of Columbia	Washington	0	North Carolina	Asheville	0
Florida	Jacksonville	25		Charlotte	10
	Key West	45	North Dakota	Bismarck	−30
Georgia	Atlanta	10	Ohio	Akron	0
	Savannah	20		Cincinnati	0
Idaho	Boise	−10		Columbus	−10
Illinois	Cairo	0	Oklahoma	Oklahoma City	0
	Chicago	−10	Oregon	Portland	10
Indiana	Indianapolis	−10	Pennsylvania	Erie	−5
Iowa	Des Moines	−15		Harrisburg	0
	Sioux City	−20		Philadelphia	0
Kansas	Topeka	−10	Rhode Island	Providence	0
Kentucky	Louisville	0	South Carolina	Charleston	15
Louisiana	New Orleans	20	South Dakota	Huron	−20
Maine	Portland	− 5	Tennessee	Knoxville	0
Maryland	Baltimore	0	Texas	Abilene	15
Massachusetts	Boston	0		Austin	20
Michigan	Detroit	−10		Brownsville	30
	Escanaba	−15		Corpus Christi	20
	Sault Ste.Marie	−20		Dallas	0
Minnesota	Duluth	−20		Houston	20
	Minneapolis	−20	Utah	Salt Lake City	−10
Mississippi	Vicksburg	10	Vermont	Burlington	−10
Missouri	Kansas City	−10	Virginia	Lynchburg	5
	St. Louis	0		Norfolk	15
Montana	Billings	−25	Washington	Seattle	15
	Helena	−20		Spokane	−15
	Miles city	−35	West Virginia	Parkersburg	−10
Nebraska	Lincoln	−10	Wisconsin	Green Bay	−20
	Valentine	−25		Madison	−15
Nevada	Reno	− 5	Wyoming	Cheyenne	−15

not included on the list of winter outside design temperatures, check with a local weather bureau. Using an outside design temperature from the nearest locality on the list can be misleading because even closely located cities can differ widely in weather conditions as a result of different altitudes, the effects of large bodies of water, and other variables.

INSIDE DESIGN TEMPERATURE

The desired inside design temperature will depend upon the intended use of the space. In a large structure, such as a hotel or a hospital, there will be more than one inside design temperature because there is more than one type of space usage. For example, hospital kitchens will generally have an inside design temperature of about 66°F. The wards, on the other hand, will range from 72 to 74°F.

Residences will generally have a single inside design temperature for the entire structure, with 70 or 71°F the most commonly used temperatures.

DESIGN TEMPERATURE DIFFERENCE

The *design temperature difference* is the variation in degrees Fahrenheit between the outside and inside design temperatures. It is used in the heat transmission loss formula (see below) and is a crucial factor in heating calculations.

Be careful that you obtain the degree *difference* between the two temperatures; do not simply subtract the smaller figure from the larger one. For example, if the outside and inside design temperatures are −20 and 70°F, respectively, the design temperature difference will be 90° (20° below zero plus 70° above zero).

DETERMINING COEFFICIENTS OF HEAT TRANSMISSION

The *overall coefficient of heat transmission*, or U-value, as it is commonly designated, is a specific value used for determining

91

the amount of heat lost from various types of construction. It represents the time rate of heat flow and is expressed in Btu per hour per square foot of surface per degree Fahrenheit temperature difference between air on the inside and air on the outside of a structural section. Furthermore, the U-value is the reciprocal of the total thermal resistance value (R-value) of each element of the structural section and may be expressed as:

$$U = \frac{I}{R_t}$$

Professional societies such as the American Society of Heating, Refrigerating, and Air-Conditioning Engineers (ASHRAE) have already determined the U-values for a wide variety of floor, ceiling, wall, window, and door construction. Tables of these U-values are made available through ASHRAE publications (e.g., the *ASHRAE 1972 Handbook of Fundamentals*) found in many libraries. Many manufacturers of heating equipment also provide tables of U-values in their literature. When calculating heat loss, you have the option of selecting U-values from the tables provided by manufacturers and certain professional associations or computing them yourself. The latter method, if done correctly, is the more precise one.

CALCULATING NET AREA

Having determined the design temperature difference and computed (or selected) an overall heat transmission coefficient for each construction material or combinations of materials, you are now ready to calculate the net area of each surface exposed to the outside or adjacent to an unheated or partially heated space.

The best procedure for calculating surface area is to work from a building plan. If one is not available, you will have to make your own measurements. Measurements for calculating net area are taken from *inside* surfaces (i.e., inside room measurements). You will not be concerned with structural surfaces (e.g, walls, ceilings, floors) between rooms and spaces heated at the same temperature because no heat transmission occurs where temperatures are constant.

Calculating the total area for each surface should be done as follows:

1. Multiply room length by room width to determine floor and ceiling area.
2. Multiply room length (or width) by room height to determine the outside wall area for each room.
3. Multiply door width by door height to determine the surface area for each door.
4. Multiply window width by window height to determine the surface area for each window.

Whether you use room length or room width in calculating the outside wall area will depend upon which wall surface is exposed to the outside. In some cases (e.g., corner rooms), both are used and require at least two separate calculations (i.e., room length × room height, and room width × room height).

Add the calculated surface area of each outside wall (see Step 2 in "Using Coefficients of Heat Transmission") to obtain the *gross wall area* for the structure. Subtract the sum of all door and window surface areas from the gross wall area. The result will be the *net wall area* for the structure. Multiply the net wall area by the heat loss in Btu per hour per square foot to calculate the heat loss through the walls.

HEAT TRANSMISSION LOSS FORMULA

The heat loss (expressed in Btu) of a given space is determined by multiplying the coefficient of heat transmission by the area in square feet by the design temperature difference (i.e., the difference between the indoor and the outdoor design temperatures). Because a heating system must supply an amount of heat equal to the amount of heat lost in order to maintain a constant indoor design temperature, heat loss is approximately equal to heat required. This may be expressed by the following formula:

$$H_t = AU (t_i - t_o)$$

Where: H_t = heat loss transmitted through a structural section (e.g., roof, floor, ceiling) expressed in Btu per hour, representing both the heat lost and the heat required.

A = area of structural components in square feet.
U = overall coefficient of heat transmission.
T_o = outdoor design temperature.
T_i = indoor design temperature.

COMPUTING TOTAL HEAT LOSS

The commonly accepted procedure for calculating the total heat loss from a structure is to calculate the heat loss for each room or space *separately* and then add the totals.

The heat loss calculation worksheet you use should contain a column in which each room or space can be identified separately. Under the identification are listed those structural sections through which heat transmission losses occur. When applicable, these will include all or most of the following:

1. Walls.
2. Glass.
3. Ceiling.
4. Floor.
5. Door(s).

In addition to the five structural sections listed above, each identified room or space should also include a line for heat loss by air infiltration. Fig. 4-1 illustrates how your worksheet should appear at this point.

TYPE OF ROOM OR SPACE	TYPE OF STRUCTURAL SECTION	
BEDROOM NO. 1	WALLS GLASS AREA CEILING FLOOR DOOR AREA INFILTRATION	

Fig. 4-1. Tabulation worksheet.

Now that you have identified the room or space, listed the structural sections through which heat loss occurs, and calculated the rate of air infiltration, you must determine the net surface area and the *U*-value for each structural section. The design temperature difference must also be determined and entered for each structural section. Except where surfaces are exposed to partially heated spaces (e.g., a garage, attic, or basement), each structural section will have the same design temperature difference. An example of how your worksheet should look at this point is illustrated in Fig. 4-2.

TYPE OF ROOM OR SPACE	TYPE OF STRUCTURAL SECTION	NET AREA/ AIR VOLUME	COEFFICIENT	TEMP. DIFF.	HEAT LOSS (BTU/h)
BEDROOM	*WALLS*	*235 SQ.FT.*	*0.26*	*90*	*5599*
NO.1	*GLASS AREA*	*45 SQ.FT.*	*0.45*	*90*	*1823*
	CEILING	*300 SQ.FT.*	*0.17*	*40*	*2040*
	FLOOR	—	—		
	DOOR AREA	—			
	INFILTRATION	*1800 CFH*	*0.018*	*90*	*2916*

Fig. 4-2. Tabulation worksheet.

Each identified room or space should also include the calculated air infiltration heat loss (see "Infiltration Heat Loss" below). The amount of heat loss due to air infiltration will depend upon the size, type, and number of windows and doors and other variables.

The bedroom given as an example in Fig. 4-1 has two exposed walls (8 ft. × 15 ft. and 8 ft. × 20 ft.). The number of air changes suggested for a room with two exposed walls is 1½ changes per hour.

The volume of air infiltration for the bedroom can be calculated as follows:

$$\frac{\text{Vol.}}{\text{of air}} = \frac{\text{no.}}{\text{of air}} \times \frac{\text{ceiling area} \times \text{ceiling height}}{2}$$

$$= 1.5 \times \frac{300 \text{ sq. ft.} \times 8 \text{ ft.}}{2}$$

Knowing that the volume of air infiltration is 1800 cu. ft./hr., the heat loss in Btuh can be calculated as follows:

$$\text{Heat loss} = 0.018 \times Q(t_i - t_o)$$
$$= 0.018 \times 1800 \times 90$$
$$= 2916 \text{ Btuh}$$

The *total* heat loss for bedroom No. 1 (as it would be designated on the worksheet) is 12,378 Btuh. If there is a door in one of the outside walls, its heat loss would also have to be calculated and entered on the worksheet. Furthermore, the door area (in square feet) would have to be subtracted from the surface area for the walls, because the latter is always a *net* figure.

After calculating the heat loss for each room or space in the structure, the results are added to obtain the total heat loss (in Btuh) for the structure.

Loss in Doors and Windows

It is not easy to compute the coefficient of heat transmission (U-value) for a door because door construction varies considerably in size, thickness, and material. A fairly accurate rule-of-thumb method is to use the U-value for a comparable size *single-pane* glass window. If you wish to compute the coefficient of heat transmission for a door, the necessary data can be found in ASHRAE publications. The same source can be used for computing the coefficients of heat transmission of windows.

Loss in Basements

The amount of heat lost from a basement will depend upon a number of different variables. An important consideration, for example, is ground temperature, which will function as the "outdoor" design temperature in the heat loss transmission formula (see above). The ground temperature will vary with geographical location (Table 4-2).

The usual method for calculating heat loss transmission through basement walls is to view them as being divided into two sections (Fig. 4-3). The upper section extends from the frost line to the basement ceiling and includes those portions of the wall exposed to outdoor air temperatures. The heat loss for this section of the

Table 4-2. Ground Temperatures below the Frost Line

State	City	Ground Temperature Commonly Used	State	City	Ground Temperature Commonly Used
Alabama	Birmingham	66	Nevada	Reno	52
Arizona	Tucson	60	New Hampshire	Concord	47
Arkansas	Little Rock	65	New Jersey	Atlantic City	57
California	San Francisco	62	New Mexico	Albuquerque	57
	Los Angeles	67	New York	Albany	48
Colorado	Denver	48		New York City	52
Connecticut	New Haven	52	North Carolina	Greensboro	62
Dist. of Columbia	Washington	57	North Dakota	Bismarck	42
Florida	Jacksonville	70	Ohio	Cleveland	52
	Key West	78		Cincinnati	57
Georgia	Atlanta	65	Oklahoma	Oklahoma City	62
Idaho	Boise	52	Oregon	Portland	52
Illinois	Cairo	60	Pennsylvania	Pittsburgh	52
	Chicago	52		Philadelphia	52
	Peoria	55	Rhode Island	Providence	52
Indiana	Indianapolis	55	South Carolina	Greenville	67
Iowa	Des Moines	52	South Dakota	Huron	47
Kentucky	Louisville	57	Tennessee	Knoxville	61
Louisiana	New Orleans	72	Texas	Abilene	62
Maine	Portland	45		Dallas	67
Maryland	Baltimore	57		Corpus Christi	72
Massachusetts	Boston	48	Utah	Salt Lake City	52
Michigan	Detroit	48	Vermont	Burlington	46
Minnesota	Duluth	41	Virginia	Richmond	57
	Minneapolis	44	Washington	Seattle	52
Mississippi	Vicksburg	67	West Virginia	Parkersburg	52
Missouri	Kansas City	57	Wisconsin	Green Bay	44
Montana	Billings	42		Madison	47
Nebraska	Lincoln	52	Wyoming	Cheyenne	42

basement wall is calculated in the same way that other surfaces exposed to the outside air are calculated. The section of the wall extending from the frost line to the basement floor and the floor itself can be calculated on the basis of the ground temperature. The ground temperature is substituted for the outside design temperature when calculating heat loss.

Loss in Slab Construction

The heat loss for houses and small buildings constructed on a concrete slab at or near grade level is computed on the basis of heat loss per foot of exposed edge. For example, a concrete slab

97

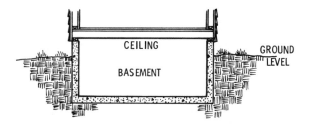

Fig. 4-3. Basement cross section.

measuring 20 ft. × 25 ft. would have an exposed edge of 90 ft. This represents the measurement completely around the perimeter of the exposed edge of the slab. The heat loss will depend upon the thickness of insulation along the exposed edge and the outside design temperature *range*. This type of information is available from ASHRAE publications. To obtain the heat loss in Btuh, simply multiply the total length of exposed edge by the heat loss in Btuh per lineal foot. For example, a 90-ft. exposed edge with 2-in.-edge insulation at an outdoor design temperature of 35° would be calculated as follows:

$$90 \text{ lin. ft.} \times 45 \text{ Btuh/lin. ft.} = 4050 \text{ Btuh}$$

Infiltration Heat Loss

During the heating season, a portion of heat loss is due to the infiltration of cooler outside air into the interior of the structure through cracks around doors and windows and other openings that are not a part of the ventilating system. The *amount* of air entering the structure by infiltration is important in estimating the requirements of the heating system, but the *composition* of this air is equally important.

A pound of air is composed of both dry air and moisture particles, which are combined (*not* mixed) so that each retains its individual characteristics. The distinction between these two basic components of air is important because each is involved with a different type of heat: dry air with specific heat, and moisture content with latent heat.

The heating system must be designed with the capability of

warming the cooler infiltrated *dry air* to the temperature of the air inside the structure. The amount of heat required to do this is referred to as the *sensible heat loss,* and is expressed in Btuh. The two methods used for calculating heat loss by air infiltration are: (1) the crack method and (2) the air-change method.

The Crack Method. The crack method is the most accurate means of calculating heat loss by infiltration because it is based on actual air leakage through cracks around windows and doors and takes into consideration the expected wind velocities in the area in which the structure is located. The air-change method (see below) does not consider wind velocities, which makes it a less accurate means of calculation.

Calculating heat loss by air infiltration with the crack method involves the following basic steps:

1. Determine the type of window or door (see Table 4-3).
2. Determine the wind velocity and find the air leakage from Table 4-3.
3. Calculate the lineal feet of crack.
4. Determine the design temperature difference.

The data obtained in these four steps are used in the following formula:

$$H = 0.018 \times Q(t_i - t_o) \times L$$

where H = heat loss, or heat required to raise the temperature of air leaking into the structure to the level of the indoor temperature (t_i) expressed in Btu per hour.

Q = volume of air entering the structure expressed in cubic feet per hour (Step 2 above).

t_i = indoor temperature.

t_o = outdoor temperature.

0.018 = specific heat of air (0.240) times density of outdoor air (approximately 0.075).

L = lineal feet of crack.

Determine the infiltration heat loss per hour through the crack

Table 4-3. Infiltration Rate through Various Types of Windows

Expressed in cubic feet per foot of crack per hour. The Infiltration rate through cracks around closed doors is generally estimated at twice that calculated for a window

Type of Window	Remarks	Wind Velocity, Miles per Hour					
		5	10	15	20	25	30
Double-Hung Wood Sash Windows (Unlocked)	Around frame in masonry wall—not calked	3	8	14	20	27	35
	Around frame in masonry wall—calked	1	2	3	4	5	6
	Around frame in wood-frame construction	2	6	11	17	23	30
	Total for average window, non-weather-stripped, $1/16$-in. crack and $3/64$-in. clearance. Includes wood frame leakage	7	21	39	59	80	104
	Ditto, weather-stripped	4	13	24	36	49	63
	Total for poorly fitted window, non-weather-stripped, $3/32$-in. crack and $3/32$-in. clearance. Includes wood frame leakage	27	69	111	154	199	249
	Ditto, weather-stripped	6	19	34	51	71	92
Double-Hung Metal Windows	Non-weather-stripped, locked	20	45	70	96	125	154
	Non-weather-stripped, unlocked	20	47	74	104	137	170
	Weather-stripped, unlocked	6	19	32	46	60	76
Rolled Section Steel Sash Windows	Industrial pivoted, $1/16$-in. crack	52	108	176	244	304	372
	Architectural projected, $1/32$-in. crack	15	36	62	86	112	139
	Architectural projected, $3/64$-in. crack	20	52	88	116	152	182
	Residential casement, $1/64$-in. crack	6	18	33	47	60	74
	Residential casement, $1/32$-in. crack	14	32	52	76	100	128
	Heavy casement section, projected, $1/64$-in. crack	3	10	18	26	36	48
	Heavy casement section, projected, $1/32$-in. crack	8	24	38	54	72	92
Hollow Metal, Vertically Pivoted Window		30	88	145	186	221	242

Courtesy *ASHRAE 1960 Guide*

of a 3 ft. × 5 ft. average double-hung, non-weather-stripped, wood window based on a wind velocity of 20 mph. The indoor temperature is 70°F, and the outdoor temperature 20°F.

The air leakage for a window of this type at a wind velocity of 20 mph is 59 cu. ft. per foot of crack per hour. This will be the value of Q in the air infiltration formula. The lineal feet of the crack is (2 × 5) plus (3 × 3), or 19 ft. (the value of L in the formula). t_i = 70°F, and t_o = 20°F. Substituting these data in the air infiltration formula gives the following results:

$$
\begin{aligned}
H &= 0.018 \times Q(t_i - t_o) \times L \\
&= 0.018 \times 59(70 - 20) \times 19 \\
&= 1.062 \times 50 \times 19 \\
&= 1008.9 \text{ Btuh} \\
&= 1009 \text{ Btuh}
\end{aligned}
$$

Table 4-4 represents a typical wall infiltration chart for a number of different types of wall construction. As is the case with air infiltration through windows, it is also based on wind velocity (see Table 4-3).

Table 4-4. Infiltration through Various Types of Wall Construction

Type of Wall	Wind Velocity, Miles per Hour					
	5	10	15	20	25	30
Brick Wall 8½ in. Plain Plastered	2 0.02	4 0.04	8 0.07	12 0.11	19 0.16	23 0.24
Plain 13 in. Plastered	1 0.01	4 0.01	7 0.03	12 0.04	16 0.07	21 0.10
Frame Wall, Lath and Plaster	0.03	0.07	0.13	0.18	0.23	0.26

Courtesy *ASHRAE 1960 Guide*

Air-Change Method. In the air-change method, the amount of air leakage (i.e., infiltration) is calculated on the basis of an assumed number of air changes per hour per room. The number of air changes will depend upon the *type* of room and the number of walls exposed to the outdoors. Table 4-5 is an example of a typical air-change chart used in the air-change method.

Table 4-5. Fresh Air Requirements

Type of Building or Room	Minimum Air Changes per Hour	Cubic Feet of Air per Minute per Occupant
Attic spaces (for cooling)	12–15	
Boiler room	15–20	
Churches, auditoriums	8	20–30
College classrooms		25–30
Dining rooms (hotel)	5	
Engine rooms	4–6	
Factory buildings (ordinary manufacturing)	2–4	
Factory buildings (extreme fumes or moisture)	10–15	
Foundries	15–20	
Galvanizing plants	20–30	
Garages (repair)	20–30	
Garages (storage)	4–6	
Homes (night cooling)	9–17	
Hospitals (general)		40–50
Hospitals (children's)		35–40
Hospitals (contagious diseases)		80–90
Kitchens (hotel)	10–20	
Kitchens (restaurant)	10–20	
Libraries (public)	4	
Laundries	10–15	
Mills (paper)	15–20	
Mills (textile—general buildings)	4	
Mills (textile—dyehouses)	15–20	
Offices (public)	3	
Offices (private)	4	
Pickling plants	10–15	
Pump rooms	5	
Schools (grade)		15–25
Schools (high)		30–35
Restaurants	8–12	
Shops (machine)	5	
Shops (paint)	15–20	
Shops (railroad)	5	
Shops (woodworking)	5	
Substations (electric)	5–10	
Theaters		10–15
Turbine rooms (electric)	5–10	
Warehouses	2	
Waiting rooms (public)	4	

Ventilation Heat Loss

In larger structures (e.g., commercial or industrial buildings), ventilation is necessary for both health and comfort. In houses, the amount of infiltrated air is generally sufficient to supply the ventilation requirements.

When ventilation is provided, the heating system warms the outside air used for this purpose (i.e., ventilation) to the inside design temperature. The heat required will be equivalent to the heat lost by ventilation and can be calculated with the following formula:

$$H = 0.018 \times Q(t_i - t_o)$$

where H = heat loss by ventilation.
       Q = volume of ventilation air.
      0.018 = specific heat of air (0.240) times density of outdoor air (approximately 0.075).

THE AVERAGE VALUE METHOD

Heat loss from a structure can also be calculated by using average values for four basic factors. These four factors and their assigned values are:

1. Wall factor (0.32 Btu).
2. Contents factor (0.02 Btu).
3. Glass factor (1 Btu).
4. Radiation factor (240 Btu).

Each of these four factors is based on a 1°F temperature difference and must be multiplied by the *design* temperature difference (see "Design Temperature Difference" in this chapter) obtained for the structure and its geographical location.

The average value for each factor represents the amount of heat (in Btu) that will pass through one square foot of the material in one hour. Thus 0.32 Btu will pass through *a square foot* of ordinary brick wall or through a wall where the average frame construction is used *in one hour*. It should be stressed that these are *average* values useful in a simple and quick way of calculating heat loss. The wall factor of 0.32 Btu varies for different kinds of walls, and heat loss calculations based on the use of average values will be only approximately accurate. More precise heat loss calculations can be made by using coefficients of heat transmission (*U*-factors).

The glass factor (1 Btu) is used to represent the amount of heat that will pass through one square foot of glass in an hour.

The radiation factor (240 Btu) was originally used to represent the amount of heat in Btu given off by an ordinary free-standing, cast-iron radiator on the basis of the square-foot area of heating surface per hour under average conditions. The average value of the radiation factor was based on repeated test observations that the amount of heat given off by ordinary cast-iron radiators per degree difference in temperature between the steam (or water) in the radiator and the air surrounding the radiator is about 1.6 Btu per square foot of heating surface. Taking this as a basis, a steam radiator under 2½ lb. pressure corresponds approximately to 220°F; when surrounded by air having a 70°F inside air temperature, it gives off heat at a rate of 240 Btu (220 − 70 × 1.6 = 240 Btu).

Radiators became so varied in design (with corresponding variations in the sizes of heating surface areas) that it became necessary to regard 240 Btu as a standard value.

Hot-water heating systems are planned on the basis of the Btu per hour capacity of each emitting unit. In steam heating systems, the radiation to be used is selected on the basis of the square foot EDR (equivalent direct radiation) capacity of each unit. One square foot of EDR equals 240 Btuh. The EDR capacity of a space is expressed in square feet and may be determined by dividing the total heat required (expressed in Btuh) by 240. For example, a room requiring 12,000 Btuh to heat it would have a steam radiation requirement of 50 sq. ft. EDR capacity.

The steps involved in determining heat loss with the average value method are:

1. Decide upon the desired inside air temperature.
2. Obtain the winter outside design temperature of the location of the structure from published lists or a local weather bureau.
3. Determine the design temperature difference (the difference between the temperatures found in Steps 1 and 2).
4. Calculate the volume of the structure or space.
5. Calculate the total area (in square feet) of window glass.
6. Calculate the total area of wall surface exposed to the outdoors.

7. Deduct the total area of window glass (Step 5) from the total area of wall surface exposed to the outdoors (Step 6).

8. Multiply the volume of the structure or space (Step 4) by the contents factor (0.02) to determine the Btu required to raise the temperature of the volume an amount equal to the design temperature difference (Step 3).

9. Multiply the *net* wall surface area (Step 7) by the wall factor (0.32) to determine heat loss through the walls.

10. Multiply the total window glass surface area (Step 5) by the glass factor (1) to determine heat loss through glass.

11. Add the figures obtained in Steps 8 to 10 to obtain the total heat loss per hour for the structure or space (this is equal to the total heat *required* per hour).

The eleven steps outlined above might be more clearly understood if shown in a sample problem.

Let's assume that you must calculate the heat loss for a space 14 ft. wide × 14 ft. long × 9 ft. high, and that 60 sq. ft. of window glass is equally divided between two exposed walls. The inside desired temperature is 70°, and the winter outside design temperature is −10°F. Your procedure (corresponding step by step to those outlined above) is as follows:

1. 70°F inside desired temperature.
2. − 10°F winter outside design temperature.
3. 80°F design temperature difference.
4. 1764 cu. ft. volume (14 × 14 × 9).
5. 60 sq. ft. of glass.
6. 252 sq. ft. of wall surface exposed (2 sides) to the outdoors (14 + 14 × 9 = 252).
7. 192 sq. ft. of *net* wall surface exposed (252 − 60 = 192).
8. 2822.4 Btu (1764 × 0.02 × 80).
9. 4915.2 Btu (192 × 0.32 × 80).
10. 4800 Btu (60 × 1 × 80).
11. 12,537.6 Btu (2822.4 = 4915.2 = 4800).

HEAT LOSS TABULATION FORMS

Manufacturers of heating equipment will often provide heat loss tabulation forms and instructions for their use. Lists of the most commonly used U-values (or factors) are included with the forms. Figs. 4-4 and 4-5 show examples of these forms and instructions obtained from Amana Refrigeration, Inc.

ESTIMATING FUEL REQUIRE-
MENTS AND HEATING COSTS

Estimating fuel requirements and heating costs is *not* an exact science. It involves too many variables to be anything other than an estimation. Moreover, there are four different formulas used for making these calculations. Depending on which formula is used, it is possible to arrive at four different estimations of fuel requirements and heating costs for a specific situation. It is no wonder, then, that two competent engineers can submit estimates that will differ as widely as 30 percent or more. These facts are not offered to discourage anyone but to present the true picture. Any attempt to calculate fuel requirements and heating costs will not produce *precise* figures—only an estimate. The problem is to make this estimate as close an approximation to the real situation as possible.

The four formulas used in calculating fuel requirements and heating costs are:

1. Heat loss formula.
2. Corrected heat loss formula.
3. NEMA formula.
4. Degree-day formula.

If one formula is used to calculate the fuel requirements and heating costs with one type of fuel (e.g., oil) and another formula is used for a second type (e.g., natural gas), the results are practically worthless for comparison purposes. For example, the NEMA formula (created by the National Electric Manufacturers Association) will present the use of electric energy much more favorably if the results are compared with calculations for oil or natural gas based on the heat loss formula. A *true* comparison is only possible when two fuels are both calculated with the *same* formula. Each of these formulas is described in the following sections.

The Heat Loss Formula

The *heat loss formula* results in higher percentages of total requirements because it does not take into consideration internal heat gains obtained from appliances, sunlight, the body heat of the occupants, electric lights, and other sources. The corrected heat loss formula includes these factors (see the following section).

The heat loss formula will include the following data:

1. Heat loss expressed in Btu.
2. Total hours in the heating season.
3. Average winter temperature difference.
4. Btu per unit of fuel.
5. Efficiency of utilization.
6. Difference between inside and outside design temperatures.

The product of the first three times (1–3) is divided by the product of the last three (4–6). In other words, heat loss × total hours × average winter temperature difference ÷ Btu per unit of fuel × efficiency of utilization × (inside design temperature − outside design temperature).

The method for calculating heat loss is described in Chapter 3 and is expressed in Btu per square foot per hour per degree Fahrenheit design temperature difference.

The total hours in the heating season will depend on the location of the house or building. If it is located in a southern state, the heating season will be much shorter than if it is located in a colder climate. Assuming that the heating season begins October 1 and ends May 1, the total number of days for which heat may be required is 212. This figure is multiplied by 24 (hours) to obtain the total hours in the heating season (5088).

The average winter temperature difference is found by subtracting the average low temperature from the average high temperature for the location in which the house or building is situated.

The Btu per unit of fuel is determined for each type, and this information can usually be obtained from the local distributor of the fuel.

Most heating equipment will burn oil or gas at an 80 percent

107

Fig. 4-4. Residential heat loss tabulation form.

RESIDENTIAL HEAT LOSS TABULATION

Job Name: _____

Location: _____

Date _____

Computed By _____

Inside Design Temp. _____
Outside Design Temp. _____
Temp. Difference _____
Temp. Difference +10 _____

Room	Living Room			Dining Room			Kitchen			Bedroom 1			Bedroom 2			Bath			Bathroom			Rooms Btu Totals
	Factor	Ar. or Ln. Ft.	BTU		Ar. or Ln. Ft.	BTU		Ar. or Ln. Ft.	BTU		Ar. or Ln. Ft.	BTU		Ar. or Ln. Ft.	BTU		Ar. or Ln. Ft.	BTU		Ar. or Ln. Ft.	BTU	
1. Room Size (LxWxH)																						
2. Linear Ft. Exp. Wall																						
3. Floor Area																						
4. Windows																						
5. Doors																						
6. Exp. Wall																						
7. Ceiling																						
a																						
b																						
8. Infiltration c																						
d																						
9. SUB TOTALS																						
10. Sub-Total for T.D.																						
11. Floor Btu for T.D.																						
12. Partition Btu for T.D.																						
13. Room Total Btu Loss																						

14. ___ + ___ % Duct Loss

15. Grand Total Btu Load _____

108

% Duct Loss Schedule

	1 Story	1½ Story	2 Story
Ducts Insul.	10	8	6
Ducts No Insul.	30	25	20

Windows — U Factors

Single Pane	1.13
Double Pane, Sealed	.57
Storm Windows, Tight	.45
Storm Windows, Loose	.75
Glass Block	.50

T.D. = TEMPERATURE DIFFERENCE

Exp. Walls — U Factors

Frame Construction With:	Clapboard or Shingle	Brick Venr.	Stone Venr.
No Insulation	.25	.28	.30
¾" Insl. Board	.19	.21	.22
2" Insl. Batts	.11	.12	.12
3⁵⁄₈" Insl. Batts	.09	.10	.10

Wood Doors — U Factors

Nom. Thickness	No Storm	Storm Door
1"	.69	.35
1½"	.52	.30
2"	.46	.28
2½"	.38	.25

Ceilings (Vent: Attic Space Above)

No Insulation	.50
2" Insulation	.10
4" Insulation	.09
6" Insulation	.05

Infiltration (factor times lineal ft. of crack)

		Weather Stripped	Not Weather Stripped
Double Hung Window	(a)	.22	.35
Pivoting Windows	(b)	.30	.50
Doors	(c)	.60	1.00
Fixed Window	(d)	.13	.13

CONCRETE FLOOR ON GROUND
Btu Per Hr. Per Lineal Ft. of Edge

Outside Design Temp.	25	20	15	10	5	0	−5	−10	−15	−20	−25	−30
1" Vert. Insul. Extending Down 18" Below Surface	55	62	68	75	80	85	90	95	100	105	110	115
1" L Type Insul. Extending 12" Down and 12" Under	50	57	63	70	75	80	85	90	95	100	105	110
2" L Type Insul. Extending 12" Down and 12" Under	40	45	50	55	60	65	70	75	80	85	90	95
No edge insulation and no heat in slab	This construction not recommended.											

Double Wood Floors or Partition

With Air Space

	10	20	30	40	50
Temp. Diff.					
Factor No Insul.	3.4	6.8	10.2	13.6	17.0
Factor 2" Insul.	1	2	3	4	5

Courtesy Amana Refrigeration, Inc.

109

Fig. 4-5. Residential heating lead estimate form.

Customer .... Address ....

Buyer .... Installation by ....

Estimate Number .... Estimate by .... Date ....

Equipment Selected ....; Model ....; Size ....

Direction House Faces ....; Gross Floor Area ....; Gross Inside Volume .... cu. ft.

Design Conditions:

Dry-Bulb Temperature (F)

Outside ....

ITEM	AREA (Sq. Ft.) Except Item 4; Linear Ft.	FACTOR (Circle the factors applicable.) DESIGN DRY-BULB TEMPERATURE (F)									BTU/HR (Area x Factor)
		−30	−20	−10	0	10	20	25	30	35	
1. WINDOWS (total of all windows)											
Single-glass		113	102	90	79	68	57	51	45	40	
Double glass or glass block		50	45	40	35	30	25	23	20	18	
2. WALLS											
No insulation (brick veneer, frame, stucco, masonry)		25	23	20	18	15	13	11	10	9	
1 in. insulation or 25/32 in. insulating sheathing		20	18	16	14	12	10	9	8	7	
2 in. or more insulation		10	9	8	7	6	5	5	4	4	

3. ROOFS

Pitched or flat with vented air space and:

No insulation	31	28	25	22	19	16	14	12	11
2 in. insulation	10	9	8	7	6	5	5	4	4
4 in. insulation	7	6	6	5	4	4	3	3	2

Flat with no air space and:

No insulation	50	45	40	35	30	25	23	20	18
$25/32$ in. insulation	25	23	20	18	15	13	11	10	9
1½ in. insulation	15	14	12	11	9	8	7	6	5
3 in. insulation	10	9	8	7	6	5	5	4	4

4. FLOORS

Over—(sq. ft. area):

Basement	0	0	0	0	0	0	0	0	0
Enclosed crawl space	23	20	18	16	14	11	10	9	8
Open crawl space	34	31	27	24	20	17	15	14	12

On ground—(linear feet):

No edge insulation	81	73	65	57	49	41	37	32	28
1 in. edge insulation	68	61	54	48	41	34	31	27	24
2 in. edge insulation	55	50	44	39	33	28	25	22	19

5. OUTSIDE AIR

Total Floor Area (sq. ft.)	14	13	11	10	8	7	6	6	5

6. TOTAL (Sum of 1 through 5)

INSTRUCTIONS FOR FIGS. 4-4 AND 4-5

Step 1. Room Size—list length, width, and height.

Step 2. Linear Feet of Exposed Walls—total the length of all walls exposed to outdoors and enter the sum in the space provided.

Step 3. Floor Area—multiply length of room by width.

Step 4. Btu loss—windows
Factor Column—enter proper U factor from "Windows" table.
Area Column—enter total of window area(s), width $\times$ height.
Btu Column—enter Btu loss for each degree of temperature difference obtained from factor $\times$ area.

Step 5. Factor Column—enter proper U factor from "Doors" table.

BTU LOSS DOORS

Area Column—enter net door area obtained from width $\times$ height minus area of glass in door.

Btu Column—enter Btu loss for each degree of temperature difference obtained from factor $\times$ area.

Step 6. Btu Loss—exposed Wall
Factor Column—enter proper U factor from "Exposed Walls" table.
Area Column—enter net wall area obtained from linear feet exposed wall (Step 2) $\times$ ceiling height minus window and door area.
Btu Column—enter Btu loss for each degree of temperature difference obtained from factor $\times$ area.

Step 7. Btu loss—ceiling
Factor Column—enter proper U factor from "Ceilings" table.
Area Column—enter ceiling area, length $\times$ width, usually the same as floor area (Step 3).
Btu Columns—enter Btu loss for each degree of temperature difference obtained from factor $\times$ area.

Step 8. Btu Loss—infiltration
In each class of infiltration: a, b, c, d take appropriate factor from "infiltration" table $\times$ the linear feet of crack to obtain Btu per degree of temperature difference.

Step 9. Subtotal Btu is the sum of Btu values, Steps 4 through 8 inclusive.

Step 10. Multiply Btu from Step 9 by temperature difference taken from upper-right corner of form.

Step 11. Btu Loss—floor
Factor from appropriate "Floors and Partitions" table $\times$ area of floor = Btu loss. If concrete floor on ground, use linear feet, Step 2, in place of area of floor. NOTE: Basement assumed 60°.

Step 12. Partition Btu
Factor from "Floors and Partitions" table $\times$ area of partition (length $\times$ height) gives partition Btu loss for temperature difference. A partition is a wall exposed to an unheated area or room.

Step 13. Room total Btu loss is the sum of Btu values, Steps 10 through 12 inclusive.

REPEAT THE STEPS FOR EACH ROOM IN THE HOME AND ADD (STEP 14) TO OBTAIN THE GRAND TOTAL.

combustion efficiency. Electric energy is generally considered to be used at 100 percent efficiency.

The outside design temperature is the lowest temperature experienced in a locality over a three- to four-year period. These outside design temperatures are available for a large number of localities throughout the United States. If the house or building is located in a small town or rural area, the nearest known outside design temperature is used. This inside design temperature is the temperature to be maintained on the interior of the house or building. The difference between the two design temperatures is used in the heat loss formula.

Table 4-6 illustrates the use of the heat loss formula for three types of fuel/energy:

1. Oil.
2. Natural gas.
3. Electricity.

Table 4-6. Applying Heat Loss Formula to Three Types of Fuels

(1) No. 2 oil, 140,000 Btu/gal, 80% efficiency, 35,000 Btuh loss, 73°F inside temperature:

$$\frac{35,000 \times 5088}{140,000 \times 0.80} \times \frac{73 - 41}{70 - 0} = 727 \text{ gallons}$$

(2) Natural gas, 100,000 Btu per therm, 80% efficiency:

$$\frac{35,000 \times 5088}{100,000 \times 0.80} \times \frac{73 - 41}{70 - 0} = 1018 \text{ therms}$$

(3) Electricity, 3413 Btu/Wh, 100% efficiency:

$$\frac{35,000 \times 5088}{3413} \times \frac{73 - 41}{70 - 0} = 23,852 \text{ kWh}$$

The Corrected Heat Loss Formula

The *corrected heat loss formula* was devised in 1965 by Warren S. Harris and Calvin H. Fitch of the University of Illinois. It takes into consideration internal heat gains that have shown a marked increase over the past forty- to fifty years. These internal heat gains make a significant contribution to the total required heat of a house or building, a factor that makes the original 65°F heat base established some forty-odd years ago too low for making a correct estimate. Furthermore, an inside average temperature of 73°F is probably more correct than the 70°F temperature previously used.

According to Harris and Fitch, the degree-day base (for all areas except along the Pacific Coast) must be corrected by the percentage given in Table 4-7. The calculated heat loss should then be reduced by 3.5 percent for each 1000 ft. the house or building is located above sea level. If the heat loss is below 1000 Btu per degree temperature difference, the 65°F degree-day base should be further reduced to the figures given in Table 4-8.

Table 4-7. Reduction in Standard Degree Days for All Areas Except the Pacific Coast

Degrees Days at 65°F Base	Reduce by This Percentage	New Total
10,000	3.15	9685
9500	3.20	9196
9000	3.22	8710
8500	3.23	8226
8000	3.25	7740
7500	3.80	7215
7000	3.85	6720
6500	4.35	6218
6000	4.55	5727
5500	4.60	5247
5000	4.65	4768
4500	4.70	4289
4000	5.30	3788
3500	5.35	3313
3000	6.15	2816
2500	7.15	2321
2000	8.00	1840
1500	9.50	1358
1000	11.90	881
500	15.10	425

Table 4-8. Reduction in Degree-Day Base When Calculated Heat Loss (Btu per Degree Temperature Difference) is Less Than 1000

Calculated Heat Loss	Revised DD Base, °F
200	55
300	60
400	61
500	62
600	63
700	64
800	64
900	64
1000	65

Courtesy National Oil Fuel Institute

Table 4-9 illustrates the use of the corrected heat loss formula for determining comparative fuel requirements for a No. 2 oil and natural gas.

Table 4-9. Using Corrected Heat Loss Formula for Determining Comparative Fuel Requirements

Assume 500 Btuh per degree temperature difference (35,000 Btu at 70°F difference); degree days with 65°F base, 5542; average indoor temperature 73°F.

Correction factor from Table 4-7 is 4.60%.

From Table 4-5, revised degree day base is 62°F.

Reduction in base is (65 − 62) 3°F.

Multiply 0.0460 × 3 = 0.1380 and deduct from 1.0000 = 0.8620.

Multiply 5542 (dd at 65° base) by 0.8620 = 4557.

EXAMPLE:

No. 2 Oil —
   35 × 0.00304 × 4557 = 485 gallons per year

Natural Gas —
   35 × 0.00429 × 4557 = 684 therms per year

Courtesy National Oil Fuel Institute

The Degree-Day Formula

The *degree-day formula* was devised some forty-odd years ago by the American Gas Association and other groups and has since been revised (see pages 627–28 of the 1970 *ASHRAE*

115

Guide) to reflect internal heat gains and the levels of insulation. The correction factors involved in the revision are given in Tables 4-10 and 4-11. These correction factors should also be applied to the corrected heat loss formula (see the previous section).

Table 4-10. Unit Fuel Consumption per Degree Day per 1000 Btu Design Heat Loss

Fuel	Utilization Efficiency		
	60%	70%	80%
Gas in therms	0.00572	0.00490	0.00429
Oil in gallons (141,000 Btu)	0.00405	0.00347	0.00304

Courtesy National Oil Fuel Institute

Table 4-11. Correction Factors for Outdoor Design Temperatures*

	Outside Design Temp., °F				
	− 20	− 10	0	+ 10	+ 20
Correction factor	0.778	0.875	1.000	1.167	1.400

*To be applied to Table 4-1 when design temperature is higher or lower than 0°F.

Courtesy National Oil Fuel Institute

The degree-day formula is based on the assumption that heat for the interior of a house or building will be obtained from sources other than the heating system (e.g., sunlight and body heat of the occupants) until the outside temperature declines to 65°F. At this point the heating system begins to operate. The consumption of fuel will be directly proportionate to the difference between the 65°F base temperature and the mean outdoor temperature. In other words, three times as much fuel will be used when the mean outdoor temperature is 35°F than when it is 55°F. The mean outdoor temperature can be determined by taking the sum of the highest and lowest outside temperatures during a 24-hour period, beginning at midnight, and dividing it by 2. Each degree in temperature below 54°F is regarded as one "degree day."

The degree-day formula is applied by dividing the heat loss figure by 1000 and multiplying the result by the figure for the unit fuel consumption per degree day per 1000 Btu design heat loss, which, in turn, is multiplied by the total number of degree days in the heating season (calculated on a 65°F base) and then by the correction factors given in Table 4-10 and 4-11. The application of this formula is illustrated in Table 4-12.

Table 4-12. Application of Degree-Day Formula

35,000 Btuh loss; outside design temperature 0; no correction factor needed; degree days, 5542.

No 2 oil, 140,000 Btu per therm, 80% efficiency —
$35 \times 0.00304 \times 5542 = 590$ gallons per year

Natural gas, 100,000 Btu per therm, 80% efficiency —
$35 \times 0.00429 \times 5542 = 832$ therms per year

Electricity, required Btu/year $= 832$ (therms) $\times$
100,000 (Btu/therm) $\times$ 0.80 (efficiency) $= 66,560,000$

$$\frac{66,560,000}{3413} = 19,502 \text{ kWh}$$

Courtesy National Oil Fuel Institute

The NEMA Formula

The *NEMA formula* was created by the National Electric Manufacturers Association. A constant of 18.5 is commonly used in the formula and reflects the percentage of heat loss that must be replaced by the heating system during a 24-hour period. In other words, the heating system will provide heat 77 percent of the time during a 24-hour period (24 $\times$ 18.5), while other heat sources will supply heat the remaining 23 percent.

The NEMA formula consists of dividing the heat loss by 3413 (the number of Btu per kilowatt-hour), and multiplying this figure by the total annual number of degree days, which figure is then multiplied by the constant (usually 18.5). The resulting product is then divided by the difference between the indoor and outdoor temperatures. The application of the NEMA Formula is illustrated in Table 4-13.

A major criticism of this formula by distributors of oil and gas heating fuels is that use of the heat loss formula gives a much

117

Table 4-13. Application of NEMA Formula

kWh = heat loss ÷ 3413
$\times$ annual degree days
$\times$ constant (usually 18.5)

Divided by
difference between indoor and outdoor design temperature

EXAMPLE: $\dfrac{35{,}000 \text{ Btuh}}{3413} = 1030$

$\dfrac{1030 \times 5542 \times 18.5}{70 - 0} = 15{,}086 \text{ kWh}$

higher kilowatt-hour consumption rate than use of the NEMA formula.

Other Heating Costs

Energy requirements for nonheating purposes are an important addition factor to be considered when estimating total fuel costs for a house or building. This nonheating energy is used to operate oil and gas burners, automatic washers and dryers, water heaters, stoves, refrigerators, and the variety of other appliances and work-saving devices deemed so necessary for modern living.

This nonheating energy is commonly electricity or gas. Electricity is used to operate thermostats and other automatic controls found in heating and cooling equipment, burners, water heaters, home laundry equipment, refrigerators, and blowers for air conditioners, to mention only a few. Much of this equipment (e.g., water heaters and home laundry equipment) is also designed to be operated in conjunction with gas. Each of these appliances consumes energy that will account for a certain percentage of the homeowner's utility bill. This must be accounted for when estimating the total fuel requirement and heating cost.

The most common method of determining which appliance is the most suitable one for a particular set of circumstances is by making a comparison of their energy use. For example, the estimated usage for home laundry equipment (either a washer or a dryer) is 1 hour per week per occupant. In other words, a dryer would be operated 5 hours per week on the average in a household containing five people. For purposes of comparison, an electric clothes dryer draws 5.0 kWh and a gas clothes dryer uses approximately 3.5 therms of gas per month. Although these figures appear insignificant, they gain importance when the total nonheating energy use for all the appliances is added together. The cost per fuel unit will depend upon the local utility rates (see the following section). Some appliances, such as water heaters, will have to be compared on a basis that includes a number of different energy requirements met under a variety of environmental conditions (see Tables 4-14 and 4-15).

Table 4-14. Usage Estimates for Gas Water Heating*

Size of Unit	Hot-Water Requirements	Northern Localities 40°F	North Central Localities 50°F	South Central Localities 60°F	Southern Localities 70°F
0-BR	30 Gallon/Day	12.8	11.2	10.0	8.5
1-BR	40 "	16.4	14.7	12.9	10.9
2-BR	50 "	19.7	17.5	15.2	13.2
3-BR	60 "	22.9	20.5	17.7	15.3
4-BR	70 "	26.6	23.6	20.6	17.6
5-BR	80 "	30.2	27.0	23.5	20.2

*Estimates are given in therms per dwelling per month. Where wholesale electricity is to be used for lighting, refrigeration, cooking, and domestic hot water, the monthly demand for all purposes may be estimated at 2.65 watts per kWh.

Courtesy National Oil Fuel Institute

Table 4-15. Usage Estimates for Electric Water Heating*

Size of Unit	Hot-Water Requirements	Northern Localities 40°F	North Central Localities 50°F	South Central Localities 60°F	Southern Localities 70°F
0-BR	30 Gallon/Day	210	185	165	140
1-BR	40 "	280	250	220	185
2-BR	50 "	350	310	270	235
3-BR	60 "	420	375	325	280
4-BR	70 "	490	435	380	325
5-BR	80 "	560	500	435	375

*Estimates are given in kWH per dwelling per month.

Courtesy National Oil Fuel Institute

The problem for the homeowner is to determine which type of appliance burns which type of fuel most efficiently and at the lowest cost. Probably the poorest sources of information are the distributor of the particular appliance and the rate correspondent of the utility. They will, of course, be biased in favor of their own product. This is as it should be because this is the reason they are in business. If they made a habit of recommending the products of their competitors, they would soon be out of business.

Quite often, consumer magazines give the energy use rates of different appliances and are quite frank in their comparison. This source is probably the most objective one to be found. However, knowing the rate at which an appliance consumes energy is only half the solution. It is also necessary to determine the utility rates.

DETERMINING UTILITY RATES

No estimation of heating costs is complete until the utility rates for the various fuels available are determined and included in the estimate. Determining utility rates depends on so many variables that we can calculate only an estimated cost, not a precise figure. The problem is to arrive at as accurate an estimation as possible.

Among the many variables involved in determining utility rates are the following:

1. Differing rate structures.
2. Fuel/energy adjustment factors.
3. Variations in energy block sizes and prices.
4. Rate structure qualification procedures.
5. Community utility taxes.
6. Maximum demand charges.

Differing rate structures exist for different facilities or usage. The problem is to determine which rate structure applies to the particular situation. This information can usually be obtained by contacting the local utility and consulting with the utility rate correspondent.

Although this is a simplified explanation, a typical utility bill is computed by adding the demand charge, the energy or commod-

Table 4-16. Determining Utility Rates

DEMAND CHARGE

300 × $0.60 = $180.00
50 × $0.45 = $ 22.50
                  $202.50

Total annual demand charges — 8 × $202.50 = $1620

COMMODITY CHARGES

Therms		Rate		Amount
3,000	×	5.0¢	=	$150.00
2,000	×	4.5¢	=	90.00
5,000				$240.00

Total commodity charges — 8 × $240.00 = $1920.00
Fuel adjustment (40,000 × $0.01819)   727.60
                                          $2647.60

Average gas cost —

$$\frac{\$1620 + \$2647.60}{40,000} = 10.6¢/\text{therm}$$

ity charge, and the fuel adjustment charge, and dividing by the total amount of electricity or gas used during the billing period. This produces the average electric or gas cost. The procedure is illustrated in Table 4-16.

CHAPTER 5

Heating Fuels

Only the principal characteristics of those heating fuels used for domestic heating purposes will be considered in any great detail in this chapter. Descriptions of *firing methods* for the more commonly used heating fuels are found in Chapter 1 (Oil Burners), Chapter 2 (Gas Burners), Chapter 3 (Coal-Firing Methods) of Volume 2, and Chapter 3 (Stoves, Fireplaces, and Chimneys) of Volume 3.

Although electricity is used as a source of power for electric-fired furnaces, its properties are such that it cannot be regarded as a heating fuel in the sense that the term is used here; therefore it is not included in this chapter. It is, however, described more fully in this volume in Chapter 2 (Heating Fundamentals), Chapter 9 (Electrical Heating Systems), and Chapter 14 (Electric-Fired Furnaces).

Heating fuels can be classified as solid, liquid, or gaseous, depending upon their physical state. Examples of solid fuels are coal and coke. Fuel oils are classified as liquid fuels; gaseous fuels include natural gas, manufactured gas, liquefied petroleum gas, and related types.

NATURAL GAS

Natural gas is a generic term commonly applied to those gases found in or near deposits of crude petroleum. It is *not* connected with the production of oil and should always be regarded as a separate and independent entity. Natural gas is piped under pressure from the gas fields to the consumer centers, and these pipelines now serve a considerable portion of the United States.

Natural gas is the richest of the gases and contains from 80 to 95 percent methane with small percentages of the other hydrocarbons. The heating value of natural gas varies from 1000 to 1200 Btu/cu. ft. with the majority of gases averaging about 100 Btu/cu. ft. The caloric value will depend upon the locality.

Natural gas can be divided into three basic types:

1. Associated gas.
2. Nonassociated gas.
3. Dissolved gas.

Associated gas is a free (undissolved) natural gas found in close contact with crude petroleum. *Nonassociated gas* is also a free gas but is not found in contact with the crude petroleum deposit. As the name suggests, *dissolved gas* is found in solution in the crude petroleum.

MANUFACTURED GAS

A *manufactured gas* is any gas made by a manufacturing process. Raw materials used in the production of manufactured gas include coal, oil, coke, natural gas, or one of the other manufactured gases. For example, coal gas is made by distilling bituminous coal in either retorts of by-product coke ovens. Often several of these raw materials are used together as a base for the production of a manufactured gas.

The types of manufactured gases commercially available include:

1. Coal gas or by-product coke-oven gas.
2. Oil gas.
3. Blue water gas.
4. Carbureted water gas.
5. Producer gas.
6. Reformed natural gas.
7. Liquefied petroleum gas.

Most of these gases have low calorific values (generally between 500 and 100 Btu/cu. ft.) and are produced primarily for industrial use. Producer gas and liquefied petroleum gas are also used for domestic purposes. The former may be used alone or in combination with other gases. Of all the manufactured gases, liquefied petroleum gas enjoys the widest application in domestic heating and cooking.

LIQUEFIED PETROLEUM GAS

Liquefied petroleum (LP) gas is a hydrocarbon mixture extracted primarily from "wet" natural gas and sold commercially as propane butane, bottled gas, or under a variety of different brand names. The terms "day" and "wet" natural gas refer to the gasoline content per 1000 cu. ft. It is regarded as "dry" if it contains less than 0.1 gallon of gasoline per 1000 cu. ft. and "wet" if it contains more than 0.1 gallon.

Propane is used extensively for domestic heating purposes. It contains 2516 Btu/cu. ft. (per gal.), of about two and a half times the Btu content of methane (natural gas). Table 5-1 compares the typical properties of propane and butane. Unlike natural gas, propane is heavier than air. An undetected leak can be quite dangerous because the escaping gas will accumulate in layers at floor level. An explosion can be set off if the gas reaches the level of the pilot flame in the furnace.

Propane is frequently delivered in bulk and stored in large, stationary tanks holding from 100 to 1000 gallons. It is then piped into the home or building in much the same manner as natural

Table 5-1. Typical Properties of LP Gas

Property	Butane	Propane
Btu per cu. ft. 60°F	3280	2516
Btu per lb.	21,221	21,591
Btu per gal.	102,032	91,547
Cu. ft. per lb.	6.506	8.58
Cu. ft. per gal.	31.26	36.69
Lb. per gal.	4.81	4.24

Courtesy National LP-Gas Association

gas. For small usage (e.g., trailers of homes that use only enough gas for one or two appliances), the LP gas is delivered to 5- to 25-gallon cylinders. The suggested clearance for LP gas cylinders and storage tanks are shown in Fig. 5-1. The size of the storage tank required by an installation depends upon the weather zone in which it is located (Fig. 5-2). Recommend tank sizes are given in Fig. 5-2.

FUEL OILS

Fuel Oils are hydrocarbon mixtures obtained from crude petroleum by refining processes. They may be divided in the following six classes or grades:

1. *No. 1 fuel oil* for vaporizing pot-type and other burners designed for this fuel.
2. *No. 2 fuel oil* for general-purpose domestic heating not requiring a lighter No. 1 oil.
3. *No. 3 fuel oil* (obsolete since 1948).
4. *No. 4 fuel oil* for installations not equipped for preheating.
5. *No. 5 fuel oil* for installations equipped for preheating.
6. *No. 6 fuel oil* for burner installations equipped for preheating with a high-viscosity fuel.

This classification of fuel oils is based on a number of characteristics, including: (1) viscosity, (2) flash point, (3) pour point, (4) ash content, (5) carbon residues, and (6) water and sediment content. All these characteristics are important to consider when selecting a fuel oil. For example, the water content will determine its suitability for outdoor storage. A low viscosity indicates a fuel oil with good flow characteristics.

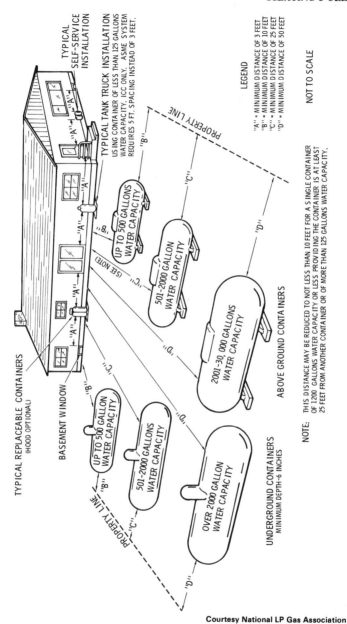

TYPICAL
SELF-SERVICE
INSTALLATION

TYPICAL TANK TRUCK INSTALLATION
USING CONTAINER OF LESS THAN 125 GALLONS
WATER CAPACITY, ICC ONLY. ASME SYSTEM
REQUIRES 5 FT. SPACING INSTEAD OF 3 FEET.

LEGEND

"A" = MINIMUM DISTANCE OF 3 FEET
"B" = MINIMUM DISTANCE OF 10 FEET
"C" = MINIMUM DISTANCE OF 25 FEET
"D" = MINIMUM DISTANCE OF 50 FEET

NOT TO SCALE

TYPICAL REPLACEABLE CONTAINERS
(HOOD OPTIONAL)

BASEMENT WINDOW

UP TO 500 GALLONS
WATER CAPACITY

501-2000 GALLON
WATER CAPACITY

2001-30, 000 GALLONS
WATER CAPACITY

ABOVE GROUND CONTAINERS

PROPERTY LINE

(SEE NOTE)

UP TO 500 GALLON
WATER CAPACITY

501-2000 GALLONS
WATER CAPACITY

OVER 2000 GALLON
WATER CAPACITY

UNDERGROUND CONTAINERS
MINIMUM DEPTH-6 INCHES

PROPERTY LINE

NOTE: THIS DISTANCE MAY BE REDUCED TO NOT LESS THAN 10 FEET FOR A SINGLE CONTAINER
OF 1200 GALLONS WATER CAPACITY OR LESS PROVIDING THE CONTAINER IS AT LEAST
25 FEET FROM ANOTHER CONTAINER OR OF MORE THAN 125 GALLONS WATER CAPACITY.

Courtesy National LP Gas Association

Fig. 5-1. Suggested clearances for LP gas cylinders and storage tanks.

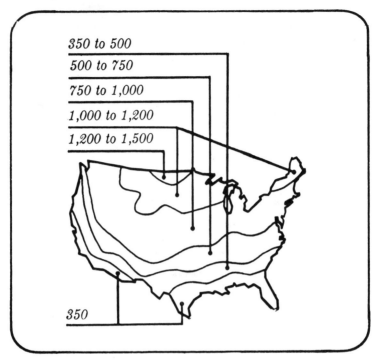

350 to 500

500 to 750

750 to 1,000

1,000 to 1,200

1,200 to 1,500

350

Fig. 5-2. The six basic weather zones and the size of tank required to ensure an adequate supply of gas.

No. 1 fuel oil has the lowest viscosity (1.4–2.2). This is a distillate oil used in atomizing and similar burners because it can be easily broken down into small droplets.

No. 2 fuel oil is a distillate oil used in domestic oil burners that do not require preheating. It is slightly heavier than a No. 1 fuel oil. No. 4 fuel oil is slightly heavier than No. 2 and will work in some oil burners without preheating.

No. 2 and No. 4 fuel oils contain the highest ash content (0.10 to 0.50% by weight, respectively). The other grades contain little or no ash. Although slagging (i.e., the formation of noncombustible deposits) is commonly associated with solid fuels such as coal, this problem also occurs in oil burning equipment (see "Ash, Slag, and Clinker Formation" later in this chapter).

No. 5 and No. 6 (or *Bunker C*) fuel oils are heavier than the other grades and require preheating. In automatically operated oil burners, the preheating of the fuel oil must be completed before it is delivered for combustion. When the fuel oil has reached a suitable atomizing temperature, the preheating is considered completed.

PS 300 and *PS 400* are heating oils used on the West Coast. They roughly correspond to No. 5 and No. 6 fuel oils, respectively. *PS 300* is the lighter of the two and is used primarily for domestic heating purposes. *PS 400* is heavier, requires preheating, and is used for industrial purposes.

The grade of fuel oil to be used in an oil burner is specified by the manufacturer. This should also be stipulated on the label of the Underwriters' Laboratories, Inc., and Underwriters' Laboratories of Canada. Table 5-2 compares the Btu content of the principal fuel oils.

Table 5-2. Btu Content of Principal Fuel Oils

Grade or Type	Unit	Btu
No. 1 Oil	Gallon	137,400
No. 2 Oil	Gallon	139,600
No. 3 Oil	Gallon	141,800
No. 4 Oil	Gallon	145,100
No. 5 Oil	Gallon	148,800
No. 6 Oil	Gallon	152,400
Natural Gas	Cu. ft.	950–1150
Propane	Cu. ft.	2550
Butane	Cu. ft.	3200

Courtesy Honeywell Tradeline Controls

COAL

Coal is a solid fuel created from deposits of ancient vegetation that have undergone a series of metamorphic changes resulting from pressure, heat, submersion, and other natural formative processes occurring over a long period of time. Because of the different combinations of these processes acting on the vegetation and the widely differing forms of vegetation involved, coal is a complex and nonuniform substance. The formation of peat represents a pre-coal stage of development. After increased pressure and time, lignite begins to form. The next stage is semibituminous, followed by bituminous, and then anthracite. Each of these stages is characterized by an increase in the hardness of the coal. Anthracite is the hardest of the coals. Graphite represents a post-coal developmental stage and cannot be used for heating purposes.

Coal is often classified according to the degree of metamorphic change to which it has been subjected into the following categories:

1. Anthracite.
2. Bituminous.
3. Semibituminous.
4. Lignite.

Anthracite is clean, hard coal that burns with little or no luminous flame or smoke. It is difficult to ignite but burnes with a uniform, low flame once the fire is started. Anthracite contains approximately 14,400 Btu/lb. and is used for both domestic and industrial heating purposes. Its major disadvantage as a heating fuel is its cost.

Anthracite coal is divided by size into a number of different grades. Each of these grades (e.g., egg size, buckwheat size, pea size) is suitable for a specific-size firepot. They are described in greater detail in Chapter 3 of Volume 2 (Coal-Firing Methods).

Bituminous coal is softer than anthracite and burns with a smokey, yellow flame. A great amount of smoke will result if it is improperly fired. The term "bituminous coal" actually covers a whole range of coals, many of which have widely differing combustion characteristics. Some of the coals belonging to this classification are hard, whereas others are soft.

The available heat for bituminous coal ranges from a low of 11,000 Btu/lb. (*Indiana* bituminous) to a high of 14,100 Btu/lb. (*Pocahontas* bituminous). The heat value of the latter approximates that of anthracite (about 14,400 Btu/lb.); however, unlike anthracite coal, it is available in far greater supply, a factor that makes it a very economical solid fuel to use.

Semibituminous coal is a soft coal that ignites slowly and burns with a medium-length flame. Because it provides so little smoke, it is sometimes referred to as *smokeless coal.*

Lignite (sometime referred to as *brown coal*) ignites slowly, produces very little smoke, and contains a high degree of moisture. In structure, it is midway between peat and bituminous coal. Lignite contains approximately half the available heat of anthracite (or about 7400 Btu/lb.), and burns with a long flame. Its fire is almost smokeless, and it does not coke, a characteristic it shares with anthracite.

Lignite is considered a low-grade fuel, and its calorific value is low when compared with the other coals. Moreover, it is difficult to handle and store.

COKE

Coke is the infusible, solid residue remaining after the distillation of certain bituminous coal or as a by-product of petroleum distillation. Coke may also be obtained from petroleum residue, pitch, and other materials representing the residue of destructive distillation.

Coke will ignite more quickly than anthracite but less readily than bituminous coal. It burns rapidly with little draft. As a result, all openings or leaks into the ash pit must be closed tightly when the coke is being burned.

Since less coke is burned per hour per square foot of grate than coal, a larger grate is required and a deep firepot is necessary to accommodate the thick bed of coal. Since coke contains very little hydrogen, the quick-flaming combustion that characterizes coal is not produced, but the fire is nearer even and regular.

The best size of coke recommended for general use, for small firepots where the full depth is not over 20 in., is that which passes over a 1-in. screen and through a 1½-in. screen. For large firepots where the fuel can be fired over 20 in. deep, coke that passes over a 1-in. screen and through a 3-in. screen can be used, but a coke of uniform size is always more satisfactory.

BRIQUETTES

Briquettes are a solid fuel prepared from coal dust and fines (i.e., finely crushed or powdered coal) by adding a binder and holding it under pressure.

The binder is commonly pitch or coal tar, but other substances and materials are also used. A briquette made with a pitch binder has the highest calorific value, exceeding that of any coal. The ash content is generally less than that of coal. Briquettes have been made from lignite coal dust and fines without a binder (only pressure being used).

COAL OIL

Coal oil is a heating fuel formerly derived from coal tar. Formerly it enjoyed widespread use, but it has now been largely replaced by petroleum-based products. It is also called *paraffin oil* or *kerosene*. Today, coal oil (kerosene) is obtained from petroleum through a refining process.

WOOD AS FUEL

A good, well-seasoned hardwood will provide approximately half as much heat value per pound as does good coal (Table 5-3). Wood is easy to ignite, burns with little smoke, and leaves comparatively little ash. On the other hand, it requires a larger storage space (commonly outdoors), and more labor is involved in this preparation.

Wood is commonly sold by the cord for heating purposes. A *standard cord* measures 4 ft. high by 8 ft. wide and contains wood cut to 4-ft. lengths for a total of 128 cu. ft. Only about 70 percent of a standard cord actually represents the wood content,

Table 5-3. Typical Firewoods

Name of Wood	Type of Firewood	Combustion Characteristics
Ash, white	Hardwood	Good firewood
Beech	Hardwood	Good firewood
Birch, yellow	Hardwood	Good firewood
Chestnut	Hardwood	Excessive sparking (can be dangerous)
Cottonwood	Hardwood	Good firewood
Elm, white	Hardwood	Difficult to split, but burns well
Hickory	Hardwood	Slow, steady fire; best firewood
Maple, sugar	Hardwood	Good firewood
Maple, red	Hardwood	Good firewood
Oak, red	Hardwood	Slow, steady fire
Oak, white	Hardwood	Slow, steady fire
Pine, yellow	Softwood	Quick, hot fire; smokier than hardwood
Pine, white	Softwood	Quick, hot fire; smokier than hardwood
Walnut, black	Hardwood	Good firewood, but difficult to find

however, because of the existence of air spaces between the wood.

The heat value of wood depends upon the *type* of wood being burned. Heat values per cord for a number of different types of wood are given in Tables 5-4 and 5-5. The data used to compile the tables was obtained from *Use of Wood for Fuel* (U.S. Department of Agriculture Bulletin No. 753). The wood heat values are determined for cords containing approximately 90 cu. ft. of solid wood (i.e., about 70 percent of a standard cord). Note that the heat values for both green and "dry" wood (12% moisture content) are given in Tables 5-4 and 5-5. It is important to remember that the amount of moisture in a wood will affect its burning characteristics. A green wood will burn much more slowly than a drier one, and its fire will be more difficult to start.

All wood can be divided into either hardwood or softwood. Contrary to popular belief, the basic difference between the two groups is *not* the hardness of the wood. In other words, a wood that can be classified as a softwood is not necessarily softer than a hardwood. These terms do not refer to the physical properties of the wood but to its classification as a coniferous (needle or cone-bearing) or deciduous (broad-leafed) tree. A softwood, for example, comes from a coniferous tree, whereas a hardwood is obtained from a deciduous tree. You will find that some soft-

Table 5-4. Heat Value per Cord (million Btu) of Wood with 12 % Moisture Content*

Name of Wood	Heat Value	Equivalent Coal Heat Value
Ash, white	28.3	1.09
Beech	31.1	1.20
Birch, yellow	30.4	1.17
Chestnut	20.7	0.80
Cottonwood	19.4	0.75
Elm, white	24.2	0.93
Hickory	35.3	1.36
Maple, sugar	30.4	1.17
Maple, red	26.3	1.01
Oak, red	30.4	1.17
Oak, white	32.5	1.25
Pine, yellow	26.0	1.00
Pine, white	18.1	0.70

*The equivalent coal heat values are based on 1 ton (2000 lb.) of anthracite coal with a heat value of 13,000 Btu/lb.

Table 5-5. Heat Value per Cord (million Btu) of Green Woods*

Name of Wood	Heat Value	Equivalent Coal Heat Value
Ash, white	26.0	1.00
Beech	27.1	1.04
Birch, yellow	27.2	1.05
Chestnut	19.2	0.75
Cottonwood	18.0	0.69
Elm, white	22.2	0.85
Hickory	29.0	1.12
Maple, sugar	27.4	1.05
Maple, red	23.7	0.91
Oak, red	27.5	1.06
Oak, white	28.7	1.10
Pine, yellow	23.7	0.91
Pine, white	17.3	0.67

*The equivalent coal heat values are based on 1 ton (2000 lb.) of anthracite coal with a heat value of 13,000 Btu/lb.

wood trees produce wood that is harder than the wood obtained from some hardwood trees.

Chimney or flue fires are always a hazard when using wood as a heating fuel. Resins and soot collect in these areas over time and can be ignited if there is a flareup of the fire. The possibility of this happening can be eliminated or greatly reduced by cleaning the chimney or flue from time to time.

Softwoods (e.g., white or yellow pine) contain greater amounts of resin than do hardwoods. This tends to give softwoods flammability, but it also results in more smoke.

ASH, SLAG, AND CLINKER FORMATION

Ash is the noncombustible mineral residue that remains in the furnace or boiler after the fuel has been thoroughly burned. During the combustion process, the combustible portion of the fuel is consumed and the noncombustible ash remains in place. As the process continues, the ash residue is subjected to a certain degree of shrinkage. This shrinkage results in portions of the ash fusing together and forming more or less fluid globules, or *slag*. This process is referred to as *slagging*.

Under the proper temperature conditions, the fluid slag globules can solidify and form *clinkers* in the fuel bed or deposits on the heating surfaces of the furnace or boiler. Clinkers can obstruct the necessary air flow to the fire and result in its reduced efficiency or extinction. These deposits on the heating surfaces will reduce their heating capacity and must be removed. Suggestions for removing clinkers are given in Chapter 3 of Volume 2 (Coal-Firing Methods).

SOOT

Soot is a fine powder consisting primarily of carbon produced by the combustion process. Because of its extreme light weight, it frequently rises with the smoke from the fire and coats the interior walls of the chimney and flue. Although the heat loss from the insulating effect of a soot layer is small (generally under 6 percent), it can cause a considerable rise in the stack temperature. Soot accumulation can also clog the flues, thereby reducing the draft and resulting in improper combustion. Soot may be blasted loose from the walls of the chimney or flue with a jet of compressed air, or it may be sucked out with a vacuum cleaner. Another method is to use a brush to remove the accumulated soot layer from the walls.

CHAPTER 6

Warm-Air Heating Systems

Air is the medium used for conveying heat to the various rooms and spaces within a structure heated by a warm-air furnace. It is also the principal criterion for distinguishing warm-air heating systems from other types in use.

The warm-air furnace is a self-contained and self-enclosed heating unit, which is usually (but not always) centrally located in the structure. Depending upon the design, any one of several fuels can be used to fire the furnace. Cool air enters the furnace and is heated as it comes in contact with the hot metal heating surfaces. As the air becomes warmer, it also becomes lighter, which causes it to rise. The warmer, lighter air continues to rise until it is either discharged directly into a room (as in the so-called pipeless gravity system) or carried through a duct system to warm-air outlets located at some distance from the furnace. After the warm air surrenders its heat, it becomes cooler and heavier. Its increased weight causes it to fall back to the furnace where it is reheated and repeats the cycle. This is a very simplified description of the operating principles involved in warm-air heating, and it especially typifies those involved in gravity heating systems.

CLASSIFYING WARM-AIR HEATING SYSTEMS

A *warm-air heating system* is one in which the air is heated in a furnace and circulated through the rest of the structure either by gravity or motor-driven centrifugal fans. If the former is the case, then the system is commonly referred to as a *gravity warm-air heating system*. Any system in which air circulation depends *primarily* on mechanical means for its motive force is called a *forced-warm-air heating system*. The stress on the word "primarily" is intentional because some gravity warm-air systems will use fans to supplement gravity flow, and this may prove confusing at first. In any event, one of the oldest forms of classifying warm-air heating systems has been on the basis of which method of air circulation is used: gravity or forced air.

Forced-warm-air heating systems are often classified according to the duct arrangement used. The two basic types of duct arrangements used are:

1. Perimeter duct systems.
2. Extended plenum duct systems.

A *perimeter duct system* is one in which the supply outlets are located around the perimeter (i.e., the outer edge) of the structure close to the floor of the outside wall or on the floor itself. The return grilles are generally placed near the ceiling on the inside wall. The two basic perimeter duct systems are:

1. The perimeter-loop duct system (Fig. 6-1).
2. The radial-type perimeter duct system (Fig. 6-2).

An *external plenum system* (Fig. 6-3) consists of a large rectangular duct that extends straight out from the furnace plenum in a straight line down the center of the basement, attic, or ceiling.

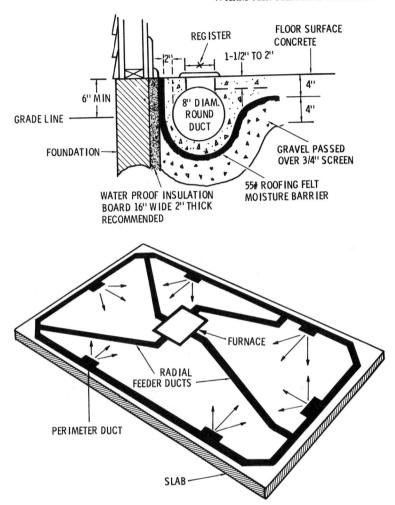

Fig. 6-1. Perimeter-loop duct system in concrete slab construction.

Round supply ducts connect the plenum to the heat-emitting units.

These and other modifications of duct arrangements are described in considerable detail in Chapter 7 of Volume 2 (Ducts and Duct Systems).

139

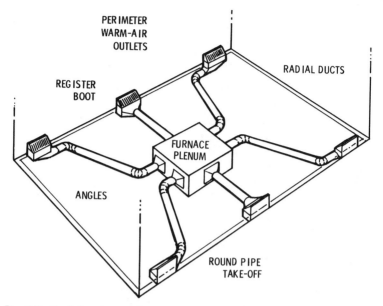

Fig. 6-2. Radial perimeter duct system.

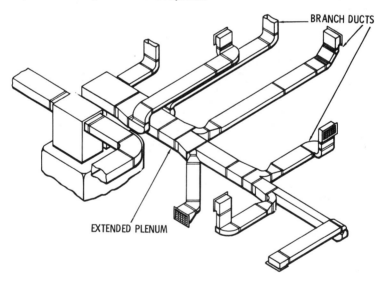

Fig. 6-3. Extended plenum system.

GRAVITY WARM-AIR HEATING SYSTEMS

A *gravity warm-air heating system* (Fig. 6-4) consists of a properly designed furnace (with casing and smoke pipe), a warm-air supply delivery system, and a cool-air return system.

In a central heating system, the warm air is delivered to the rooms and spaces being heated through a system of air ducts, the exception being a gravity floor furnace or a pipeless furnace. Those ducts located in the basement and extending (or leading) from the furnace to the basement ceiling are referred to as *leaders*. The *stacks* are warm-air ducts or pipes that connect to the leaders and extend vertically within the walls and up through the

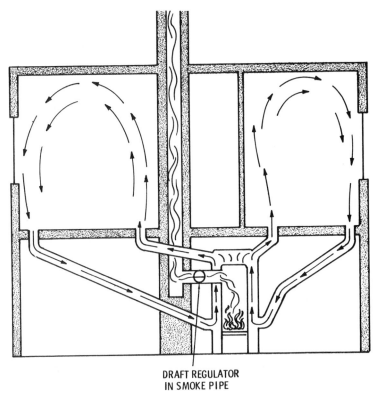

DRAFT REGULATOR
IN SMOKE PIPE

Fig. 6-4. Operating principle of a gravity warm-air heating system.

building. The warm air enters a room through registers located on the walls or floor (the former will generally connect to a stack, the latter to a leader). The cooler room air exits a room by means of return grilles and travels back through return air ducts to the furnace where it is reheated and recirculated.

A gravity warm-air heating system differs from the forced-warm-air type by relying primarily on gravity to effect air circulation. The operating principle of the gravity warm-air heating system is based on the fact that the weight per unit volume of air decreases as its temperature increases or increases as its temperature decreases. As the air is heated in the furnace, it expands and becomes lighter. Because it is lighter, it is displaced by the cooler, heavier air entering the furnace. The warmer, lighter air moves through the leaders and up through the wall stacks where it enters the rooms through the registers. As the air cools, it becomes heavier and returns through the return air grilles and ducts to the furnace—hence the name *gravity* warm-air heating systems.

Air circulation in a gravity warm-air heating system depends upon the difference in temperature between the rising warm air and the cooler air that is falling and returning to the furnace for reheating. The greater the difference in temperature, the faster the movement of the air; however, natural conditions will place upper limits on the speed of the air movement, making it advisable to equip a gravity-type central furnace with a fan as an integral part of its construction. Such an integral fan should be powerful enough to overcome internal duct resistance to air flow and is *absolutely* necessary if air filters are used in the heating system. The motive force of air circulation in a standard gravity system is not strong enough to overcome the resistance of filters.

Planning a Gravity Warm-Air Heating System

A number of different trade and professional associations will provide information for planning a gravity warm-air heating system. A very useful source of information is the most recent edition of the National Warm Air Heating and Air Conditioning Association's *Gravity Code and Manual for the Design and Installation of Gravity Warm Air Heating Systems* (Manual 5).

The following recommendations for planning such a system are offered as a basic planning guide:

1. Calculate the heat loss from each room in the structure, and add these together for the *total heat loss*.
2. Plan the leaders (i.e., the horizontal ducts or pipes leading from the furnace) so that none are longer than 10 ft. in length.
3. Keep the number of elbows in the leader to a minimum.
4. Locate the warm-air outlets (registers) and return air grilles as close to the floor in each room as possible.
5. Locate the warm-air outlets (registers) on the inside walls of each room and in that part of the room closest to the furnace.
6. Select warm-air and return air ducts according to their Btu carrying capacities.
7. Select a furnace capable of delivering in Btuh a register delivery *equal* to the total heat loss calculated for the structure.
8. Locate the furnace in the *lowest* part of the structure and as near to the center as possible.

FORCED-WARM-AIR HEATING SYSTEMS

A *forced-warm-air heating system* (Fig. 6-5) consists of a furnace equipped with a blower, the necessary controls, an air-duct system, and an adequate number of suitably located warm-air registers and cold-air returns. The warm air travels from the furnace through the supply ducts to the room registers. The cold air returns through the return to the furnace, where it is reheated and recirculated. The return-air ducts are joined together before reaching the furnace, a feature distinguishing the duct arrangement in this type of heating system from that found in heating systems relying upon the gravity flow principle of air circulation (Fig. 6-6).

Planning a Forced-Warm-Air Heating System

ASHRAE publications (see, for example, the *ASHRAE 1960 Guide*) summarize design procedures for planning several types of forced-warm-air heating systems. The following six elements are common to each planning procedure:

143

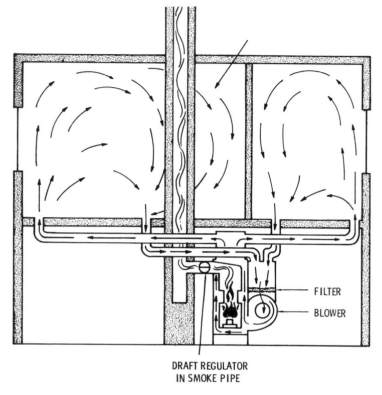

FILTER

BLOWER

DRAFT REGULATOR
IN SMOKE PIPE

Fig. 6-5. Operating principles of a forced-warm-air heating system.

1. Calculate the heat loss for each room or space, and from this data determine the total heat loss for the structure.
2. Determine the required furnace-bonnet capacity from the total heat loss of the structure.
3. Determine the location of the diffusers (warm-air outlets) on the building plan.
4. Calculate the required Btuh delivery of each diffuser.
5. Locate the position of the feeder ducts on the building plan.
6. Calculate the size of the feeder ducts on the basis of the total Btuh delivery it must supply.

Other elements in the design of a forced-warm-air heating system will be specifically oriented to the characteristics of each

144

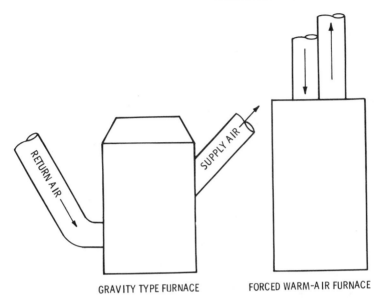

Fig. 6-6. Duct arrangement for gravity and forced-warm-air furnaces.

system. For example, it is recommended that the maximum length of an extended plenum (nonexistent in a forced-warm-air perimeter system) not exceed 35 ft. The locations and types of warm-air outlets and return-air inlets selected for the system will also depend upon the type of forced-warm-air heating system used in the structure.

The following publications are recommended as being the most useful and authoritative sources of information for installing forced-warm-air heating systems. In each case, the most current edition should be used.

1. *Warm Air Perimeter Heating* (National Warm Air Heating and Air Conditioning Association. Manual 4).
2. *Four-Inch Pipe Warm Air Perimeter Heating* (National Warm Air Heating and Air Conditioning Association. Manual 10).
3. *Standards for the Installation of Residence Type Warm Air Heating and Air Conditioning Systems* (National Fire Protection Association. No. 90A).

145

Proprietary systems always should be installed according to the manufacturer's instructions. Any variation from these instructions increases the probability of error and future problems in the system.

PERIMETER-LOOP, WARM-AIR HEATING SYSTEMS

The *perimeter-loop, warm-air heating system* (Fig. 6-1) was originally designed for use in residences built on a concrete slab rather than over a basement. The proven success of this duct arrangement in providing efficient and economical heating has resulted in its installation in all types of construction. This is rapidly becoming one of the most popular forms of warm-air heating.

In the perimeter-loop system, round ducts are imbedded in the concrete slab or suspended beneath the floor. The air is heated in a warm-air furnace equipped with a blower and forced through the ducts leading from the furnace to a continuous duct extending around the outer perimeter of the structure. The registers (diffusers) through which the heated air enters the various rooms and spaces are located along this outer perimeter duct. The most efficient operation can be achieved by placing these warm-air outlets on the floor next to the wall and *beneath a window* (Fig. 6-7). By placing the warm-air outlets here, window drafts are eliminated and the colder outside walls are kept warmed. Research shows that up to 80 percent of the heat loss from a structure can occur at these locations.

The National Warm Air Heating and Air Conditioning Association's publication *Warm Air Perimeter Heating* (Manual 4) is probably the most authoritative and useful source of information about this subject. (Ask for the most recent edition because the manual has undergone a number of revisions.)

CEILING PANEL SYSTEMS

A *ceiling panel system* is a forced-warm-air heating system in which the heated air is delivered through ducts to an enclosed

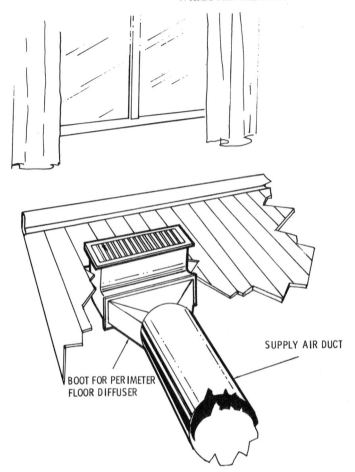

Fig. 6-7. Floor location for a warm-air outlet.

SUPPLY AIR DUCT

BOOT FOR PERIMETER
FLOOR DIFFUSER

space above a false ceiling. There are no air supply outlets in the
ceiling, and the heated air is therefore blocked from direct penetration of the occupied spaces below. Because its downward path
is blocked, the heated air spreads over the entire surface area of
the false ceiling. The ceiling eventually absorbs the heat, and the
cooler air is returned to the furnace for reheating and recirculation. The heat is transferred from the ceiling to the occupied
spaces by radiation. A more appropriate name for this type of

panel heating system might be "ceiling *space* panel system." Other ceiling panel systems are described in Chapter 1 of Volume 3 (Radiant Heating).

A ceiling panel system of this type does not differ from the standard forced-warm-air heating system except in the design of the heat-emitting unit. Instead of using compact (individual heat-emitting units) a ceiling panel system uses a portion of the structure itself. As a result, considerable experience is necessary to install this type of heating system. It is recommended only for new construction, and never as a conversion from an existing heating system.

ZONING A FORCED-WARM-AIR HEATING SYSTEM

The design of some residences, particularly the larger split-level homes, often causes balancing problems for the heating system. As a result, certain rooms will remain much colder than the rest of the house. This is a problem that cannot be corrected simply by turning up the heat because the rest of the house will become too hot. The most effective solution is to divide the heating system into two separate zones, each controlled by its own thermostat. A motorized damper is installed in the existing duct system and is regulated by thermostats to obtain the balance.

BALANCING A WARM-AIR HEATING SYSTEM

It is not always necessary to go to the expense of zoning to ensure that each room receives enough heat. Sometimes a heating system can be balanced by reducing the air delivery to rooms or sections of the structure that require less heat. This results in automatically diverting more air and heat to those areas that require it.

The procedure for balancing a warm-air heating system is as follows:

1. Pick a day for balancing the system when the outdoor dry-bulb temperature is 40°F or below.
2. Open the dampers in all warm-air outlets as wide as possible (Fig. 6-8). The same holds true if the dampers are in the supply ducts (Fig. 6-9).
3. Leave the thermostat at one setting for at least 3 hours, and make certain the furnace blower is running.
4. Check the temperatures in all rooms. This can be done with thermometers (if you are certain they register equally), or simply by making your own judgment.
5. Leave the dampers in the coldest rooms wide open, and adjust the dampers in the warmer rooms to obtain the desired balance. After each adjustment, allow the system to stabilize for at least 30 minutes before checking the temperature or making the next adjustment.

When balancing a heating system, do not expect immediate results. A little patience makes it well worth the effort.

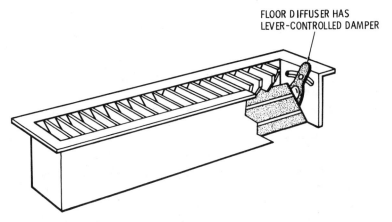

FLOOR DIFFUSER HAS
LEVER-CONTROLLED DAMPER

Fig. 6-8. Adjusting floor diffuser to full open position.

WARM-AIR FURNACES

The warm-air furnace is a self-contained heating unit designed to supply warm air to the interior of a structure. Warm-air furna-

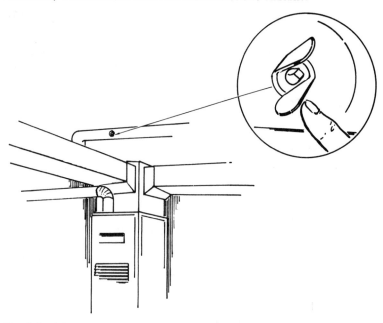

Fig. 6-9. Adjusting supply dampers to full operating position.

ces used in central heating systems usually employ a system of
ducts to distribute the air to the various rooms and spaces within
the structure.

These furnaces can be classified according to the method of air
circulation into the following two basic categories: (1) *gravity*
warm-air furnaces, and (2) *forced*-warm-air furnaces. A distinct
advantage of the forced-warm-air furnace is that its blower can
move the air in any direction. As a result, this type of furnace can
be located anywhere in the structure. Furthermore, the ducts do
not have to be located above the heating unit, as is the case with
gravity warm-air furnaces. These advantages of the forced-
warm-air furnace have contributed to its tremendous popularity
over gravity-type furnaces. Both types are described in considerable detail in Chapter 10 (Furnace Fundamentals).

Other criteria used for classifying warm-air furnaces include:
(1) the type of fuel used (e.g., gas, oil, coal, or electricity), (2) the
method of air distribution (ducts or pipeless), and (3) the method
of firing (automatic or manual). Specific chapters in this book

deal individually with some of these criteria (e.g., Chapter 11, Gas-Fired Furnaces; or Chapter 12, Oil-Fired Furnaces).

CONTROL COMPONENTS

The controls used in a warm-air heating system will depend upon a number of factors, including: (1) the method of air circulation used (e.g., forced or gravity), (2) the size of the structure, and (3) the method used for supplying the fuel (automatic or manual) to the fire. In other words, for all intents and purposes, a control system is custom designed to fulfill the requirements of a specific heating system.

All gravity and forced-warm-air furnaces in central heating systems are controlled by a thermostat located in one of the rooms. If the heating system is zoned, two or more room thermostats will be used. In forced-warm-air systems equipped with automatic burners or strokers, the thermostat will also control the operation of these units.

Each furnace should be equipped with a high-limit control to shut off the air or gas burned when plenum air temperatures exceed the furnace manufacturer's design limits. The high-limit control is automatic and will switch on the burner again as soon as the air temperature in the plenum has returned to normal (i.e., reached a level below the manufacturer's design limits). The high-limit control is frequently designed to operate in conjunction with the fan (blower) switch in forced warm-air furnaces.

Other controls used in warm-air heating systems are described in Volume 2—Chapter 4 (Thermostats and Humidistats), Chapter 5 (Gas and Oil Controls), and Chapter 6 (Other Automatic Controls).

DUCTS AND DUCT SIZING

Ducts are passageways used for conveying air from a furnace or cooling unit to the rooms and spaces within a structure. The ductwork is also used to return the air to its source for recirculation.

The air is distributed to the rooms or spaces by means of grilles, registers, and diffusers. The size and location of the supply-air outlets and the return-air inlets are determined by such design factors as: (1) use (heating and cooling), (2) air velocity, (3) throw, (4) drop, and (5) desired distribution pattern. These and other aspects of air outlet and air inlet design and selection are considered more thoroughly in Chapter 7 of Volume 2 (Ducts and Duct Systems).

Designing and sizing a duct system is a compromise between the requirements and limitations of both the structure and the heating and cooling system. Duct sizing will be affected by a number of variables, including: (1) the architectural design of the structure, (2) space limitations, (3) the required air supply, (4) the allowable duct air velocities, (5) the desired noise level, and (6) the capacity of the blower. Adequately sized ducts represent the best possible compromise among these variables.

Accuracy in estimating the resistance to the flow of air through the duct system is important in the selection of blowers for application to such systems. Resistance should be kept as low as possible in the interest of economy. However, underestimating the resistance will result in failure of the blower to deliver the required volume of air. The various calculation methods used for sizing ducts are given in Chapter 7 of Volume 2 (Ducts and Duct Systems).

A number of precautions should be taken in the design of a duct system. For example, careful study should be made of the building drawings with consideration given to the construction of duct locations and clearances. Other recommendations that should be considered when designing a duct system can be summarized as follows:

1. Keep all duct runs as short as possible, bearing in mind that the air flow should be conducted as directly as possible between its source and delivery points, with the fewest possible changes in direction.
2. Select locations of duct outlets so as to ensure proper air distribution.
3. Provide ducts with cross-sectional areas that will permit air to flow at suitable velocities.

4. Design for moderate velocities in all ventilating work to avoid waste of power and reduce noise.

COOLING WITH A WARM-AIR HEATING SYSTEM

The simplest method of cooling a house in the summer is to use the blower of a forced-warm-air furnace. Most furnaces have a manual blower switch that can be used to circulate the air.

If you plan to install a central air conditioning system capable of maintaining comfortable indoor temperatures during the warmer months, you must first examine the existing furnace and duct system and determine what changes (if any) must be made to handle the cooling load.

Forced-warm-air furnaces can often be converted to year-round air conditioning by installing a cooling coil in the furnace supply duct and connecting it to a compact compressor-condenser unit located outdoors. The existing duct system may be inadequate for air conditioning, but this can be modified to handle the cool air by increasing the size of certain ducts or by installing additional ducts.

Some furnaces (particularly gravity warm-air furnaces) cannot be fitted with cooling coils, *but* the existing ductwork is adequately sized for cooling. If this should be the case, the furnace can be bypassed with the cooling coils and air handling equipment installed in a short length of duct that feeds into the main supply- and return-air ducts at a point near the furnace. Dampers must be placed in the ducts between the furnace and the cooling coil to prevent the warm and cool air from mixing.

If both the existing furnace and ductwork cannot be used, it will be necessary to install a separate and independent air conditioning system. Adding one or more window units is the most common method of accomplishing this. Slightly more expensive is the use of a through-the-wall unit, which is installed by cutting a hole in the wall. The compressor-condenser section is located on the outside to reduce the noise level. Neither window units nor through-the-wall units use ductwork to distribute the cool air.

The most expensive separate and independent air conditioning systems are the single-package-unit systems and the very popular split systems. Both use ductwork to distribute the cool air. A single-package-unit system is limited to the installation of the cooling unit in the attic or basement, with the former location offering the fewest design problems. The advantage of a split system is that the cooling unit can be located anywhere inside the structure. In both systems, the preferred location of the compressor-condenser section is outdoors.

AIR CLEANING

The air cleaning equipment used in forced-warm-air heating and cooling systems commonly takes the form of washable or disposable air filters, or electronic air cleaners.

Washable and disposable air filters are installed in the return-air plenum of a forced-warm-air furnace as shown in Fig. 6-10. These are dry filters consisting of cellulose fibers, steel wood, or some other suitable material set in a wire frame. These filters are effective only when dust concentrations are limited to 4 grains

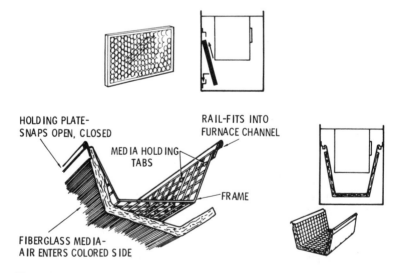

HOLDING PLATE-
SNAPS OPEN, CLOSED

RAIL-FITS INTO
FURNACE CHANNEL

MEDIA HOLDING
TABS

FRAME

FIBERGLASS MEDIA-
AIR ENTERS COLORED SIDE

Fig. 6-10. Examples of washable and throwaway air filters used in warm-air furnaces.

per 1000 cu. ft. of air or lower. Consequently, they are restricted in use to residences or small buildings. Air filters are not used in gravity warm-air heating systems because they impede the rate of air flow.

Electronic air filters are designed for return-air-duct installation at the furnace, air handler (in the duct) or air conditioning unit. They generally operate on the electrostatic precipitation principle and are capable of removing up to 95 percent of all airborne particles (e.g., dust, tobacco, and smoke). The dust particles are given an electric charge when they pass through an ionizing field and are collected when they subsequently pass between collector plates having an opposite charge. Typical installations are illustrated in Fig. 6-11.

The various types of air filters and electronic air cleaners used in cleaning and filtering the air are described in Chapter 14 of Volume 3 (Air Cleaners and Filters).

HUMIDIFIERS AND DEHUMIDIFIERS

Air that has a very low level of humidity is too dry for comfort and can cause damage to walls, interior woods, and furnishings. This results from the fact that dry air absorbs moisture from other sources, including the human body. The moisture-robbing effect of dry air is particularly noticeable on the sensitive nasal and throat membranes. Moisture can be added to the air by humidification, and the device used to add moisture is called a *humidifier*.

The humidifiers used in heating and cooling systems are designed to maintain the relative humidity within the comfort zone. They are available in a number of different types, each based on a different operating principle. Pan-type humidifiers, for example, contain a reservoir of water that evaporates into the warm air flowing through the supply duct. The water level in the reservoir is controlled by a float control. The desired humidity level is determined by a humidity control setting. When the relative humidity drops below the humidity control setting, a humidifier fan is actuated and air is blown over the water in the reservoir. Moisture is then picked up by the air and blown into the

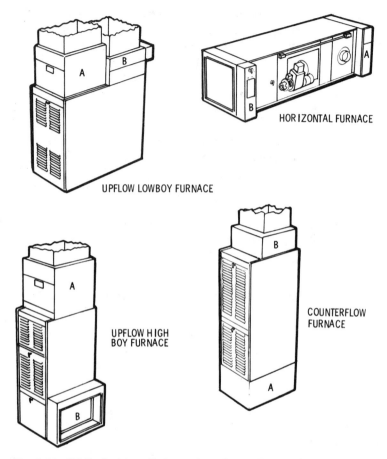

HORIZONTAL FURNACE

UPFLOW LOWBOY FURNACE

UPFLOW HIGH
BOY FURNACE

COUNTERFLOW
FURNACE

Fig. 6-11. (A) Typical installations of cooling coils and (B) electronic air
cleaners on forced-warm-air furnaces.

space to be humidified. Other types of humidifiers are: (1) spray-
type air washers and (2) steam-type humidifiers (either air or
electrically operated). Typical humidifier installations are illustrated in Fig. 6-12.

Sometimes air will have a humidity level that is too high. The
excess moisture resulting from this condition can also cause damage to walls, interior woods, and furnishings, as well as prove
very uncomfortable for the occupants. Excess moisture can be

removed from the air, and the device used to remove the moisture is called a *dehumidifier.*

Dehumidifiers operate either on the cooling or absorption method. The former method accomplishes dehumidification by an air washer with a water-spray temperature lower than the dew point of the air passing through the unit. Condensation occurs, and both latent and sensible heat are removed.

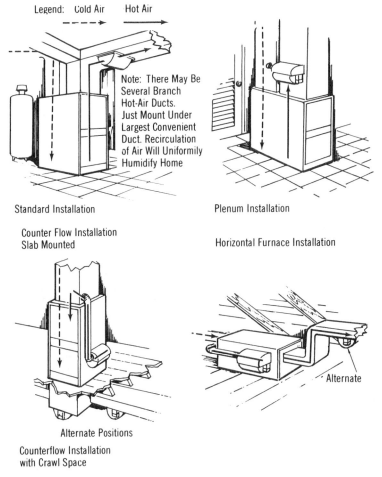

Fig. 6-12. **Typical installation of power humidifiers on forced-warm-air furnaces.**

In the absorption method of dehumidification, sorbent materials are used for removing moisture from the air. Air to be dehumidified is drawn or blown through a screened bed of dry-solid absorbent, and the water vapor in the air is caught and retained in the pores of the absorbent.

Both humidifiers and dehumidifiers are described in greater detail in Chapter 13 of Volume 3.

ADVANTAGES OF A WARM-AIR HEATING SYSTEM

By present comfort standards, gravity warm-air heating offers no special advantage and too many disadvantages to be recommended as a heating system in the types of structures popular with the public today. The architectural design of these structures *necessitates* the use of a forced-warm-air heating system. Because the gravity system lacks a blower, air circulation (and therefore heat distribution) depends upon the temperature difference between the rising warm air and the descending cold air. Unfortunately, many of our houses and buildings are designed in such a way that a gravity system would require a considerable temperature difference to obtain the desired rate of air movement.

If you design the structure *around* a gravity heating system, then quite different results are possible. For example, a residence built by a gentleman named Wendell Thomas in western North Carolina contains a gravity-type heating system using a simple wood-burning stove that produces indoor temperatures ranging between 60 and 75°F year-round. The north and west sides of the structure were set into the ground. No windows were placed in the north wall, and the windows in the other walls were two or three panes thick.

As shown in Fig. 6-13, the warm air heated by the stove rises and gives off its heat to the room. As it loses its heat, it cools and descends along the inside surfaces of the exterior walls, passes through metal air vents in the floor, and then enters the basement where it is warmed somewhat by the slightly warmer indoor temperatures. It then rises through metal air vents placed in the

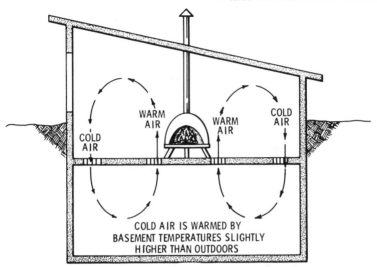

Fig. 6-13. A gravity-type system.

floor around the stove where it is reheated and recirculated. (Construction and operation details of this heating system can be obtained by sending 10 cents and a self-addressed envelope to *The Mother Earth News*, Box 957, Des Moines, IA 50304. Ask for Reprint No. 39, "The Self-Heating, Self-Cooling House.")

In a structure designed to take into consideration the operating principles of a gravity heating system, the operating costs will be lower than any other type system.

The advantages of a forced-warm-air heating system are as follows:

1. The installation costs are lower than those for hot-water or steam heating systems.
2. Heat delivery is generally quicker than other systems.
3. Heat delivery can be shut off immediately.
4. Air cleaning and filtering can be cheaply and easily provided.
5. The humidity level of the air can be easily controlled.
6. Air conditioning can be added without great difficulty or cost.

7. The furnace (because of its blower) can be placed in a number of different locations in the structure.

DISADVANTAGES OF A WARM-AIR HEATING SYSTEM

One of the principal disadvantages of a *gravity* warm-air heating system is that the furnace *must* be centrally located at the lowest point in the structure. This, of course, severely limits the design flexibility of the structure itself. Other disadvantages of gravity warm-air heating systems include:

1. Slow response to heat demand from the thermostat.
2. Slow air movement.
3. Inadequate or nonexistent air filtering.

Forced-warm-air heating systems depend upon blowers to circulate the air through a network of sheet-metal ducts. Both the operation of the blower and the expansion and contraction of the ducts as the air temperature rises or falls can cause distracting noise.

Warm-air outlets can frequently create decorating problems. In order to operate effectively, the registers should never be blocked by furniture or carpeting. This requirement, of course, restricts furniture arrangement and the placement of carpet.

The blower in a forced-warm-air heating system will create a certain amount of air turbulence. This rapid agitation of the air causes dust particles to circulate and deposit on walls, furniture, and other surfaces. Ordinary filters will remove most of the larger particles but are not very effective against smaller ones. An electronic air filter is recommended for the latter.

Warm-air heating systems require separate hot-water heaters to supply the hot water necessary for household use. This requirement for additional equipment adds to the initial installation costs for the system (in a hot-water heating system, the hot water for household use is supplied by the boiler).

Forced-warm-air heating systems supply convected heat through forced-air movement. The heat is supplied in bursts (rather than a continuous, sustained flow) when the room thermostat calls for it. This results in room temperatures varying up and down several degrees. The variation in temperature can be considerably reduced by setting the fan switch for continuous blower operation.

CHAPTER 7

Hot-Water
Heating Systems

Hot-water heating systems use water as the medium for conveying and transmitting the heat to the various rooms and spaces within a structure. The motive force for the water in these systems is based either on the gravity flow principle or forced circulation. The latter type (referred to as *hydronic*, or *forced hot-water, heating*) is the most commonly used method.

CLASSIFYING HOT-WATER
HEATING SYSTEMS

Hot-water heating systems can be classified in a number of different ways, depending on the criteria used. Three broad classification categories based on the following criteria are generally recognized:

1. Type of water circulation.
2. Piping arrangement.
3. Supply water temperature.

In all hot-water heating systems, the water is circulated either by forcing it through the line or by allowing it to flow naturally. The latter is referred to as a *gravity hot-water heating system* because circulation results from the difference in weight (specific gravity) of the water due to temperature differences (heavy when cold, light when hot). In a *forced hot-water heating system*, the accelerated circulation of the water can result from several commonly employed methods, including: (1) using high pressures, (2) superheating the circulating water and condensing the steam, (3) introducing steam or air into the main riser pipe, (4) using a combination of pumps and local boosters, and (5) using pumps alone.

The four principal piping arrangements used in hot-water systems are as follows:

1. One-pipe system.
2. Two-pipe direct-return system.
3. Two-pipe reverse-return system.
4. Series-loop system.

These piping arrangements are described and illustrated in several sections of this chapter (see, for example, "One-Pipe System" and "Two-Pipe Direct-Return System."

If a hot-water heating system uses supply water temperatures above 250°F, it is classified as a *high-temperature system*. High-temperature systems are used in large heating installations such as commercial or industrial buildings. A *low-temperature system* is one having a supply water temperature below 250°F and is used in residences and small buildings.

ONE-PIPE SYSTEM

A *one-pipe system* (Fig. 7-1) is one in which a single main pipe is used to carry the hot water throughout the system. In other words, the same pipe that carries the hot water to the heat-

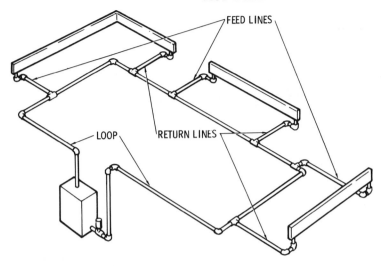

Fig. 7-1. One-pipe, hot-water heating system.

emitting units (i.e., radiators and convectors) in the various rooms and spaces within the structure also returns the cooler water to the boiler for reheating. Each heat-emitting unit is connected to the supply main by two separate branch pipes (a separate feed and return line). (See Fig. 7-2.)

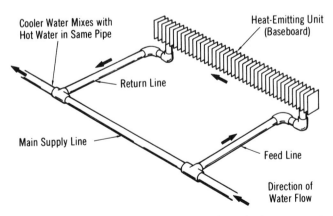

Fig. 7-2. Heat-emitting unit with two separate branch pipes connected to the main supply line.

165

The hot water flows from the boiler or heat exchanger (if a steam boiler is the heat source) to the first heat-emitting unit, through it to the second unit, and so on through each of the heat-emitting units in the one-pipe system until it exits the last one and returns to the boiler or heat exchanger.

One-pipe systems may be operated on either forced or gravity circulation. Special care must be taken to design the system for the temperature drop found in the heat-emitting units farthest from the boiler. This is particularly true of one-pipe systems designed for gravity circulation.

A principal advantage of a one-pipe system is that one or more heat-emitting units can be shut off without interfering with the flow of water to other units. This is *not* true of series-loop systems in which the units are connected in series and form a part of the supply line.

In some large one-pipe systems, zoning is possible by providing for more than one piping circuit from the boiler (Fig. 7-3). In such cases, each piping circuit is equipped with its own thermostat and circulating pump. Sometimes these circuits are erroneously referred to as "loops" and are confused with the piping arrangement in a series-loop system.

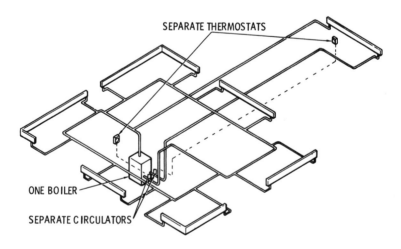

SEPARATE THERMOSTATS

ONE BOILER

SEPARATE CIRCULATORS

Fig. 7-3. One-pipe, hot-water heating system with more than one piping circuit for zoning.

SERIES-LOOP SYSTEM

In the *series-loop system* (Fig. 7-4), the heat-emitting units form a part of the piping circuit, i.e., *loop*, which carries the hot water from the boiler around the rooms and spaces within the structure and back to the boiler again for reheating. In other words, there are no branch pipes connecting the main supply pipe to the heat-emitting units as in the one-pipe system. In the series-loop system, the hot water flows from the boiler through a length of the main supply pipe to the first heat-emitting unit in the circuit (loop). It then flows through the unit to a length of the main supply pipe connected at the opposite end, flows through this pipe to the second heat-emitting unit in the circuit, and so on until the entire circuit is completed.

The series-loop system is cheaper and easier to install than other piping arrangements because it eliminates the need for branch pipes and reduces the amount of pipe used in the main circuit to the comparatively short lengths connecting the heat-emitting units. There is no need for one continuous length of main pipe.

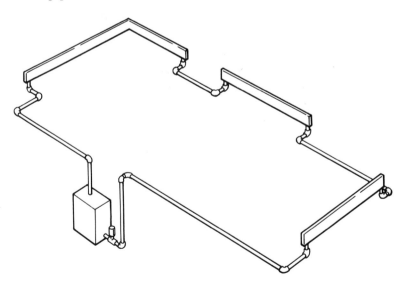

Fig. 7-4. Series-loop, hot-water heating system.

Because the heat-emitting units are connected in series and constitute a part of the main supply line, the same hot-water supply passes through each unit in succession. As a result, the heat-emitting unit closest to the boiler receives the hottest water, whereas units farther away receive water several degrees cooler. Furthermore, individual units cannot be shut off (unless there is a special bypass piping arrangement) without obstructing the flow of water to units farther along the line.

TWO-PIPE, DIRECT-RETURN SYSTEM

In a *two-pipe, direct-return system* (Fig. 7-5) hot water returns directly to the boiler from each heat-emitting unit. In other words, the hot-water supply and return mains are separate pipes. The heat-emitting units are connected to the supply and return lines by separate branches (Fig. 7-6).

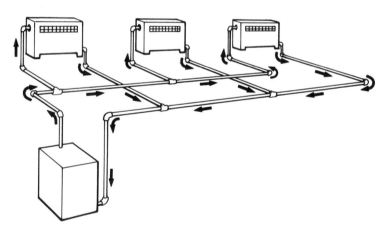

Fig. 7-5. Two-pipe, direct-return, hot-water heating system. The water from each heat-emitting unit returns directly to the boiler without passing through other units in the system.

Each heat-emitting unit represents the midpoint in a complete circuit within a two-pipe, direct-return system. The boiler completes the circuit. The farther a heat-emitting unit is located from the boiler, the greater the length of the piping in the circuit.

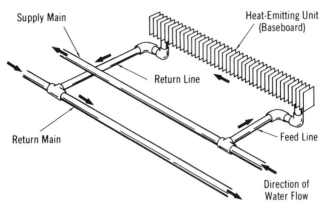

Supply Main

Heat-Emitting Unit
(Baseboard)

Return Line

Return Main

Feed Line

Direction of
Water Flow

Fig. 7-6. Heat-emitting unit with feed line connected to the supply main and return line connected to the return main.

Although this factor causes problems in balancing the water supply among the various circuits, balancing can be achieved in a number of ways, including: (1) balancing cocks, (2) proper pipe sizing, or (3) using a reverse-return system.

TWO-PIPE, REVERSE-RETURN SYSTEM

A *two-pipe, reverse-return system* (Fig. 7-7) achieves a balance in the water supply by creating circuits to the radiators of approximately equal length. Instead of allowing the return supply of water to proceed directly to the boiler from each radiator, the return main carries the water in the opposite direction for a predetermined distance before turning back to the boiler. The first radiator has the shortest supply main but the longest return main. For the farthest radiator, the reverse is true. Regardless of the position of a radiator in the system, the total length of pipe within the circuit of which it forms a part will be essentially the same as that of any other circuit.

COMBINATION PIPE SYSTEMS

Sometimes two or three different pipe arrangements will be combined in a single heating system. For example, it is not unus-

169

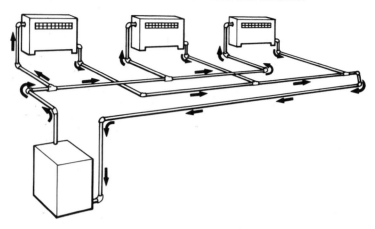

Fig. 7-7. Two-pipe, reverse-return, hot-water heating system.

ual to find a two-pipe, reverse-return system combined with a series-loop system. Combination pipe systems are usually found in commercial or industrial buildings and are fitted to specific design needs.

ZONING A TWO-PIPE SYSTEM

Balancing problems for two-pipe hydronic heating systems in larger houses and buildings can be solved by splitting the existing system into two or more separate zones independently controlled by their own thermostats. This can be accomplished by installing a two-position valve in the hot-water supply line. The valve is actuated by its own thermostat (Fig. 7-8).

RADIANT PANEL HEATING

Forced hot water can also be used in a radiant panel heating system (Fig. 7-9). In this heating system, no radiators or convectors are used. The hot water circulates through pipes concealed in the floor, ceiling, or walls, which function as heat-emitting surfaces. Radiant baseboard systems are also used. The principles of

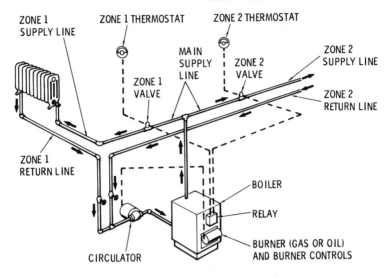

Fig. 7-8. Zoning a two-pipe forced hot-water heating system. Thermostatically actuated zone valves are installed on the main water supply line.

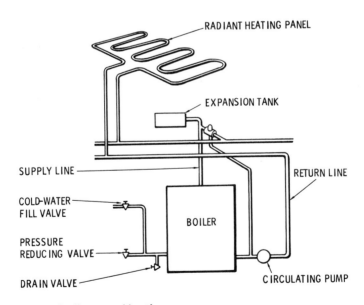

Fig. 7-9. Radiant panel heating.

171

radiant heating are described in greater detail in Chapter 11 of Volume 2 (Radiant Heating).

OTHER APPLICATIONS

It is possible to use the boiler of a hydronic heating system to supply heat for such purposes as snow melting, heating a swimming pool, providing domestic hot water for household use, and other applications. Separate circuits are created for each of these purposes and are controlled by their own thermostats. They are designed to tap into the main heating circuit from which they receive their supply of hot water.

Hot water for household use can be obtained by means of a heat exchanger or special coil (Fig. 7-10) inserted into the boiler. Cold water is run through this coil, heated as the boiler water gets hot, and then distributed to the various hot-water faucets.

In the snow-melting system illustrated in Figs. 7-11 and 7-12, the circuit is filled with a water and antifreeze solution. The snow melts as soon as it touches the surface.

GRAVITY HOT-WATER HEATING SYSTEMS

All hot-water heating systems are either gravity or forced circulation types. The latter generally use pumps to force the hot

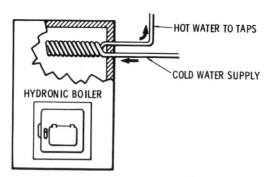

Fig. 7-10. Using the heat exchanger principle to provide domestic hot water without a special tank.

Fig. 7-11. Copper tubing imbedded in concrete driveway is connected to the hydronic system that heats the house. A small pump is used to circulate the liquid through the tubing.

water through the pipes (see "Forced Hot-Water Heating Systems" in this chapter). The circulation of hot water in gravity hot-water systems, on the other hand, results from the difference in weight (specific gravity) due to temperature differences. In

173

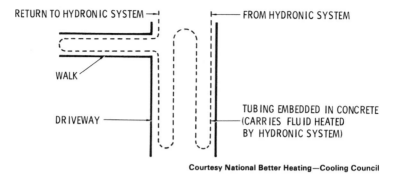

RETURN TO HYDRONIC SYSTEM

FROM HYDRONIC SYSTEM

WALK

DRIVEWAY

TUBING EMBEDDED IN CONCRETE
(CARRIES FLUID HEATED
BY HYDRONIC SYSTEM)

Courtesy National Better Heating—Cooling Council

Fig. 7-12. Hydronic snow-melting system.

other words, the motive force is due to the difference in density of the water at different temperatures—heavy when cold, light when hot. For this reason, gravity systems are also referred to as *thermal, or natural, hot-water heating systems.*

In order that water may transmit heat from the boiler (heating unit) to the radiators or other heat-emitting devices, there must be a constant movement (circulation) of the water from the heater to the radiators and back again. As mentioned above, this circulation in gravity systems is due to the difference in density (weight per unit volume) of water at different temperatures. For example, 1 cu. ft. of water at, say, 68°F weighs 62.31 lb.; at 212°F it will weigh 59.82 lb. This difference in weight is 2.49 lb. $(62.31 - 59.82)$, which is available to cause circulation, as shown in Fig. 7-13.

The difference in weight is caused by the expansion of water as its temperature is increased. This principle is illustrated in Fig. 7-14. Note that the top surface *abcd* of the cubic foot *A* has expanded to *A'*, an increase in volume represented by the linear increase *aa'*.

In the two-pipe system illustrated in Fig. 7-15, the unbalanced weight between the water in the riser and downflow pipe forms a motive force that causes the water to circulate through the system as indicated by the arrows.

The two-pipe, gravity hot-water system illustrated in Fig. 7-16 is equipped with an expansion tank. In operation after the fire is started, the temperature of the water in the boiler rises and

174

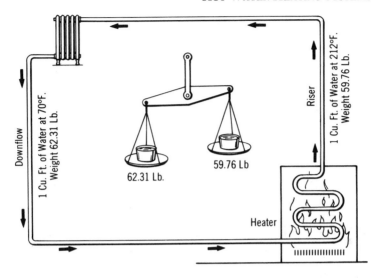

Fig. 7-13. Motive force in a gravity hot-water heating system.

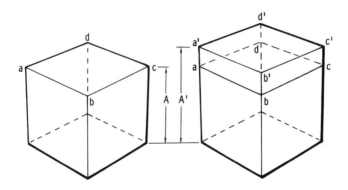

Fig. 7-14. The expansion of water with rise in temperature.

expands. This disturbs the equilibrium of the system, causing the colder and heavier water in the downflow pipe to flow downward, pushing the warmer and lighter water in the riser upward, thus starting circulation.

A circulation, or bypass, pipe is provided to form a continuous

175

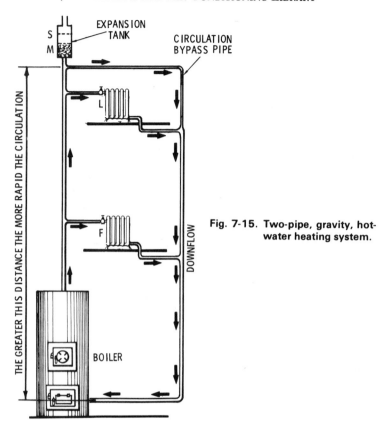

Fig. 7-15. Two-pipe, gravity, hot-water heating system.

path for the flowing water in the event that all the radiators are shut off. In the absence of this provision, all the water in the upflow pipe would be pushed up into the expansion tank followed by steam.

The expansion tank provides two essential functions for this system. As the water is heated, it expands and the tank provides space for this increase in volume. Thus, as it is heated, the water will expand from, say, elevation M to elevation S in the expansion tank (Fig. 7-15). The higher the elevation of the tank, the greater the pressure that may be brought on the water in the boiler (due to the head). As a result, the system may be worked at higher temperatures when desired.

176

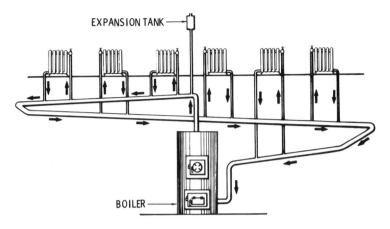

EXPANSION TANK

BOILER

Fig. 7-16. Circuit two-pipe, gravity, hot-water heating system.

Sometimes a special piping circuit can be built into a heating system (see "Two-Pipe, Direct-Return System" and "Two-Pipe, Reverse-Return System" in this chapter). In the two-pipe, gravity hot-water heating system illustrated in Fig. 7-16, a single main is taken from the boiler to a high point under the basement ceiling and then is pitched along its run as much as possible to the return inlet of the boiler. The piping circuit formed by this main consists of a closed loop of extra-large pipe in the basement having a pitch of not less than ½ in. per 10 ft. of run. The main supplies all the risers to the radiators above. This arrangement is not as efficient as the two-pipe system illustrated in Fig. 7-15, and for this reason the circuit main should be of very liberal size to reduce friction to a minimum.

In the so-called one-pipe system shown in Fig. 7-17, special distribution tees are employed to deflect part of the water from the main into the radiators while letting the balance flow through the main to the next radiator.

This is called a *one-pipe system* because one pipe serves both inlet and outlet of each radiator. In this system, there must be an upflow side *L* and a downflow side *F*. Thus, there are in reality two main pipes, but in function only *one* main pipe. The downflow may be considered as a continuation of the upflow instead of as a second pipe. The major objection to this system (as well as

177

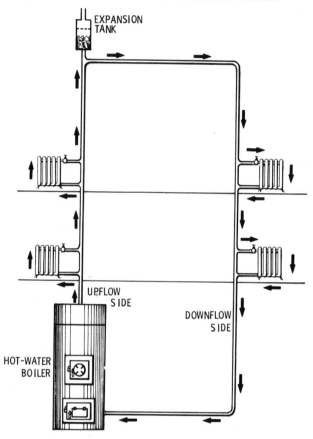

Fig. 7-17. One-pipe, gravity, hot-water heating system.

to the one illustrated in Fig. 7-16) is a lack of uniformity in heat distribution. This is primarily due to the fact that the radiators on the upflow side L are hotter than those on the downflow side F.

Distribution tees (Fig. 7-18) are used in this one-pipe system. They are provided with a baffle tongue, which deflects part of the water from the main in and out of the radiator while bypassing the balance along the line as indicated by the arrows in the illustration.

178

Another piping system used to distribute the hot water is the *overhead arrangement* illustrated in Fig. 7-19. This is practically the same as the so-called one-pipe system (Fig. 7-17), except that it has a divided circuit at the high point connecting with the downflow mains. No air vents are necessary because the arrangement is such that all air works to the top and passes off into the expansion tank. One advantage of this system is a better and more uniform distribution of the heat. The hot water enters each downflow pipe at the same temperature, equalizing the heat given off by each pair of radiators.

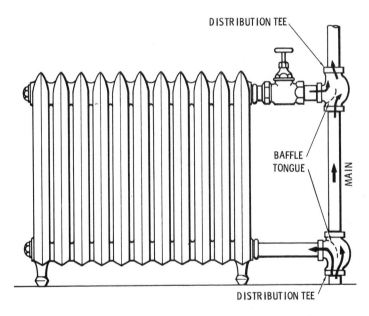

Fig. 7-18. Using distribution tees to connect a cast-iron radiator to the main in a one-pipe, hot-water heating system.

Although the overhead piping arrangement is not suited to all classes of buildings, there are many buildings, such as apartments, stores, office buildings, and hotels, where the general arrangement lends itself to this system.

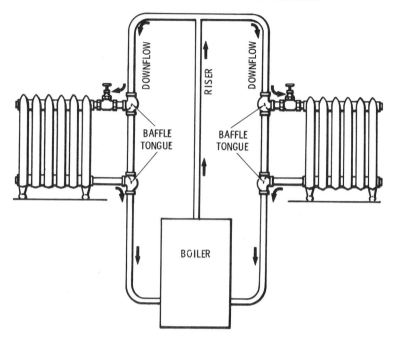

Fig. 7-19. Overhead piping arrangement in a gravity hot-water heating system.

FORCED HOT-WATER (HYDRONIC) HEATING SYSTEMS

Water as a medium for transmitting heat to radiators and other heat-emitting units gives up only sensible heat, as distinguished from steam systems that heat principally by the latent heat of evaporation. The result is that the temperature of the heat-emitting units of a steam system is relatively high as compared with hot water. In a hot-water system, latent heat is *not* given off; hence more radiating surface is needed to obtain equal heating effect.

The reasons for speeding up the flow of the water in hot-water heating systems are to increase the heating capacity of the heat-emitting units, make the system more responsive to load conditions, and allow the use of smaller pipes.

Numerous methods have been introduced to accelerate the circulation of hot water. The most commonly used methods have been based on the following techniques:

1. Introducing high pressures to gain greater temperature differences.
2. Superheating a part or all of the circulating water as it passes through the boiler and condensing the steam thus formed by mixing it with a portion of the cold circulating water of the return main.
3. Introducing steam or air into the main riser near the top of the system.
4. Forcing by pumps.
5. Combining pumps and local boosters.

The last two methods (pumps or a combination of pumps and local boosters) are generally the ones used in most modern forced hot-water heating systems. (See Fig. 7-20.)

HOT-WATER BOILERS

The boilers used in hot-water heating systems are made of cast iron or steel (Figs. 7-21 and 7-22). The cast-iron boilers generally display a greater resistance to the corrosive effects of water than the steel ones do, although the degree of corrosion can be significantly reduced by chemically treating the water.

Most hot-water heating boilers are designed to burn coal, gas (natural or propane), or oil. Some manufacturers will provide conversion devices for switching from one type of gas to the other (Fig. 7-23). Changing from coal to oil or gas is a frequent conversion provided for by conversion burners (see Chapter 16, Boiler and Furnace Conversions).

Never purchase an uncertified boiler. All certified boilers are approved by one of several organizations that have assumed this responsibility. For example, many boilers are certified by either the Institute of Boiler and Radiator Manufacturers (cast-iron boilers) or the Steel Boiler Institute (steel boilers). They will be

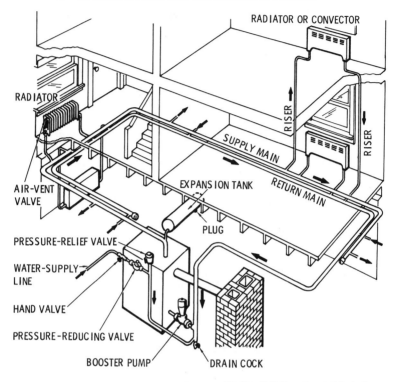

RADIATOR OR CONVECTOR

RADIATOR

RISER

RISER

SUPPLY MAIN

RETURN MAIN

AIR-VENT
VALVE

EXPANSION TANK

PLUG

PRESSURE-RELIEF VALVE

WATER-SUPPLY
LINE

HAND VALVE

PRESSURE-REDUCING VALVE

BOOSTER PUMP

DRAIN COCK

Courtesy U. S. Department of Agriculture

**Fig. 7-20. Two-pipe, forced, hot-water heating system with separate supply
and return mains.**

stamped "I.B.R." or "SBI," respectively. Fig. 7-24 illustrates some
of the certifications you will encounter on boilers.

Steam boilers can also be used in hot-water heating systems,
but a heat exchanger must be incorporated into the system to
transfer the heat from the steam to the hot water that flows
through the heat-emitting units in the rooms (see the next
section).

HEAT EXCHANGERS

Steam can also be used as a source of heat in a hot-water
heating system. If such is the case, then a heat exchanger (Fig.
7-25) must be used to convert the steam to hot water before it

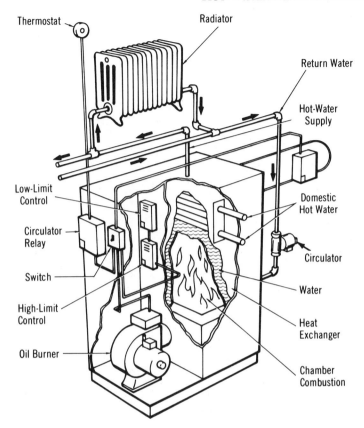

Thermostat

Radiator

Return Water

Hot-Water Supply

Low-Limit Control

Domestic Hot Water

Circulator Relay

Switch

Circulator

Water

High-Limit Control

Heat Exchanger

Oil Burner

Chamber Combustion

Fig. 7-21. Interior view of a typical boiler used in a hot-water heating system.

reaches the heat-emitting units. The heat exchanger is commonly of the steel-shell and copper-tube design. The steam flows through the shell and the water through the tubes (Fig. 7-26). The hot water returns to the steam boiler where it is reheated and changed to steam again.

CONTROL COMPONENTS

The efficient and safe operation of a hot-water heating system requires the use of a variety of different types of controls, which

Fig. 7-22. Examples of boilers used in hot-water heating systems.

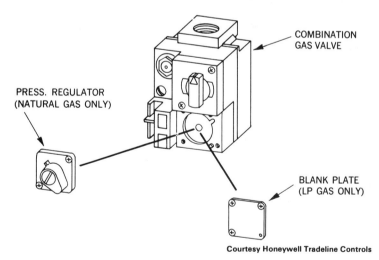

COMBINATION
GAS VALVE

PRESS. REGULATOR
(NATURAL GAS ONLY)

BLANK PLATE
(LP GAS ONLY)

Courtesy Honeywell Tradeline Controls

Fig. 7-23. Gas conversion devices used for changing from one type of gas to another.

CSA   FACTORY MUTUAL   A.G.A.   I.B.R.   ASME

BELGIUM   ENGLAND   CANADA   HOLLAND   GERMANY

Fig. 7-24. A partial list of certification symbols for boilers.

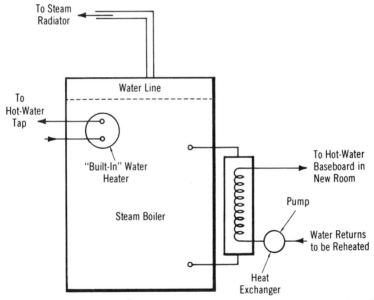

Courtesy National Better Heating-Cooling Council

Fig. 7-25. Heat exchanger used in connection with a steam boiler to provide hot water to baseboard heat-emitting units.

can be roughly divided into either system-actuating controls (e.g. the room thermostat, burner controls, and circulating pump controls) or safety controls (e.g. high-limit controls, pressure-relief valves, and pressure-reducing valves).

185

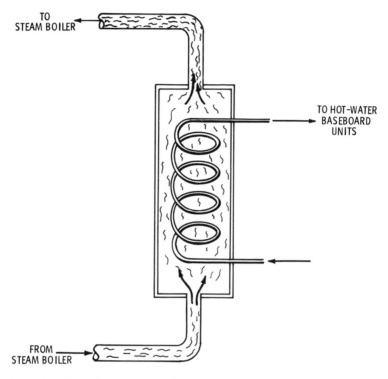

Fig. 7-26. Operating principles of a heat exchanger.

The safety controls prevent damage to the system by shutting it down when pressure and temperature levels become excessive. The *high-limit control,* or *aquastat,* is an example of such a control. It is a device designed to operate in conjunction with the circulating pumps and is located on the hot-water boiler. If the pressure or temperature of the hot water exceeds the design limits of the system, the system is shut down until conditions return to an acceptable level. Some aquastat relays provide multizone control when used with a separate circulator and a relay for each zone.

Another very important safety control for hot-water heating systems are *pressure-relief valves.* These valves are designed to open when pressure in the boiler reaches a certain level and close when the pressure returns to a safe level again. Pressure-relief valves *must* be installed in accordance with current ASME or local codes.

These and other controls (e.g., main shutoff valves, pressure-reducing valves, and drain cocks) are described in considerable detail elsewhere in this book—Chapter 15 (Boilers and Boiler Fittings) and Chapter 4 (Thermostats and Humidistats)—and in Chapter 9 of Volume 2 (Valves and Valve Installation).

PIPE AND PIPE SIZING

The selection of pipe used in hot-water heating systems depends primarily upon the following two factors: (1) the flow rate of the water and (2) the friction loss in the pipes. The *flow rate* of the water is measured in gallons per minute (gpm), and constant *friction loss* is expressed in thousandths of an inch for each foot of pipe length.

Pipes and pipe sizing methods are described in considerable detail in Chapter 8 of Volume 2 (Pipes, Pipe Fittings, and Piping Details).

EXPANSION TANKS

Expansion tanks (Figs. 7-27 and 7-28) are installed in hot-water heating systems to provide for the expansion and contraction of the water as it changes in temperature. Water expands with the rise of temperature, and the excess volume of the water flows into the expansion tank.

Another feature of the expansion tank is that the boiling point of the water can be increased by elevating the tank. In other words, increasing the *head* (i.e., the difference in elevation between two points in a body of fluid) increases the pressure. As a result, the water can be heated to a higher temperature without generating steam, which, in turn, causes the radiators or other heat-emitting devices to give off more heat.

There are both open and closed expansion tanks used in hot-water heating systems. The *open* expansion tank is used on low-pressure systems, and the *closed* tank is used on high-pressure

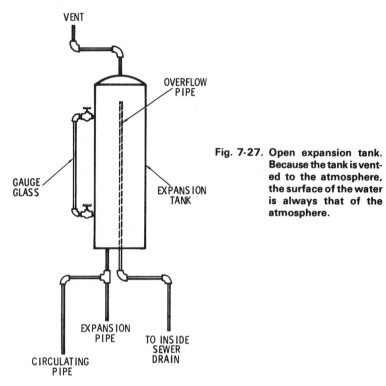

Fig. 7-27. Open expansion tank. Because the tank is vented to the atmosphere, the surface of the water is always that of the atmosphere.

systems. Air in the tank above the water forms a cushion for increasing the pressure. As the temperature of the water rises, it expands and flows into the tank, thus compressing the air and increasing the pressure.

The relation between pressure and volume changes of the air should be understood. According to Boyle's law, *at constant temperature the pressure of a gas varies inversely as its volume.* Thus, when the volume is reduced by one-half, the pressure is doubled. This is *not* gauge pressure, but absolute pressure (the pressure measured from true zero or point of no zero pressure).

In gravity hot-water heating systems, either closed or open piping arrangements can be used. In an *open* gravity system, the expansion tank is located at the *highest* point in the system (e.g., roof, attic, or top floor). (See Fig. 7-27.) The expansion tank used in this piping arrangement is an open type with an overflow pipe located at the top. Provisions can be made to return the overflow water to the boiler or to discharge it through outside runoff drains.

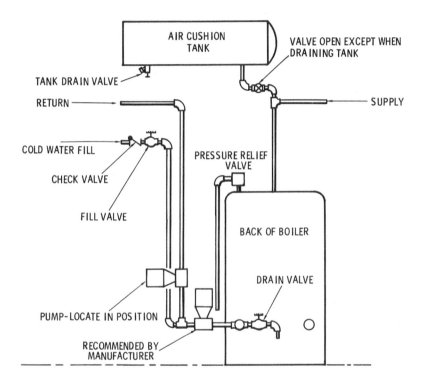

Fig. 7-28. Closed expansion tank. This type of expansion tank is sealed against free venting to the atmosphere.

In a *closed* gravity system, a closed, airtight expansion tank is located near the hot-water boiler. Higher pressures (and, consequently, higher water temperatures) result as pressure builds up in the system. Pressure-relief valves are installed on the main supply line to prevent the buildup of too much pressure.

Forced (hydronic) hot-water heating systems are closed systems with expansion tanks located near the boiler.

CIRCULATING PUMPS

In forced hot-water heating systems, circulating (booster) pumps are used to force or circulate the hot water through the pipes. Fig. 7-29 illustrates a typical location of one of these pumps on a return line.

Circulating pumps for hot-water heating systems are designed and built to handle a wide range of pumping capacities. They will vary in size from small booster pumps with a 5-gpm capacity to those capable of handling thousands of gallons per minute.

In order to select a suitable pump for a hot-water heating system (i.e., one that will produce maximum flow without overloading the pump motor), it is necessary to correctly match the oper-

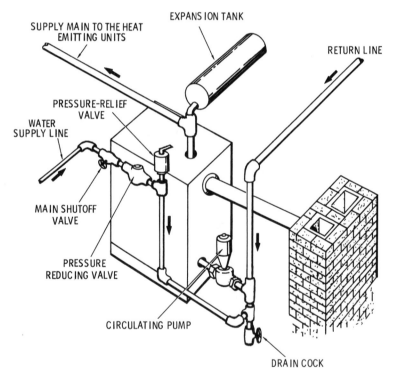

Fig. 7-29. Circulating pump is located on the return water line next to the boiler.

ating characteristics of the pump to the requirements of the heating system.

The correct location of a pump is crucial to its successful operation. Pump manufacturers will usually give detailed information in their literature on this point. Circulating pumps with mechanical seals are very common in hot-water heating systems.

DRAINAGE

There should be a provision for adequate drainage facilities in a hot-water heating system during shut-down periods. If drainage is not provided for, there is always the danger of the unheated water freezing in the pipes during extreme cold weather. Chemical "antifreeze" solutions have also been developed for use in hot-water heating systems in an attempt to prevent the water from freezing during shut-down periods.

Pitching all pipes to a central point is an important consideration when designing the system because it is essential to effective drainage.

HEAT-EMITTING UNITS

The hot water in both gravity and forced hot-water heating systems is circulated through the pipes to the radiators or other heat-emitting units from which the heat is transferred into the room.

Conventional radiators (Fig. 7-30) are often set on the floor or mounted on the wall. Many examples of these can still be found in older buildings. Some attempts have been made to conceal radiators by recessing them in the wall of the room or enclosing them (partially or entirely) in cabinets. Recessed radiators must have some sort of insulation between them and the wall. One-inch thick insulation board, a sheet of reflective insulation, or a combination of both is recommended for this purpose. If the radiator is partially or entirely covered with a cabinet, then openings should be provided at both the top and bottom of the cabinet for air circulation.

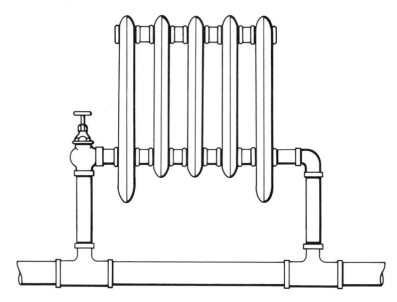

Fig. 7-30. Conventional cast-iron radiator still found in many older hot-water heating systems.

Convectors usually consist of finned tubes enclosed in a cabinet (Fig. 7-31) or baseboard unit (Fig. 7-32) with openings at the top and the bottom. The hot water circulates through the tubes, which radiate heat. Air enters the bottom of the cabinet or baseboard and exits through the openings in front.

Baseboard radiator units (Fig. 7-33) consist of long, narrow tubes directly behind a baseboard face. The hot water flows through the tubes, heats the baseboard face, and the heat is radiated from the surface into the room. Make sure that the metal cover of the baseboard unit is not too thin (an indication of a cheaply built unit). Thin covers are easily dented, and water temperature changes in the heating element can cause a thin baseboard cover to be noisy.

AIR CONDITIONING

Air conditioning can be added to a structure heated by a hot-water heating system in a number of different ways. Central air

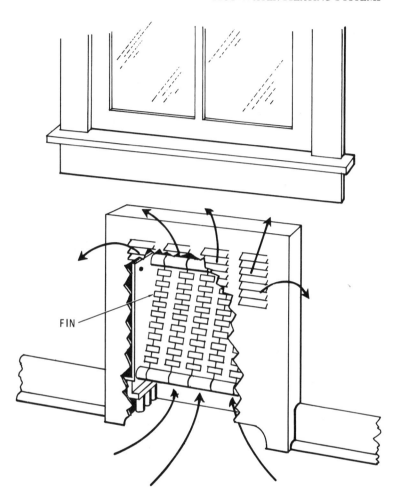

FIN

Fig. 7-31. Finned-tube connector in an upright cabinet.

conditioning can be added with the use of the so-called split-system installation. The evaporator-blower unit is installed in the attic, and the condensing unit (compressor and condenser) is located outside the house.

Chapters 8 to 11 of Volume 3 provide details for a variety of air conditioning systems, many of which are suitable for adding to houses and buildings with existing hot-water heating systems.

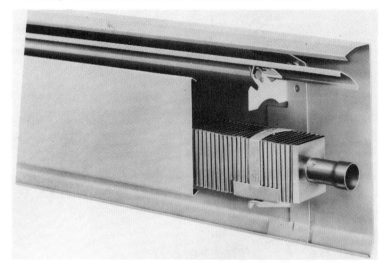

Fig. 7-32. Baseboard radiator units.

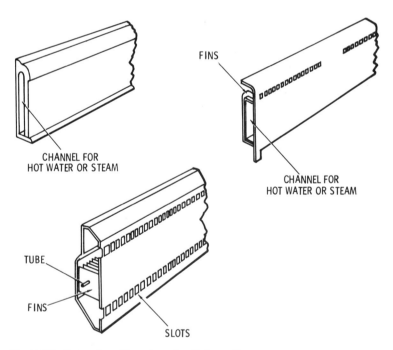

FINS

CHANNEL FOR
HOT WATER OR STEAM

CHANNEL FOR
HOT WATER OR STEAM

TUBE

FINS

SLOTS

Fig. 7-33. Examples of baseboard radiator units.

MOISTURE CONTROL

If the humidity level in a structure is too low, the dry air will absorb moisture from other sources, such as plaster, wood floors, wood furniture, fabrics (e.g., drapes and upholstery), and human bodies. As a result, the plaster tends to crack, the wood shrinks, the fabrics become dry and brittle, and the sensitive membranes in the nose and throat dry out faster than the human body can replace the moisture. These problems caused by dry air can be avoided by maintaining an adequate amount of moisture in the air.

A hot-water heating system will generally maintain an adequate humidity level without the additional assistance of a separate humidifying unit. The humidifier illustrated in Fig. 7-34 is an example of the type that could be used in a hydronic heating system should one be required. It is designed with a hot-water coil on the air inlet side of the unit. The heated boiler water circulates through this coil. A blower in the unit draws filtered room air over the coil and through a wetted rotating pad. The moist air is then circulated through the rooms by the blower. The humidifier is controlled by a room humidistat.

ELECTRICALLY HEATED SYSTEMS

Compact electric-fired boilers, operated essentially the same way as a heating plant powered by coal, oil, or gas, are also available for hot-water heating systems. These boilers are characterized, not only by their compactness, but also by their quiet operating characteristics.

The extreme compactness of an electric-fired, hot-water heating system is illustrated in Fig. 7-35. The heat exchanger, expansion tank, and operating controls are all mounted on the wall in a relatively small space.

A central heating unit (e.g., an electric-fired boiler) can be eliminated in some systems by using thermostatically controlled electric heating components in the baseboard units. Fig. 7-36 illustrates a system of this type. Note that there is no boiler. The water is circulated through the system by a pump, and a uniform

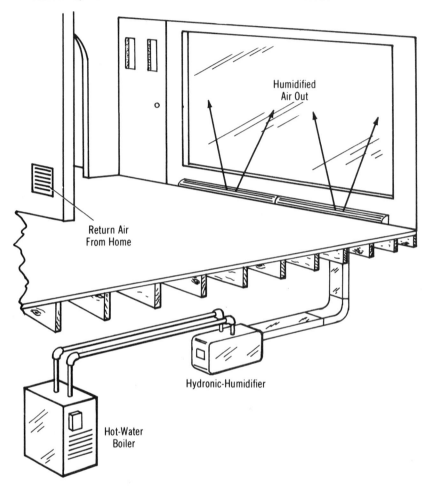

Fig. 7-34. Typical installation of a humidifier in a hydronic heating system.

temperature can be maintained by the heating element in each baseboard.

The heating system illustrated in Fig. 7-36 is a single-loop installation. It is also possible to divide the whole system into a series of sealed or closed units filled with a water and antifreeze solution operating on the gravity-flow principle. Each unit could be controlled by a thermostat, or several units by a wall thermostat. The advantage of this type of installation is that the heating capacity can be easily increased if the house or building is enlarged.

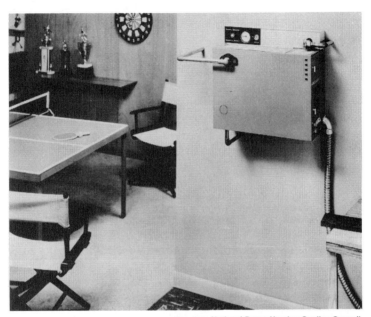

Fig. 7-35. Compact electric-fired boiler.

ADVANTAGES AND DISADVANTAGES OF HOT-WATER HEATING SYSTEMS

There are several distinct advantages to using hot-water heating as opposed to steam heating. For one thing, hot-water heating is more flexible than low-pressure (above-atmospheric) steam systems because the temperature may be widely varied. Due to the low working temperature of the water, the heat from a hot-water system is relatively mild and the room atmosphere is not robbed of any of its healthful qualities. Moreover, these systems function as reservoirs for storing heat because the radiators

197

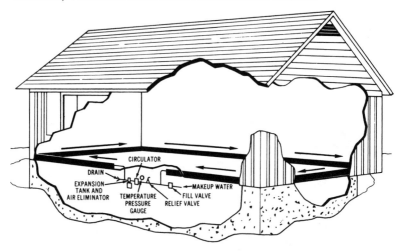

Fig. 7-36. Single-loop installation consisting of electrically heated hydronic baseboard units. Water is circulated through the entire loop by a pump.

remain warm a considerable length of time after the fire in the boiler is extinguished.

A principal disadvantage of using hot-water heating is the initial high installation cost as compared with steam and warm-air heating systems.

CHAPTER 8

Steam Heating Systems

Steam is a very effective heating medium. Until recently, this property of steam has resulted in its being the most commonly used method of heating residential, commercial, and industrial buildings. Over the past forty years or so, steam heating has been largely replaced in residences and small buildings by other heating systems that have often proven to be less expensive to install and operate or that operate at similar or greater levels of efficiency in small structures.

The basic operating principles of steam heating are relatively simple. A boiler is used to heat water until it turns to steam. When the steam forms, it rises through the pipes in the heating system to the heat-emitting units (radiators, convectors, etc.) located in the various rooms and spaces in the structure. The metal heat-emitting units, being cooler, cause the steam to condense and return to the boiler in the form of water (condensation) for reheating.

CLASSIFYING STEAM HEATING SYSTEMS

There are a number of different methods of classifying steam heating systems, but the most commonly used methods include one or more of the following features:

1. Pressure or vacuum conditions.
2. Method of condensation flow to the boiler.
3. Piping arrangement.
4. Type of piping circuit.
5. Location of condensation returns.

Steam heating systems can be divided into low-pressure and high-pressure types, depending on the operating pressure of the steam used in the system. A *low-pressure system* commonly operates at a pressure of 0 to 15 psig, whereas a *high-pressure system* uses operating pressures in excess of 15 psig.

Both *vapor* and *vacuum steam heating systems* operate at low pressures (0 to 15 psig) and under vacuum conditions. The latter system uses a vacuum pump to maintain the vacuum; the vapor system does not, relying instead on the condensation of the steam to form the vacuum.

The condensation from the heat-emitting units is returned to the boiler either by gravity or some mechanical means. When the former method is used, the system is referred to as a *gravity return system*. If mechanical means of returning the condensation are employed, the system is referred to as a *mechanical return system*. The three types of mechanical devices used to return the condensation in mechanical return systems are: (1) the vacuum return pump, (2) the condensation return pump, and (3) the alternating return trap. (Each of these devices is described in the appropriate sections of this chapter.)

Using the piping arrangement as a basis for classification, a steam heating system will be either a one-pipe or a two-pipe system. A *one-pipe system* is designed with a single main that carries the steam to the heat-emitting units and the condensation back to the boiler. In other words, it functions as both a supply and a return main. In a *two-pipe system*, there is both a supply main and a return main.

The *piping circuit* may best be described as the "path" taken by the steam to the riser (or risers). In a *divided-circuit* installation, two or more risers are provided for the steam supply. A *one-pipe-circuit* installation, on the other hand, employs a single riser from the boiler to carry the steam supply to the heat-emitting units. A *loop-circuit* installation is used when it is necessary to operate heat-emitting units at locations *below* the water level of the boiler.

A steam heating system may also be classified according to the direction of steam flow in the risers (i.e., supply mains). An *upfeed system* is designed so that the risers are *below* the heat-emitting units. In other words, the steam supply moves from the boiler up to the heat-emitting units in the rooms and spaces within the structure. An upfeed system is also sometimes referred to as an *upflow,* or *underfeed, system.* A *downfeed system* is one in which the steam supply flows down to the heat-emitting units. In this system, the supply main is located *above* the heat-emitting units.

Sometimes the location of the condensation return is used as a basis for classifying a steam heating system. If the condensation return is located below the water level in the boiler, it is referred to as a *wet return.* A *dry return* is a condensation return located above the water-level line.

GRAVITY STEAM HEATING SYSTEMS

A gravity steam heating system is characterized by the fact that the condensation is returned to the boiler from the heat-emitting units by means of gravity rather than mechanical means. Both one- and two-pipe installations are used, with the former being the oldest and most commonly used for a number of years (Figs. 8-1 and 8-2). Gravity systems are generally limited to residences and small buildings where the heat-emitting units can be located at least 24 in. above the water-level line of the steam boiler.

Gravity steam heating systems are the cheapest and easiest to install because they are adaptable to most types of structures, but they do have a number of disadvantages inherent to the system. The principal disadvantages of gravity steam heating systems are:

1. The return lines in *one-pipe* gravity systems must be large enough to overcome the resistance offered by the steam flowing up from the boiler in the opposite direction.

2. There is the possibility of water hammer developing in one-

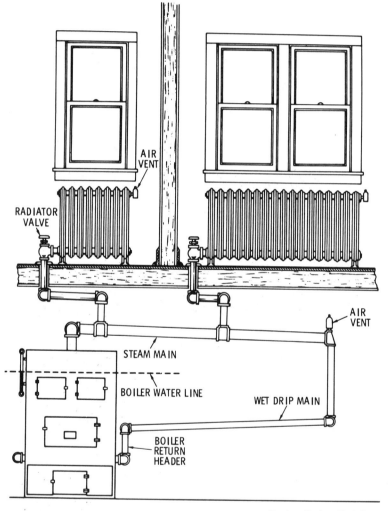

Fig. 8-1. One-pipe, gravity, steam heating system.

pipe systems because the steam and condensation must flow in opposite directions in *the same pipe.*

3. Air valves (which are required) sometimes malfunction by either spurting water or failing to open. If the latter situation is the case, excess heat will build up in the system.

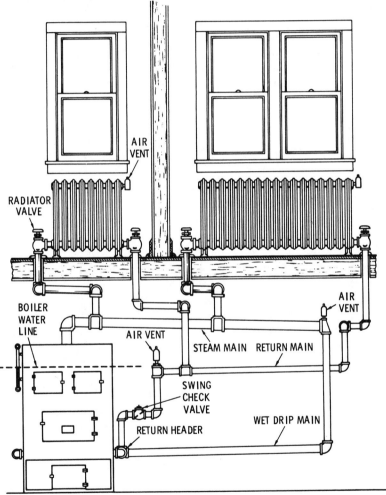

Fig. 8-2. Two-pipe, gravity, steam heating system.

203

4. Automatic control of the steam flow from the boiler results in room-temperature fluctuations.
5. Comfortable room temperatures are possible by manually regulating the valves on individual heat emitting units, but this results in some inconvenience.
6. In *two-pipe gravity systems*, the condensation return from each heat emitting unit must be separately connected to a wet return or water sealed. This is expensive.

The principal reason for the development of the two-pipe gravity system is to create a means of overcoming the resistance offered by the steam flow to the condensation returning to the boiler.

ONE-PIPE, REVERSE-FLOW SYSTEM

The *one-pipe, reverse-flow system* is the simplest and cheapest of this type to install. This system is easily distinguished by the absence of any wet or dry returns to the boiler. Supply mains from the boiler are inclined upward and connect with the room heat-emitting units, there being no other piping. This system is called "reverse flow" because the condensation flows back through the mains in a reverse direction or opposite to that of the steam flow.

The operation of a typical one-pipe, reverse-flow system is shown in Fig. 8-3. Steam from the boiler flows into the main or mains (inclined upward) through the risers to the heat-emitting units at the bottom. The steam pushes the air out of the mains, risers, and heat-emitting units and escapes through air valves placed at the other end of the heat-emitting units, as shown. The condensation forming in the units flows back to the boiler through the same piping, but in the opposite or reverse direction to the steam flow.

For satisfactory operation, every precaution should be used to install the *correct* pipe sizes, expecially for the mains. If the piping is too small, it will be necessary to carry excess pressure in the boiler to ensure proper operation. Without the excess pressure, operation of the remote heat-emitting units will be unsatisfac-

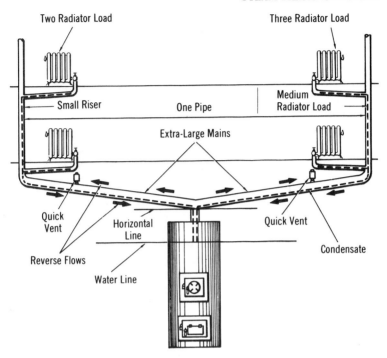

Fig. 8-3. One-pipe, reverse-flow, heating system.

tory. Satisfactory operation of these remote units can be achieved without excess boiler pressure if the pipes used in the system are the correct size.

The smaller the main, the greater the speed of the steam flow and the greater the resistance to the flow of condensation. It is very difficult for the condensation to flow back to the boiler against the onrushing steam in a long main that is in an almost horizontal position. The main should be inclined as much as conditions in the basement will permit.

Leaky radiator valves can be a problem. When the valve does not close tightly, steam will work its way into the radiator and stop the condensation from coming out. The result is that the radiator soon fills with water, and when turned on again, there is difficulty getting the condensation out. This produces gurgling, hissing, and the more violent effect known as "water hammer."

UPFEED ONE-PIPE SYSTEM

One-pipe steam heating systems can also be of the *upfeed* type. In a standard upfeed one-pipe steam heating system, both the steam and condensation travel through the same pipes, and the heat-emitting units are located *above* the supply mains (hence the name *upfeed system*).

An attempt to reduce the amount of condensation return in the pipes carrying the steam has resulted in a modification of the upfeed one-pipe system. In the modified system, the condensation is dripped at each radiator (and therefore from the main itself) into a wet return.

The upfeed one-pipe system shown in Figs. 8-4 and 8-5 consists of a main or mains branching off from the boiler steam outlet and inclined downward, instead of upward as in the previously described one-pipe system. This arrangement causes the condensation in the mains to flow in the same direction as the steam.

The mains connect with return pipes, which carry all the condensation back to the boiler. It is only in the risers that reverse flow of the condensation takes place, and accordingly they

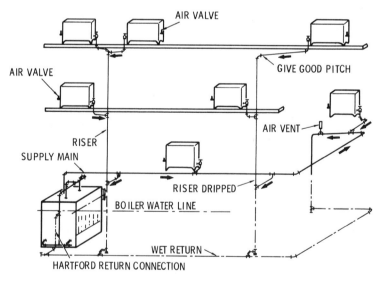

Fig. 8-4. Upfeed gravity, one-pipe, air-vent, steam heating system.

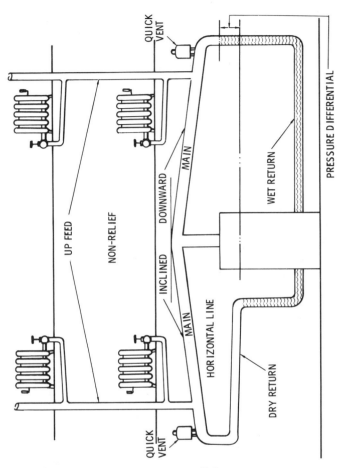

Fig. 8-5. One-pipe, under- or upfeed, nonrelief system.

should be large enough to take care of this reverse flow without undue turbulence.

Steam from the boiler flows through the main or mains, accruing condensation with it; the condensation returns (hence the name) to the boiler at a much lower level, as shown in Figs 8-4 and 8-5. After traversing the mains, steam flows through the risers and into the radiators. Condensation takes place in warming the radiators and drains through the risers and return pipes to the boiler.

207

A distinction is made between a *wet, or sealed, return* and a *dry return*. A wet return is below the water level in the boiler, whereas a dry return is above the water level. The advantage of a wet return is that it seals and prevents steam at a slightly higher pressure entering the return. Under most circumstances, a wet return is preferable to a dry return. The latter is necessary to clear doorways or other openings.

There are no disadvantages to a wet return when the system is properly installed and the valves maintained in tight condition. In most installations, these requirements are often lacking, resulting in many noninherent disadvantages. Among the installation problems that should be avoided are: (1) pipes that are too small for the job, (2) sharp turns, (3) air pockets, (4) not enough pitch to the mains, (5) not enough air valves, and (6) air valves that are too small or cheap.

UPFEED ONE-PIPE RELIEF SYSTEM

Fig. 8-6 shows an *upfeed one-pipe, relief system* applied to an eight-radiator installation. Connected to the main outlet *A* are two or more branch mains *AB* and *AC*, which supply the various risers. Steam is supplied to each riser, the condensation draining in the riser in reverse direction to the steam flow.

The condensation returns from the risers to the boiler by gravity through *drip pipes*, which are virtually continuations of the return pipe so that the condensation will flow back into the boiler.

Steam (usually at from 1 to 5 psig) passes from the boiler to mains *AB* and *AC*. These branches being slightly inclined, any condensation will drain into the drip pipes. The steam passes through the risers to the radiators, where its heat is radiated in warming the rooms, thus causing condensation. The risers being of liberal size, the condensation is carried by gravity in a direction reverse (that is, opposite) to the direction of the steam flow, drains down the drip pipes, and returns to the boiler through the low-level return pipe.

The condensation is forced back into the boiler by the pressure resulting from the greater head of the column of condensation in the drip pipes, as compared with the head of water in the boiler. Moreover, the water in the drip pipes, being at a lower temperature than the water in the boiler, is heavier, which upsets the equilibrium of the two columns.

This system is characterized by a slight difference in pressure (pressure differential) in the various parts of the system caused by frictional resistance offered by the pipe to the flow of the steam. The steam flow is variable in different parts of the system due to variable condensation and badly proportioned pipe sizes. This can be explained by Fig. 8-6. If the water level in the boiler is at D, then in operation with wet returns, the pressure difference will be balanced by the condensation standing at different levels in the different drip pipes, as at E and F, these levels being such that the difference in head and density will restore equilibrium. The effect of a wet return can be obtained with the dry return (shown at the left in Fig. 8-6) by attaching a siphon to the bottom of the drip pipe. Water from the drip falls into the loop formed by the siphon; and after it is filled, it overflows into the dry return.

The water will rise to different heights, G and H, in the legs of siphon to balance the difference in pressure at points P and P'. If the siphon were omitted and the drip pipe connected directly to

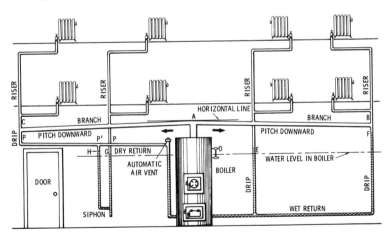

Fig. 8-6. One-pipe, under- or upfeed, relief system showing locations of dry and wet returns.

the dry return, there would be a tendency for the condensation in the dry return to back up instead of draining into the boiler because the pressure in the drip pipe at P is greater than the pressure in the dry return. In general, the pressure varies because there is a gradual reduction in pressure as the steam flows from the boiler to the remote parts of the system. This is caused by frictional resistance offered by the pipe and fittings (Fig. 8-7).

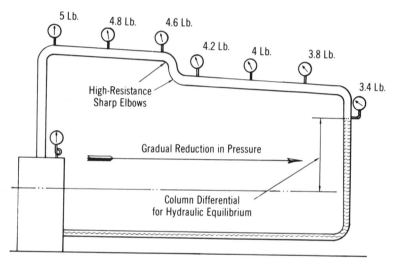

Fig. 8-7. Fractional resistance offered by the pipe and fittings.

The steam flows into the pipes only when condensation is taking place. The variation in pressure exists, on the other hand, only when the steam is flowing in the pipes. The effects of pressure variation can be explained with the aid of Fig. 8-8. With the plant in operation and condensation taking place in the radiators and draining into the drip pipes (suppose the pressure in the boiler is 5 psig—4 psig in drip 1 and 3 psig in drip 2), then to balance these pressure differences the water will rise in drip 1 to L 2.3 ft. above the water level in the boiler, because there is a pressure difference of 1 psig (5 − 4) and the weight of a column of water 2.3 ft. is 1 psig for each square inch of cross section. Similarly, for drip 2, the pressure difference is 2 psig; hence, the water will rise in this column to an elevation above the water level in the boiler equal to 2.3 × 2, or 4.6 ft., to balance the 2-psig pressure difference. Strictly speaking, these figures are correct only when the temperature of the water in the two drip pipes is the same as the temperature of the water in the boiler. For simplicity, the difference in weight or density of the "cold" water in the drips and hot water in the boiler were not considered. Under these circumstances, the heavy cold water in the drips would rise to lower elevations than those shown at *L* and *F* in Fig. 8-8.

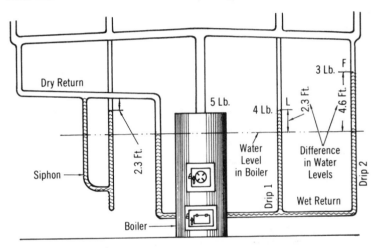

Fig. 8-8. Effects of pressure variation in the different parts of a system.

Sometimes there are problems encountered on long lines with radiators located at the end of the line and near the level of the water in the boiler. On long lines where there is considerable reduction of pressure, the water sometimes backs up into the radiator, as in Fig. 8-9, interfering with its operation.

Radiators located at elevations below the water level in the boiler may be operated by means of a steam loop (Fig. 8-10). In the steam loop, the condenser element may consist of a pipe radiator placed on the floor above the boiler. The liberal condensing surface thus provided will render the loop very active in removing the condensation, and at the same time the heat radiated from the condenser is utilized in heating. The drop leg is provided with a drain cock *D*, and the connection to the boiler with a check valve. To start the system, turn on the steam at the

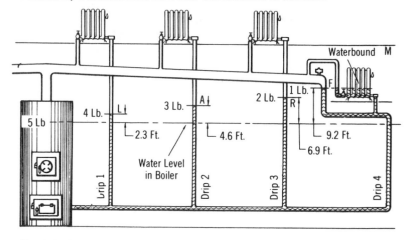

Fig. 8-9. Effects of water backing up in radiator on end of long line.

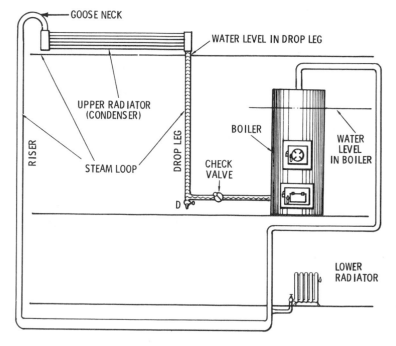

Fig. 8-10. Steam-loop method of operating radiator placed below water level in boiler.

boiler and open *D* until the steam appears. The condensation of steam in the condenser (upper radiator) will cause a rapid circulation in the riser, carrying with it the condensation from the radiator, which, in passing over the goose neck, cannot return but must gravitate through the upper radiator and drop leg past the check valve and into the boiler. The pipe at the bottom of the main riser, which acts as a receiver for the condensation from the lower radiator, should be one or two sizes larger than the pipe in the main riser.

DOWNFEED ONE-PIPE SYSTEM

The *downfeed one-pipe system* (also referred to as the *one-pipe oversystem* or simply the *downfeed system*) is characterized by having the heat-emitting units located *below* the supply mains. The condensation drips through the risers, thereby keeping the supply main relatively free of the condensation.

The downfeed one-pipe system is well suited for buildings two to five stories high because if the heat-emitting units were fed from below, the risers would have to be excessively large. Instead of a steam main encircling the basement, it is carried to the attic, forming a central riser for all the heat-emitting units. The branches in the attic correspond to the mains in the systems already described but do not carry any condensation from the heat-emitting units. These branches connect with the drops or downflow pipes that feed the heat-emitting units and drain the condensation.

As shown in Fig. 8-11, steam flows from the boiler up the central upflow pipe through the branches and down through the drops or downflow pipes. The branches being inclined, any water or condensation in the steam drains into the upflow riser. The steam passes to the heat-emitting units through the connecting pipes. Condensation forming in the units drains through these connecting pipes into the downflow pipes. Air valves or vents are provided to rid the system of air. It should be noted that there is no *reverse* flow of condensation in the drop pipes. As a result, pipe sizes can be made smaller because, with parallel flow, steam velocities can be higher than in the one-pipe system.

213

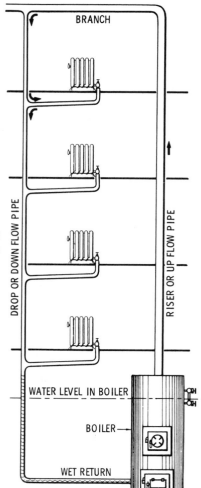

Fig. 8-11. One-pipe, overhead feed system installed in a tall building.

ONE-PIPE-CIRCUIT SYSTEM

In the *one-pipe-circuit system* the steam main is carried entirely around the basement, taken off from the boiler by an elbow at the high point, as in Fig. 8-12.

Note that the main *must* incline all the way from the high point to the low point. To allow for this inclination requires twice as much riser between high and low points as with the divided-circuit system, that is, where there are two mains taken off from a tee connection.

The one-pipe system is adapted for a rectangular building of low or moderate size. The size of the main (since all the steam flows through it) must be larger than the divided-circuit system. However, especially in large installations, a saving in piping may be made by installing a "tapered main." A tapered main is one that is reduced in size along its length by connecting lengths of different size pipes with reducers. Eccentric reducers should be used to avoid water pockets, which would interfere with the proper drainage of the condensation. The risers are connected by being tapped from the main at various points to serve the heat-emitting units.

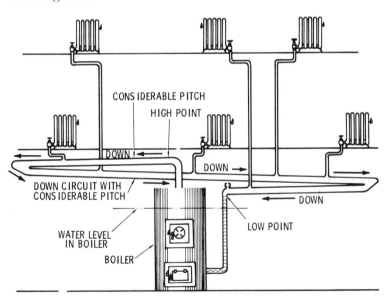

Fig. 8-12. One-pipe, under- or upfeed, nonrelief, circuit system.

In a one-pipe system, the condensation drains into the main and flows in the same direction as the steam flow, being carried to the drip pipe, and then into the boiler. Since there is no return pipe as with the relief system, the circuit arrangement is less expensive to install.

Proportioning a tapered main is very important. The amount of condensation increases from the beginning to the end of the main and near the end is considerable, depending upon the number of heat-emitting units. Allowance should be made for this, and too much tapering should be avoided.

ONE-PIPE, DIVIDED-CIRCUIT NONRELIEF SYSTEM

The *one-pipe, divided-circuit nonrelief system* differs from the one just described in that there are two mains at the high point taken off by a tee as in Fig. 8-13. These mains terminate in a U-shaped drip connection (*LF*) at the low point. Each should be vented with a quick vent as shown. Evidently each main takes care of only half the total condensation.

This steam heating system is suited to long buildings with a boiler located near the center. The end of each main is connected to a separate drip pipe connected with a common return, giving separate seals for each end.

For proper operation, these ends should not be at a lower

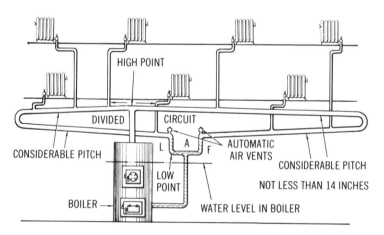

Fig. 8-13. One-pipe, under- or upfeed, nonrelief, divided-circuit system.

elevation than 14 in. above the boiler water line. The individual seals make the two halves of the divided circuit independent, which is desirable for unequal loads. Thus, there may be considerable difference between the pressure at L and F, each being what is necessary to balance the load.

ONE-PIPE-CIRCUIT SYSTEM WITH LOOP

The *one-pipe-circuit system with loop* is adapted to L-shaped buildings, a circuit being used for the main building and a loop (tapped from the circuit) servicing the wing (Fig. 8-14). The mains are installed for the proper drainage of the condensation by providing two high points—one at the beginning of the circuit and the other at the beginning of the loop—thus giving ample margin above the boiler water line for liberal pitch in both the circuit and the loop. This is clearly shown in Fig. 8-14.

The low point of the loop is higher than the low point of the circuit because the pressure at the end of the loop is less than the pressure at the end of the circuit. This requires a longer vertical drip pipe since the liquid column rises higher to balance the lower pressure.

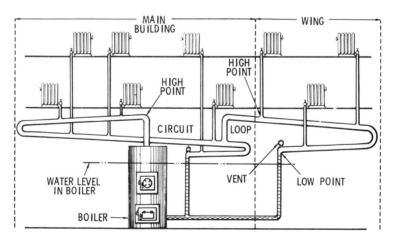

Fig. 8-14. One-pipe, under- or upfeed, nonrelief, circuit system with loop for L-shaped building.

TWO-PIPE STEAM HEATING SYSTEMS

In two-pipe systems, separate pipes are provided for the steam and the condensation; hence they may be of smaller size than in one-pipe systems where a single pipe must take care of both steam and condensation. Various piping arrangements are used in two-pipe systems (e.g., circuit, divided-circuit, and loop) to best meet the requirements of the building. Steam is supplied to the heat-emitting units through risers, and the condensation is returned through downflow or drip pipes.

TWO-PIPE, DIVIDED-CIRCUIT SYSTEM

A *two-pipe, divided-circuit system* is shown in Fig. 8-15. In this two-pipe system, steam passes from the boiler at the high point to the mains, along which risers bring the steam to the heat-emitting units. From the opposite side of each unit is connected a drop or drip pipe. Fig. 8-15 shows a wet return on the left side and a dry return on the right side.

In the wet return, the condensation is returned to the boiler by means of individual drips provided at each connection. There is a different arrangement for the dry return. The drip pipe from

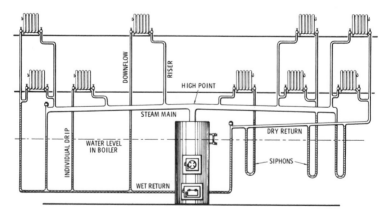

Fig. 8-15. Two-pipe, under- or upfeed, relief, divided-circuit system.

each heat-emitting unit terminates in a loop or siphon, which is tapped to the dry return. In operation, the condensation gradually fills the siphons and flows over into the dry return, passing into the boiler drip pipe.

A two-pipe system should also be provided with a check valve so placed as to allow water to pass into the boiler but prevent undue outflow. Under certain conditions, this prevents the water in the boiler from being driven out into the return system by the boiler pressure. The disadvantage of a check valve is that it sometimes gets stuck, a problem that can interfere with the operation of the system. An equalizing pipe with a Hartford connection loop may be used in place of a check valve to avoid this situation (see "Hartford Return Connection" in this chapter).

VAPOR STEAM HEATING SYSTEMS

A *vapor steam heating system* is one that commonly uses steam at approximately atmospheric pressure or slightly more, and that operates under a vacuum condition without the aid of a vacuum pump (see "Vacuum Steam Heating Systems" and Figs. 8-16 to 8-18).

The steam pressure at the boiler necessary to operate a vapor system is generally very low (often less than 1 lb.), being no more than is required to overcome the frictional resistance of the piping system. Under most operating conditions, the pressure at the vent will be zero or atmospheric.

Vapor steam heating systems may consist of various combinations of closed or open, upfeed or downfeed, and one-pipe or two-pipe arrangements, depending on the requirements of the installation. Any of these combinations will have certain advantages and disadvantages.

OPEN (ATMOSPHERIC) VAPOR SYSTEMS

A vapor system with a return line open to the atmosphere *without* a check, trap, or other device to prevent the return of air is sometimes referred to as an *open*, or *atmospheric, system.*

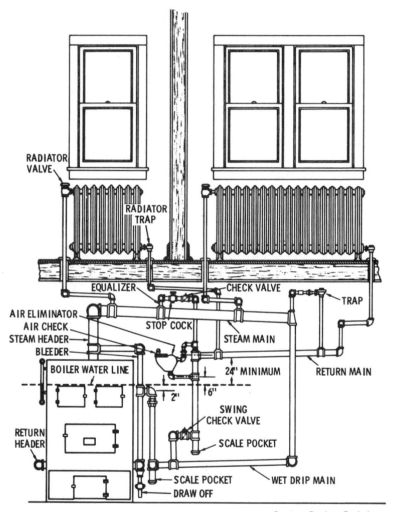

RADIATOR VALVE

RADIATOR TRAP

EQUALIZER

CHECK VALVE

TRAP

AIR ELIMINATOR
AIR CHECK
STEAM HEADER
BLEEDER

STOP COCK

STEAM MAIN

BOILER WATER LINE

24" MINIMUM

RETURN MAIN

RETURN HEADER

2"

6"

SWING CHECK VALVE

SCALE POCKET

SCALE POCKET

WET DRIP MAIN

SCALE POCKET

DRAW OFF

Fig. 8-16. A vapor steam heating system.

An open vapor system is frequently used when the steam is delivered from its source under high pressure. When this is the case, pressure-reducing valves should be installed in the system to reduce the pressure of the steam to a suitable operating level. An open vapor system is also used when there is no need to return the condensation to the boiler (i.e., when it is wasted within the system). A condensation-return pump should be used when the system design requires the return of the condensation to the boiler.

In an open vapor system, the pressure at the boiler is 1 to 5 oz., or enough to overcome the frictional resistance of the piping system. The pressure at vent is zero gauge or atmospheric. In operation, steam is maintained at about 5 oz. pressure in the boiler by the action of the automatic damper regulator. The amount of heat desired at the radiators is regulated by the degree of opening of the supply valve. Steam enters at the top of the radiator and pushes out the air through the outlet connection, which is open to the atmosphere. The condensation returns to the boiler by gravity. This system has the advantage of heat adjustment at the radiator, but the devitalizing effect in the air is somewhat greater than in the vacuum systems because the steam

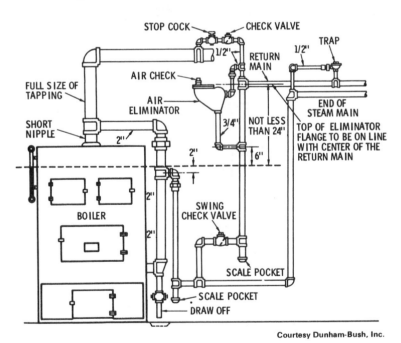

Courtesy Dunham-Bush, Inc.

Fig. 8-17. Piping of a vapor system to a boiler with only one supply tapping.

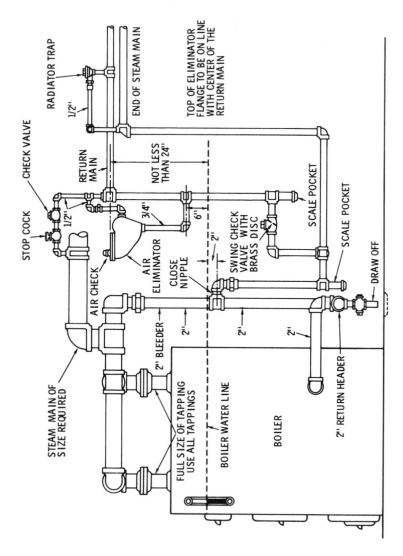

Courtesy Dunham-Bush, Inc.

Fig. 8-18. Piping of a vapor system to a boiler with two or more supply tappings.

entering the radiators is at a higher temperature than the steam of lower pressure in the vacuum system. It is, however, simple.

Fig. 8-19 illustrates a very simplified, mechanically controlled, vapor steam heating system. Though it does not present latest practice, it does present control principles very plainly.

The success of a vapor steam heating system depends upon the proper working of the automatic damper regulator in keeping the boiler pressure within proper limits. To accomplish pressure regulation, the dampers are controlled by a float working in a float chamber in communication with the water space in the boiler, as shown in Fig. 8-19.

When the pressure in the boiler is the same as that of the atmosphere (0 psig), the water level in the float chamber (Fig. 8-19) is the same as that in the boiler and the index hand points to zero.

As steam generates, the steam pressure increases and the water level in the boiler is forced downward. The latter action causes the level in the float chamber to rise until the pressure due to the difference *AB* (Fig. 8-19) of water level balances that in the boiler.

The float, in rising, connected as it is by pulleys and chains to

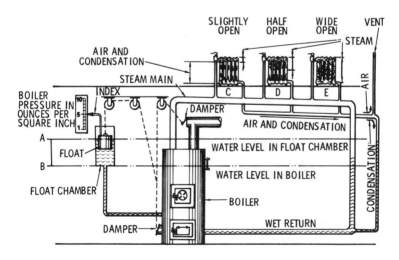

Fig. 8-19. Open atmospheric vapor system.

the dampers, closes the ash pit damper, thus checking the draft and preventing a further increase of steam pressure.

In this system, the steam feed is connected to the top of the radiators and the air and condensation is taken from the bottom because steam is lighter than either air or condensation. Accordingly, when steam is admitted, it floats on top of the air, thus driving the air out through the lower connection.

The chief feature of a vapor system is that the amount of heat given off by each radiator may be regulated by the steam valve. Thus, in Fig. 8-19, the valve of radiator C is opened just a little, which will admit only just enough steam to heat a larger portion of the radiator; with the valve wide open on E, the entire radiator is heated.

The kind of radiator used is the downflow type in which steam enters at one end at the top and the air and condensation passes out at the other end at the bottom.

As steam enters a cold radiator it forces the cool air in the radiator out through the trap into the return piping. The operation of a typical downflow radiator is shown in Figs. 8-20 to 8-22. Fig. 8-20 shows the steam entering and air passing out through thermostatic retainer valve. Fig. 8-21 shows more steam entering

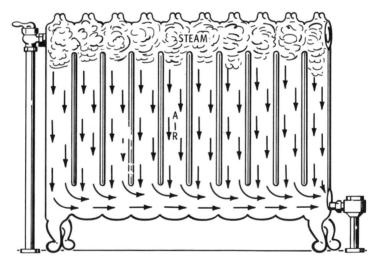

Fig. 8-20. Steam entering top of radiator and pushing air out the bottom.

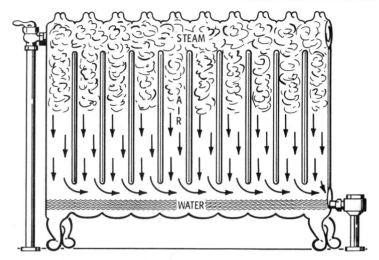

Fig. 8-21. Condensation and air leaving the radiator.

and condensation and the balance of the air passing out through the trap, the action progressing until (as in Fig. 8-22) the radiator is full of steam.

As the radiator warms up, the steam gives off heat and in so

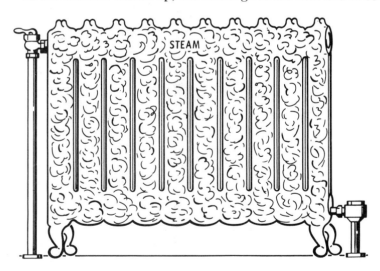

Fig. 8-22. Radiator filled with steam.

225

doing condensation takes place. The condensation, being heavier than steam, falls to the bottom of the radiator and flows to the trap through which it passes into the return piping. After the air is forced out, the steam fills the radiator and follows the condensation to the trap. The trap closes when the steam enters it because the steam is hotter than the water. This excess heat expands the valve control element, closing and holding the valve against its seat with a positive pressure, thus preventing the steam from flowing into the return piping.

The trap closes once the radiator is completely filled with steam, and heat is given off as the steam condenses. The condensation thus formed, which is cooler than the steam, flows in a steady stream to the trap which it slightly chills, causing it to open and allowing the condensation to pass out into the return piping (Fig. 8-23 and 8-24).

When properly working, the trap adjusts itself to a position corresponding to the temperature of the condensation, just as a thermometer does to the room temperature, and permits a continuous flow of condensation from the heat-emitting units (Fig. 8-25).

CLOSED VAPOR SYSTEMS

A vapor steam with a return line closed to the atmosphere is sometimes referred to as a *closed vapor* system. The condensation returns by gravity flow to a receiving device (an alternating receiver or boiler-return trap), where it is discharged into the boiler. The condensation from the alternating receiver is discharged against the boiler pressure.

Because air cannot enter a closed vapor system, a moderate vacuum is created by the condensing steam. As a result, steam is produced at lower temperatures, and the system will continue to provide heat after the boiler fire has died down.

Fig. 8-26 shows the arrangement of a typical upfeed, two-pipe vapor system with an automatic return trap. The heat-emitting units discharge their condensation through thermostatic traps to the dry return pipe. These systems operate at a few ounces pressure and above, but those with mechanical condensation return

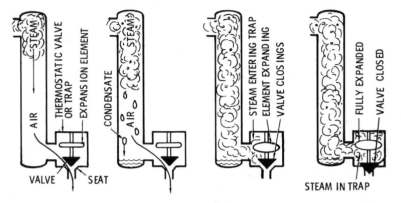

Fig. 8-23. Progressive action of thermostatic check or trap from beginning of air entrance to closing of valve by expansion of actuating element.

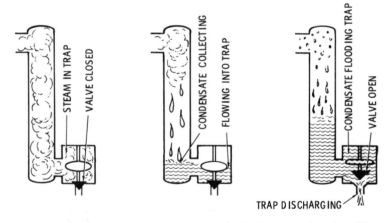

Fig. 8-24. Progressive action of thermostatic check or trap closed position to open position due to contraction of actuation element when chilled by relatively cool consensation.

devices may operate at pressures upward of 10 psi. The simplest method of venting the system consists of a ¾-in. pipe with a check valve opening outward. Most systems employ various forms of vent valves, which allow air to pass and prevent its return. A dry return is provided so that the air will easily go out the vent pipe.

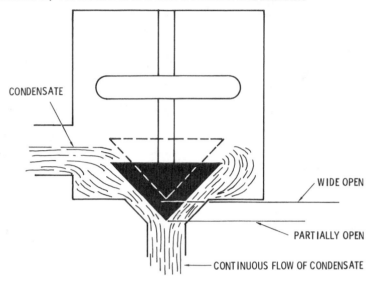

CONDENSATE

WIDE OPEN

PARTIALLY OPEN

CONTINUOUS FLOW OF CONDENSATE

Fig. 8-25. Detail of trap showing the valve in intermediate position for ideal continuation flow of condensation.

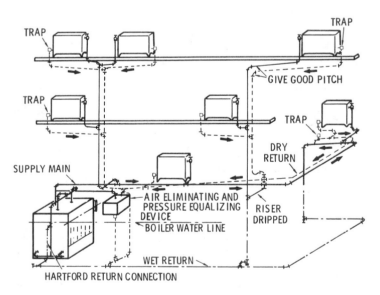

TRAP

TRAP

GIVE GOOD PITCH

TRAP

TRAP

DRY RETURN

SUPPLY MAIN

AIR ELIMINATING AND PRESSURE EQUALIZING DEVICE

BOILER WATER LINE

RISER DRIPPED

WET RETURN

HARTFORD RETURN CONNECTION

Fig. 8-26. Two-pipe, upfeed, vapor system with automatic return trap.

VACUUM STEAM HEATING SYSTEMS

A *vacuum heating system* is one that operates with steam at pressures less than that of the atmosphere. The object of such systems is to take advantage of low working temperatures of the steam at these low pressures, giving a mild form of heat such as is obtained with hot-water heating systems. Vacuum systems such as these, which operate at all times at pressures less than atmospheric, should not be confused with the combined atmospheric and vacuum systems described in the next section. There are distinct design differences between the two systems.

A distinction should be made between a vacuum system and a subatmospheric system. The latter differs from an ordinary vacuum system in that it maintains a controlled partial vacuum on both the supply and return sides of the system instead of only on the return side. In the vacuum system, steam pressure above that of the atmosphere exists in the supply mains and heat-emitting units practically at all times. The subatmospheric system is characterized by atmospheric pressure or higher existing in the steam supply piping and heat-emitting units only during severe weather (Figs. 8-27 and 8-28).

There are a number of different methods of classifying vacuum systems (e.g., one-pipe or two-pipe and vacuum pressure or subatmospheric). For the purposes of this chapter they will be classified according to the type of vacuum: natural vacuum systems and mechanical vacuum systems.

NATURAL VACUUM SYSTEMS

Any standard one- or two-pipe steam system may be converted into a *natural vacuum system* by replacing the ordinary air valve with a mercury seal or connecting thermostatic valves to the radiator return outlet on radiators and providing a damper regulator on coal-burning boilers adapted to vacuum working. The mercury-seal system is shown in Fig. 8-29.

A mercury seal is virtually a barometer, consisting, as shown in Fig. 8-30, of a tube (A) that dips just below the surface of the mercury in a cup (B). When the steam is raised in the boiler to

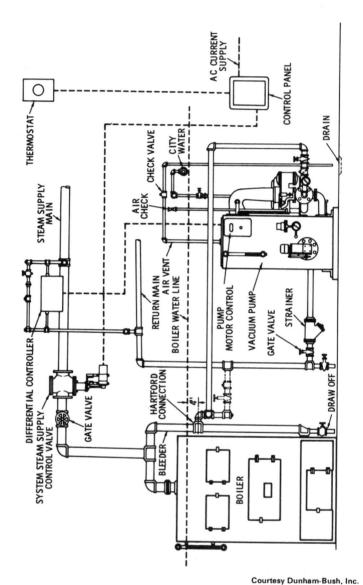

Fig. 8-27. A typical variable-vacuum, steam-heating system installation.

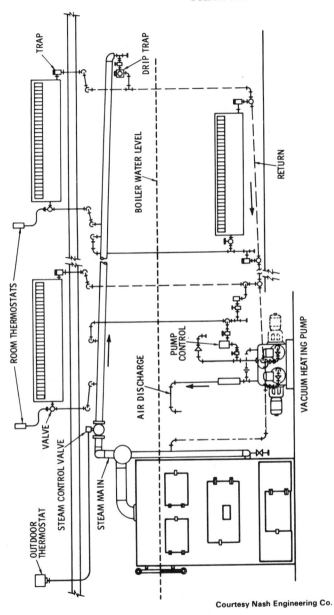

TRAP

DRIP TRAP

BOILER WATER LEVEL

RETURN

ROOM THERMOSTATS

PUMP CONTROL

VACUUM HEATING PUMP

AIR DISCHARGE

VALVE

STEAM CONTROL VALVE

STEAM MAIN

OUTDOOR THERMOSTAT

Courtesy Nash Engineering Co.

Fig. 8-28. A subatmospheric steam-heating system.

231

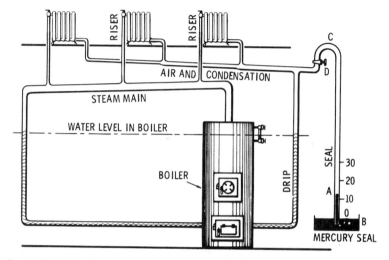

Fig. 8-29. Natural vacuum, mercury-seal system.

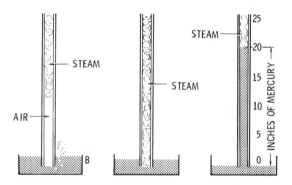

Fig. 8-30. Operation of mercury seal.

pressures above atmospheric, it drives all air out of the system, the air leaving by bubbling through the mercury in cup B.

If the fire is allowed to go out, the steam will condense and produce a vacuum, provided all pipe fitting has been carefully done and the valve stuffing boxes are tightly packed.

In Fig. 8-29 the loop at C prevents water from being carried over into the seal pipe when purging the system of air. If air should again enter the system, it can be expelled by raising the steam pressure above atmospheric. In very cold weather, the system can be operated at pressures above atmospheric by closing valve D. When fires are banked for the night, valve D may be opened and the system worked as a vacuum system. The flexibility of vacuum systems is in sharp contrast with low-pressure systems where steam disappears from the radiators as the temperature drops below 212°F. According to weather demands, the radiators may be kept at any temperature from, say, 150 to 220°F.

Another method of maintaining a natural vacuum is by using thermostatic valves instead of a mercury seal. A thermostatic valve has an expansion element that operates to close the valve when heated by hot steam and to open the valve when chilled by the relatively cold condensation.

The two kinds of thermostatic valves used are the single-unit (or retainer) valves and two-unit, or combined retainer and air-check valves, sometimes called *master thermostatic valves.*

Fig. 8-31 shows the details of a master thermostatic valve, con-

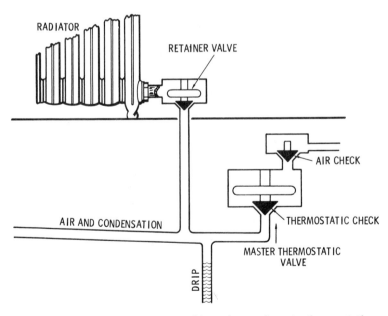

Fig. 8-31. A natural vacuum system with retainer and master thermostatic valve showing sectional views of the valves.

sisting of a thermostatic unit and an air check. The thermostatic unit has an expanding element, the air check consisting of a group seat poppet check valve that is practically airtight and will retain the vacuum within the system for a considerable length of time. The air check operates when excess pressure is generated in the boiler to purge the system of air, the check at other times remaining closed.

The thermostatic valve remains open while the system is being purged of air and condensation but closes when steam enters the valve chamber—it retains vacuum in the air line.

Fig. 8-32 shows the operation of the natural vacuum system with retainer and master thermostatic valves. Individual thermo-

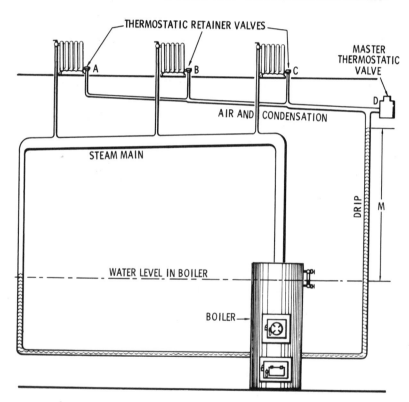

Fig. 8-32. Natural vacuum system with retainer and master thermostatic valves.

static retainer valves *A, B,* and *C* are placed in the outlet of each radiator, which pass air or water but close to steam. At the end of the air line is a master thermostatic valve (*D*), which operates when the system is purged of air by excess pressure.

The drip should be proportioned to prevent water entering the air line in case of high vacuum in such a manner that the vertical distance *M* between the water level in the boiler and the lowest point of air line is not less than 2 ft. for each inch of vacuum to be carried in the system.

The successful operation of natural vacuum systems depends largely on efficient damper regulators on coal-burning boilers (i.e., efficient draft control), for unless the fire is held in proper check, the pressure will rise and break the vacuum. This can waste fuel, for there may be sufficient heat in the boiler to supply steam to the system with a 5- or even 10-in. vacuum and hold that heat in the system for hours.

Automatic damper regulators are designed to act by pressure, temperature, or a combination of these two. Fig. 8-33 shows a regulator that acts on the pressure principle. It consists of a diaphragm connected at *B* to a lever fulcrumed at *A* and having a weight *W* free to slide along a slot between the stops.

In starting, the weight is placed on the left side of the lever as shown (Fig. 8-33). This tilts the lever (position *LF*) and opens the

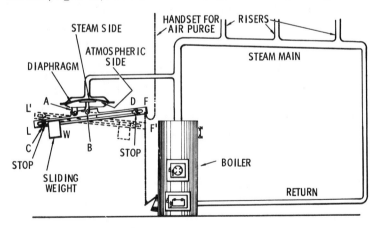

Fig. 8-33. Damper regulator for natural vacuum system operating on the pressure principle.

damper. The weight is adjusted by the stop so that sufficient pressure is produced to clear the system of air before the regulator tips to position L′F′ (shown in dotted lines) and closes the damper.

The regulator is gradually closing as the pressure comes on. When entirely closed, the weight slides to the right and remains in this position until the vacuum in the system becomes strong enough to gradually open the damper—just enough to maintain a vacuum.

In the morning, the regulator may be set to open position from the floor above by the pull chain M. This generates pressure and purges the system of any air that might have accumulated; then the regulator weight automatically goes to the vacuum side of the regulator and maintains the vacuum heat until more fuel is required or further regulation is necessary. Temperature controls or damper regulators that depend on temperature changes for their operation may also be used (Fig. 8-34). Since the temperature of steam increases with the pressure, evidently the expansion and contraction of a rod exposed to the steam can be made to operate the damper.

Fig. 8-34 shows the construction of a typical thermostatic regulator. The "expansion element" or rod is fastened at A in a closed cylindrical chamber through which steam from the boiler passes to the main. The end B is free to move, passing out of the chamber through a stuffing box. The motion of the rod is considerably magnified by the bell crank lever, which is connected to the damper by a chain attached at C.

In operation, as the pressure of the steam rises, so does its temperature, and the rod (which is made of a metal having a higher coefficient of expansion than that of the cylindrical chamber) end B, will move to the right, thus causing end C of the lever to descend, closing the damper.

When the pressure falls, the rod contracts and the spring that keeps the bell end in contact with the rod causes end C of the lever to rise and open the damper.

The lever will assume some intermediate position in actual operation, thus holding the steam at some predetermined pressure, which may be varied by means of the screw adjustment (D).

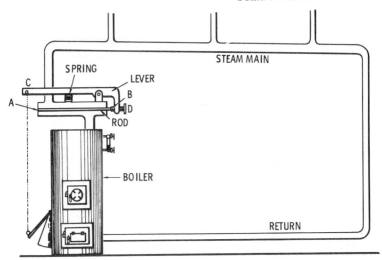

Fig. 8-34. Damper regulator for natural vacuum system operating on the temperature principle.

The major objection to regulation by temperature is that there is no provision for securing excess pressure to purge the system of air in starting. This must be done by hand control of the damper. This objection can be overcome by the method of combined pressure and temperature regulation, which employs pressure for starting and temperature for running. In starting, the thermostatic portion of the regulator is closed off from the systems during which pressure is generated sufficient (about one pound) to purge the air from the system. After purging, the regulator automatically opens a valve to the thermostatic position, which then maintains the temperature desired, its range embracing both vacuum and low-pressure operation.

MECHANICAL VACUUM SYSTEM

A *mechanical vacuum system* is one in which an ejector or pump is used to maintain the vacuum. The hookup of the ejector system is shown in Fig. 8-35. The ejector, which may be operated either by steam or water, is started before steam is turned on in the system. Thus, after the air is removed, steam will quickly fill the heat-emitting units full of steam since the air is automatically removed as fast as it accumulates.

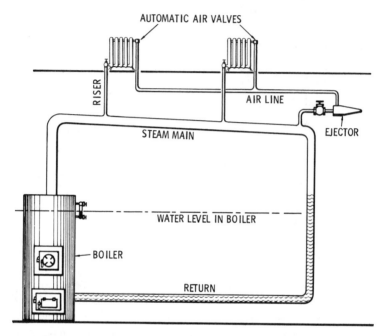

Fig. 8-35. Mechanical vacuum ejector system.

The system commonly used in exhaust heating employs an alleged or so-called vacuum pump, which ejects the air and condensation from the system. In operation, this device pumps out most of the air (or other gas) from the condenser, maintaining a partial vacuum. The pump cannot obtain a perfect vacuum because each stroke of the pump piston or plunger removes only a certain fraction of the air, depending on the percentage of clearance in the pump cylinder, resistance of valves, and so forth. Hence, theoretically an infinite number of strokes would be necessary to obtain a perfect vacuum (not considering resistance of the valves, clearance, etc.).

Condensation of steam creates the vacuum, and the pump that removes the air maintains a vacuum. A "wet" pump (that is, one that removes both air and condensation) is the type generally used. A "dry" pump removes only the air.

The essential features of a mechanical vacuum pump system are shown in Fig. 8-36. This system is of the fractional valve distribution type. In operation, air, being heavier than steam, passes off through thermostatic retainer valves to the pump. When the steam reaches the retainer valves, they close automatically to prevent the steam passing into the dry return line to the pump, thereby breaking the vacuum. The condensation is pumped from the receiver back into the boiler by a feed pump, passed on its way through a feed water heater, where it is heated by the exhaust steam from the air and feed pumps.

COMBINED ATMOSPHERIC PRESSURE AND VACUUM SYSTEMS

A combined atmospheric and vacuum system works at pressures in the boiler at from 1- to 5-oz. gauge pressure (that is, above atmospheric). This pressure is needed on coal-burning installations to operate the damper regulator.

The desired vacuum in the heat emitting units is obtained by "throttling" the supply steam with the unit feed valves. The working principle of this system is shown in the elementary sketch found in Fig. 8-37. In operation, when steam is raised in the boiler, it passes through the steam main risers and supply valves to the heat-emitting units.

The proper working of this system is obtained by an automatic device or trap that closes against the pressures of either steam or condensation and allows air, but not the steam, to pass out. The trap (Fig. 8-37) consists of three elements: (1) a diaphragm valve (point L), (2) a float valve (point A), and (3) a thermostatic valve (point F). Connection R, in Fig. 8-37, connects the steam outlet of the boiler to the diaphragm. When there is no pressure in the boiler, diaphragm valve L is held closed by the spring.

When the fire in the boiler is started and the air in the boiler expands, the diaphragm is inflated and moves the valve spring to the right (against the action of the spring), which opens valve L.

239

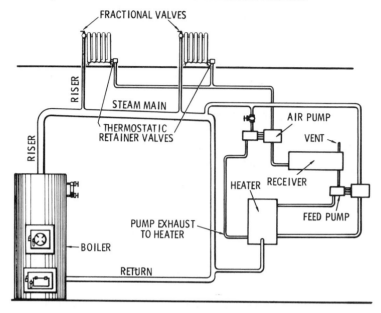

Fig. 8-36. Mechanical vacuum air-pump system as applied to fractional valve distribution.

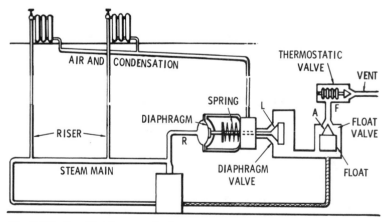

Fig. 8-37. Combined atmospheric pressure and vacuum system.

This makes a direct opening through float valve A and thermostatic valve F, thus opening the system to the atmosphere. Valve L remains open as long as there is a fraction of an ounce of pressure on the boiler. When steam forms and passes through the system, it drives all the air out of the system through the three open valves (L, A, and F).

The heat of the steam causes the expansion element of valve F to expand and close the valve; thus the system is filled only with steam.

The vacuum is obtained on the principle that the steam admitted to the radiators condenses while giving off heat through the radiator walls. This causes a tremendous reduction in volume of the steam remaining in the radiators, resulting in a pressure reduction in radiators that is less than atmospheric; i.e., a vacuum is formed. The steam condenses because of a reduction in temperature below that corresponding to the pressure of the steam.

When the radiator gives off heat in heating the room, the temperature of the steam in the radiator is lowered. This reduction in temperature causes some of the steam to condense in a sufficient amount to restore equilibrium between temperature and pressure of the steam.

The pressure falls because the temperature falls. If a closed flask containing steam and water is allowed to stand for a length of time, the atmosphere being at a lower temperature than that inside, the flash will abstract heat from the steam and water, but the heat will leave the steam quicker than the water. The result is a continuous condensation of the steam and reevaporation of the water, during which process the temperature of the whole mass and the boiling point are gradually lowered until the temperature inside the flask is the same as that outside. This process is accomplished by a gradual decrease in pressure. Figs. 8-38 to 8-40 illustrate the effect of pressure on the boiling point.

A one-pipe system may be converted into a combined system by replacing the air vent on the heat-emitting units and making the piping absolutely tight. In this conversion, a "compound" gauge recording both pounds of steam and of vacuum inches is required (Fig. 8-41).

241

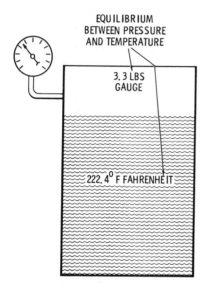

EQUILIBRIUM
BETWEEN PRESSURE
AND TEMPERATURE

3.3 LBS
GAUGE

222. 4° F FAHRENHEIT

Fig. 8-38. Equilibrium between
the steam and the water.
Equalized temperature.

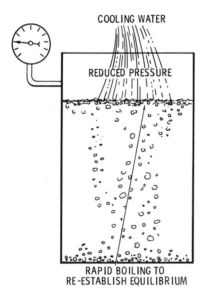

COOLING WATER

REDUCED PRESSURE

RAPID BOILING TO
RE-ESTABLISH EQUILIBRIUM

Fig. 8-39. Reducing the pressure
by applying cooling
water to the closed
vessel. Reduction of
pressure causes the
water to boil.

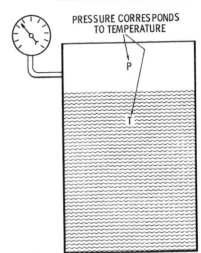

Fig. 8-40. Equilibrium reestablished between pressure and temperature.

EXHAUST-STEAM HEATING

The term *exhaust-steam heating* relates to the source of the steam rather than to its distribution. In fact, after the exhaust steam enters the heating system, its action is no different from live steam taken from a heating boiler, because it is adapted to both low-pressure and vacuum systems.

The chief difference between exhaust systems and those already described are the provisions for delivering steam from the engine to the heating system free from oil and at a constant pressure and for returning the condensation to the high-pressure boiler (Fig. 8-42). Fig. 8-43 shows the essential features of an exhaust-steam heating system having fractional control vacuum distribution.

The necessary devices between the engine and the inlet to the heating system are: (1) the oil separator, (2) the trap, (3) the back pressure valve, and (4) the pressure-regulating valve. In addition, for mechanically producing the vacuum and returning the condensation to the high-pressure boiler, an air pump, receiver with vent, and feed pump are required. A feed water heater should also be provided both for economy and to permit returning the

243

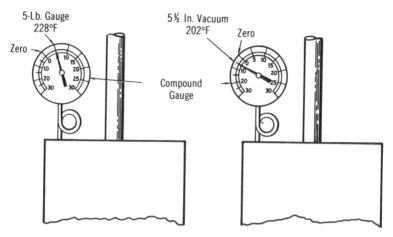

5-Lb. Gauge
228°F

Zero

5½ In. Vacuum
202°F

Zero

Compound
Gauge

Fig. 8-41. Compund gauge used on combined atmospheric-vacuum systems.

condensation and makeup feed water to the boiler at the proper temperature.

In operation, exhaust steam from the engine first passes through the heater and then through the oil separator, which frees it from the lubrication oil, the latter passing off into the oil trap. The steam now enters the heating system at *A*, its pressure being prevented from rising above a predetermined limit by the back pressure valve (regulated by weight *B*) and maintained at a predetermined constant pressure by the pressure-regulating valve (adjusted by weight *C*).

The pressure-regulating valve is, in fact, an automatic steam "makeup" valve, which admits live steam to the heating system when the exhaust is not adequate to supply the demand, thus "making up" for this deficiency and maintaining the pressure constant.

The condensation and air removed from the system at *D* by a wet pump (as distinguished from a dry pump, which removes only the air). The condensation and air are discharged into a receiver where the air passes off through a vent, the condensation being pumped by a feed pump back into the boiler after passing through a feed water heater.

There is a continued loss of water through various leaks; the

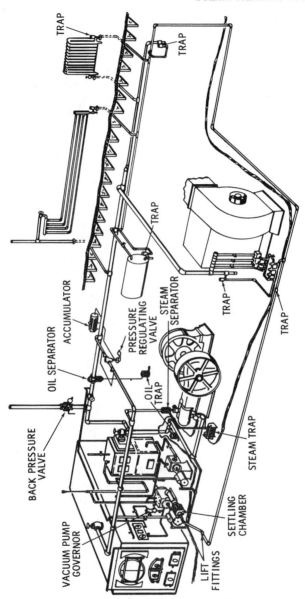

Fig. 8-42. Typical exhaust-steam heating system with vacuum distribution showing application to a power plant containing an open feed water heater.

feed pump inlet (alleged suction) is connected at E (Fig. 8-43), with the supply from the street main or other source, the amount entering the system being controlled by the "makeup" valve.

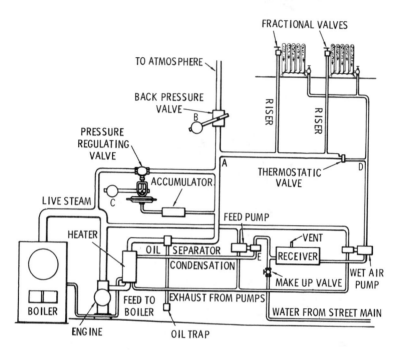

Fig. 8-43. Exhaust-steam heating system with fractional valve control and automatic makeup.

The construction of a typical regulating valve and its connections are shown in Fig. 8-44. The valve is controlled by means of governing pipe A (Fig. 8-44), connecting the diaphragm chamber to the accumulator, the latter being connected to the heating main at the point from which the pressure regulator is to be governed.

The accumulator is always half full of water, and its elevation must be such that the water line in the accumulator is level with the diaphragm so that there will not be an unbalanced column of water to exert pressure on the diaphragm.

The water is provided to protect the diaphragm from the

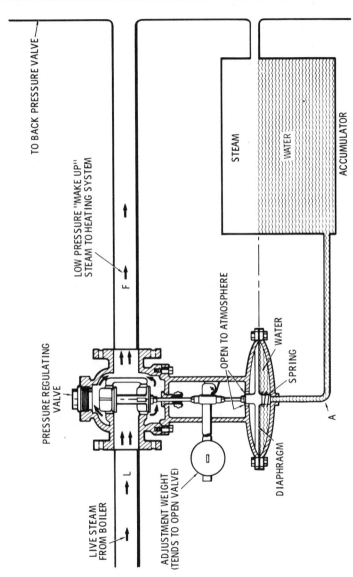

Fig. 8-44. Pressure-regulating valve and accumulator for maintaining a constant pressure in steam heating main of an exhaust-steam heating system.

steam, the pressure of the water being transmitted from the surface of the water in the accumulator to the diaphragm.

The working principle of the regulating valve and its connected devices is relatively simple. In operation, when the exhaust side F is at the predetermined pressure, this brings sufficient force against the under, or water, side of the diaphragm to overcome the downward thrust due to the adjustable weighted level and close the valve.

If the engine slows down or there is heavy demand for heat so that the exhaust steam is not adequate to supply the demand, the pressure in the exhaust side F (Fig. 8-44) will fall, and the downward thrust of the adjustment weight will overcome the opposing pressure of the water on the diaphragm and open the valve, admitting live steam from the boiler side L into the exhaust side F in sufficient quantity to restore the pressure.

The inertia of the water in the accumulator acts as a damper to prevent oversensitiveness of the valve or hunting (i.e., the behavior of any mechanism that runs unsteadily, oscillating either too far or too little in an attempt to adjust itself to momentary fluctuations of pressure or other conditions that cause this action).

The spring under the diaphragm acts to balance the downward thrust of the lever and hold the valve in closed position when the pressure is the same on both sides of the diaphragm.

The back pressure valve is used to prevent exhaust pressure exceeding a predetermined limit. This is virtually a lever safety valve designed to work at very low pressure. Some back pressure valves are so light that they will open or close with a variation of only 2 oz. The position of the weight on the lever, which is fulcrummed at F (Fig. 8-45), determines the exhaust pressure at which the valve will open.

PROPRIETARY SYSTEMS

Over the years, a number of automatic heating systems have been designed and patented by manufacturers of steam heating equipment. Because these heating systems are protected by patent, they are referred to by the name of the manufacturer. Three of the most popular of these proprietary steam heating systems are:

1. The Trane vapor system.
2. The Dunham differential system.
3. Webster moderator systems.

The *Trane vapor system*, illustrated in Fig. 8-46, is a combined atmospheric and natural vacuum system installed for residential heating. In this system, the air and water return runs in the same direction and is practically the same length as the supply main. When properly worked out, this feature gives the same effect as though each convector radiator were placed at an equal distance from the boiler. This tends to synchronize the heating effect of all the heat emitting units in the system.

When the fire is started in the boiler of the Trane system, the water becomes heated and steam is formed, which flows through the supply main and enters the radiators, displacing the air, which is heavier than the steam. The air and condensation drain from the radiators through radiator traps and return piping to a point near the boiler where the air is exhausted through the quick vents (4) and float vents (1) at the end of the steam and return mains. The water is returned to the boiler by the direct-return trap (2).

As the rooms become warm, less steam is condensed and the pressure in the boiler begins to rise. The rising pressure causes the damper regulator (3) to operate, closing and opening the drafts

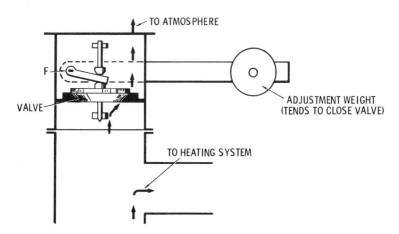

Fig. 8-45. Details of exhaust-steam heating system back valve.

249

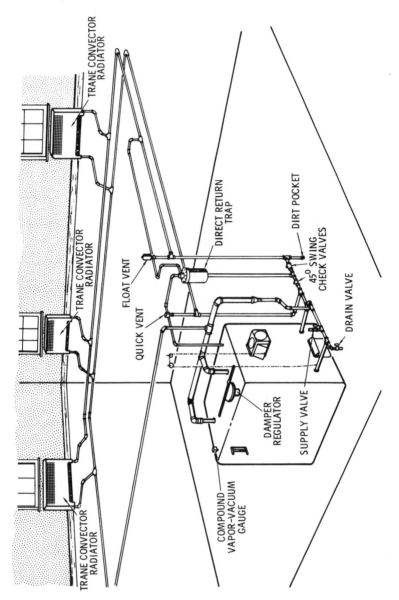

TRANE CONVECTOR RADIATOR

TRANE CONVECTOR RADIATOR

TRANE CONVECTOR RADIATOR

FLOAT VENT

QUICK VENT

DIRECT RETURN TRAP

DIRT POCKET

45° SWING CHECK VALVES

DRAIN VALVE

DAMPER REGULATOR

SUPPLY VALVE

COMPOUND VAPOR-VACUUM GAUGE

Fig. 8-46. Trane vapor heating system. A combined low-pressure and natural vacuum, steam heating system.

and maintaining just the amount of pressure necessary for proper heating.

When the first in the boiler becomes lower, condensation forms, air is prevented from entering the system, and a vacuum is created. As a result, the operation changes from atmospheric to natural vacuum, hence the name "combined atmospheric and natural vacuum system." The reduced pressure of the vacuum allows the water to boil and furnish steam to the radiators at a lower temperature, a decided economy when the fire is low.

The boiler connections for the Trane system are shown in Fig. 8-47. It is very important that the steam connection to the return trap be taken from the steam space of the boiler and *not* from the supply piping of the header. The top of the direct-return trap (2) must be placed at least 22 in. above the water line of the boiler. In *no* case should the top of the trap be less than 4 in. below the air and water return main. The dimensions from the water line to the end of the mains and the top of the return trap are the minimum allowable. Greater clearance above the water line should be employed where possible. When the ends of the steam and return mains occur in remote parts of the building away from the boiler, use the connections shown in Fig. 8-48. The ends of such mains must always be vented before dropping to the wet return. The vent pipe must be installed from the piping below the trap up to the return main when a wet return is used. If a dry return is used, the vent pipe may be omitted.

If desired, a Hartford connection may be used between the inlet to the boiler and the return connections from the steam and return mains (see "Hartford Return Connection" in this chapter).

The recommended size for vertical piping immediately below the float vents (1) and the quick vents (4) is 1¼ in. pipe at least 6 in. long. This affords a separate chamber for air and water.

The connections for a dripping steam main that rises to a higher level are illustrated in Fig. 8-49. The trap and strainer may be omitted provided the lower main is at least 18 in. above the water line and the return is connected directly into the return header of the boiler.

When the motor-operated steam valves are used on the mains, the boiler connections should be arranged as shown in Fig. 8-50. An equalizer line is required to equalize between the steam main

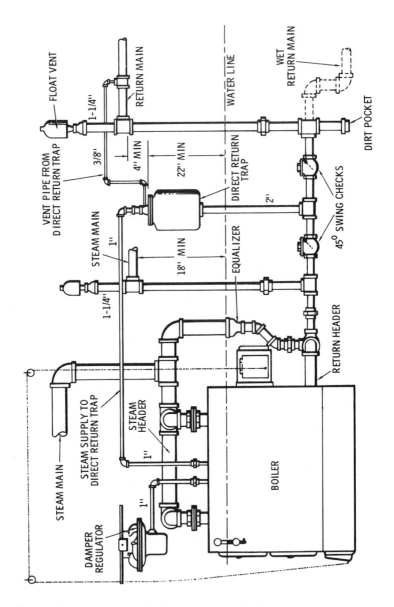

Fig. 8-47. Recommended boiler connections for Trane vapor system.

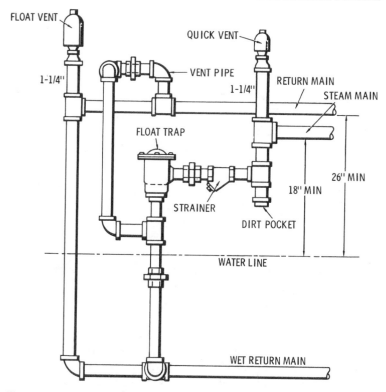

FLOAT VENT

QUICK VENT

VENT PIPE

1-1/4"

1-1/4"

RETURN MAIN

STEAM MAIN

FLOAT TRAP

26" MIN

18" MIN

STRAINER

DIRT POCKET

WATER LINE

WET RETURN MAIN

Fig. 8-48. Pipe connections for use where steam and return mains drop to wet return in remote part of building.

and the return main when a vacuum forms in the former after the motor-operated valve closes. A swing check valve prevents the flow of steam into the return main (Fig. 8-51). Fig. 8-52 shows the arrangement of boiler connections when air is eliminated from the return main through the No. 9 vent trap. Sometimes reversed circulation resulting from rapid condensation of steam will tend to create greater vacuum in the steam main than in the return main. This can be prevented by using the boiler connections illustrated in Fig. 8-49.

A typical convector radiator used in the Trane system is shown in Fig. 8-53. It is equipped with an angle valve and an angle trap having horizontal laterals below the floor. Connections for Trane convector radiators are illustrated in Fig. 8-54. The upper unit

253

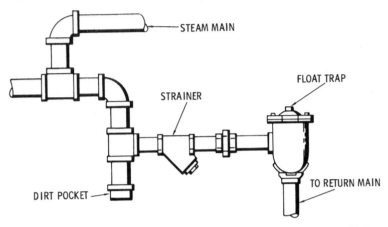

Fig. 8-49. Connections for dripping steam main where it rises to a higher level.

shows a vertical trap and a straightway valve concealed within the convector radiator. The lower unit is equipped with an angle valve and trap with a downfeed riser dripped through the angle trap.

The *Dunham differential vacuum heating system* is a simple two-pipe, power, vacuum (air-pump) return system working normally at pressures below atmospheric (subatmospheric) in the system and employing orifice supply valves on the radiators. The principal advantage of the Dunham system is that it continuously distributes heat at a variable rate equal to heat losses from the structure. This is accomplished by reducing the capacity of the system by decreasing the volume temperature of the steam.

Standard radiators, pipe fittings, and boiler connections are used in the Dunham system. The principal features of the Dunham system that distinguish it from other heating systems are:

1. The traps and valves.
2. The condensation pump.
3. The controller.

Each of these components is differentially controlled; that is, each is actuated by a pressure differential (the difference in pressure in the supply piping and the pressure return piping). The Dunham system distributes the steam proportionately to all

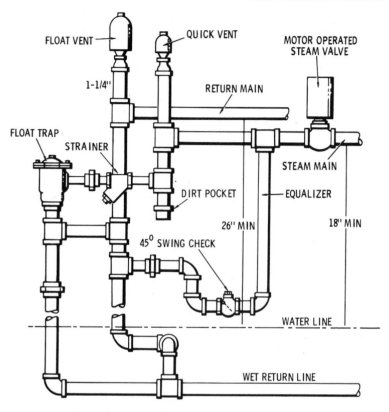

Fig. 8-50. Pipe connections when motor-operated steam valve is used on mains.

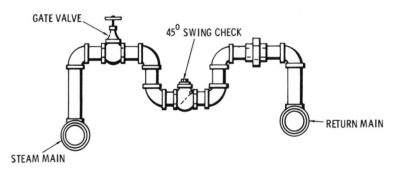

Fig. 8-51. Equalizer connection.

255

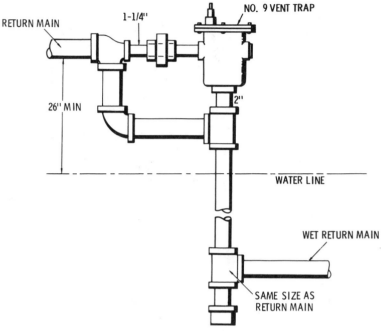

Fig. 8-52. Connection when air is eliminated from the return main through the vent trap No. 9.

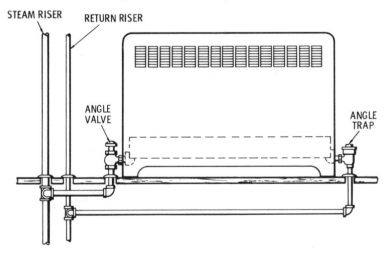

Fig. 8-53. Trane convector radiator with angle valve and angle trap having horizontal laterals below floor.

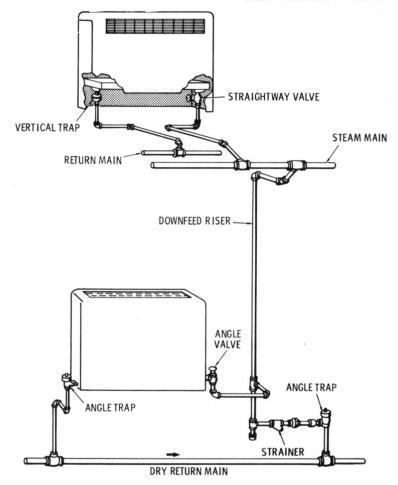

Fig. 8-54. Typical connections for Trane convector radiators.

radiators, the pressure range being from 2 psig to pressures considerably lower than atmospheric. Because of the relatively constant differential in pressures between the steam and return lines, the radiators are filled with steam.

A condensation pump is connected to the differential controller and to the supply and return piping. The controller starts the pump when the pressure difference between the supply and return piping tends to fall or disappear, and stops it when the

257

pressure differential is restored. The condensation pump is a wet air pump that handles both air and condensation.

The thermostatic radiator traps are actuated by temperature changes within the radiator. Drip traps are installed at drip points to which large volumes of condensation flow. These are combined thermostatic and float traps. A control valve regulates the admission of a *continuous* flow of steam into the heating main.

Webster moderator systems are special control systems used primarily with two-pipe vapor, vacuum, and vented return systems or in modified form with some one-pipe steam heating systems. These Webster moderator systems are controlled by the weather with an automatic outdoor thermostat. One or more hand-operated *variators* can be used to adjust the outdoor thermostat. The two moderator control systems produced by Webster are based on the following operating principles:

1. Continuous steam flow.
2. Pulsating steam flow.

The continuous steam flow arrangement is called the *Webster electronic moderator system* (Fig. 8-55) and is suitable for medium to large buildings requiring one or more control valves. This system consists of the following basic components:

1. Outdoor thermostat.
2. Main steam control valve.
3. Variator.

The steam supply may be taken from a high- or low-pressure boiler or any other source. The initial pressure should not be over 15 psig, using a reducing valve if necessary. The return piping may be either open or closed.

The moderator control regulates the pressure difference and will function equally well regardless of whether the pressure in the return piping is atmospheric or below. The main steam-control valve is adjusted automatically by the moderator control, which acts to reverse the direction of the motor, causing it to move the valve in the closing direction when *less* steam is required and in the opening direction when *more* steam is required. The outdoor thermostat automatically varies the steam flow in accordance with changes in outdoor temperature. Depending upon the outdoor temperature, the thermostat automati-

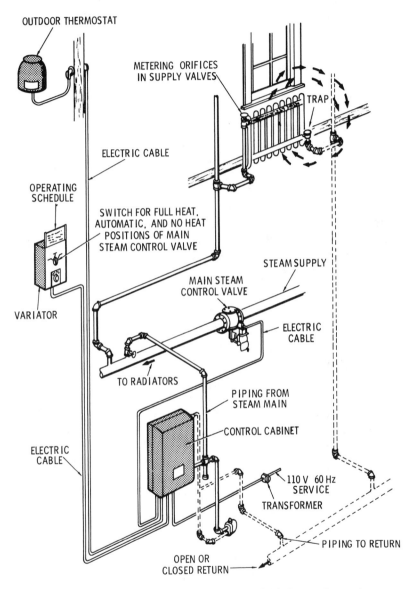

OUTDOOR THERMOSTAT

METERING ORIFICES
IN SUPPLY VALVES

TRAP

ELECTRIC CABLE

OPERATING
SCHEDULE

SWITCH FOR FULL HEAT,
AUTOMATIC, AND NO HEAT
POSITIONS OF MAIN
STEAM CONTROL VALVE

STEAM SUPPLY

MAIN STEAM
CONTROL VALVE

VARIATOR

ELECTRIC
CABLE

TO RADIATORS

PIPING FROM
STEAM MAIN

ELECTRIC
CABLE

CONTROL CABINET

110 V 60 Hz
SERVICE

TRANSFORMER

PIPING TO RETURN

OPEN OR
CLOSED RETURN

**Fig. 8-55. Typical arrangement of Webster E-5 electronic moderator
system.**

259

cally selects the position of the main steam-control valve. Its position may be advanced or reduced by the variator to give more or less steam than is called for by the outdoor temperature.

Changes in pressure difference in the heating system are automatically compensated for by a pressure-actuated mercury tube in the control cabinet. One end of the tube is connected to the steam supply mains and the other end to the return main. If the supply pressure is unduly increased, mercury rises in the tube to unbalance resistances contained therein and the main steam valve begins to close in amount sufficient to balance resistances. A reverse action takes place when the pressure difference falls below that called for by the control equipment.

The pulsating steam flow arrangement is called the *Webster moderator system*. This system comprises a central heat control of the pulsating flow type for new or existing steam or hot-water systems. It is designed for small- and medium-size buildings and for zoning of large buildings. It directly controls the operation of burner, stoker, blower, or draft damper motors. The basic components of this system are:

1. Outdoor thermostat.
2. Pressure difference controller.
3. Control cabinet.
4. Capillary tubing.

These four components work together to open and close a valve in the steam main or to start and stop the automatic firing device at the boiler, generally through a relay. Fig. 8-56 shows the general arrangement of a Webster EH-10 moderator system for a building served by a central station or street steam. A Webster EH-10 moderator system used for controlling a burner on a steam boiler is illustrated in Fig. 8-57.

The pressure difference controller maintains the correct pressure difference between supply and return piping. In combination with metering orifices, this device gives an even distribution of the steam to all radiators in the system, thereby preventing over- or underheating.

Control is accomplished by varying the length of intervals during which steam is delivered to the radiators. These intervals are longest in cold weather and shortest in mild weather. The timing

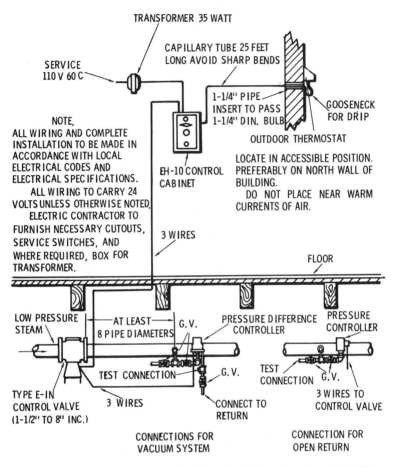

Fig. 8-56. Typical arrangement of Webster EH-10 moderator system in a building served by a central station or street steam.

is such that the longest *off* interval is comparatively short so that heat output from radiators is practically continuous.

The timing mechanism inside the control cabinet is powered by a synchronous motor, which turns a cam. Timing gears between motor and cam set the length of the operating cycle. Rising on the cam is a roller ·connected to the arm of a switch. When the roller is on the high part of the cam, the switch is in the position for opening the control valve or starting the firing

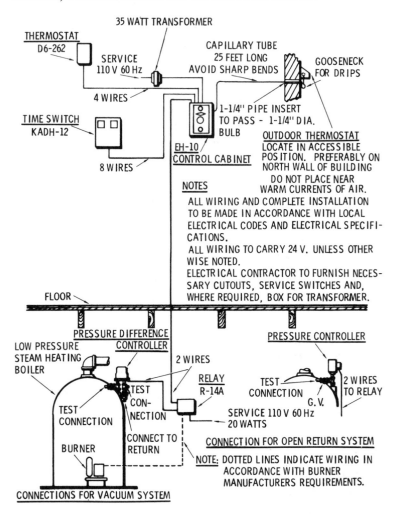

Fig. 8-57. General arrangement of the Webster EH-10 moderator system controlling a burner of a steam boiler.

equipment. When the roller is on the low part of the cam, the switch is on the position for closing the control valve or stopping the firing equipment. The length of the *on* interval is changed automatically by the outdoor thermostat or by adjusting the variator by hand. The average length of the cycle is 30 minutes.

Other gears can be furnished for cycle lengths from 12 to 60 minutes.

A variator is included in the control cabinet for manual adjustment of the rate of heat delivery to the building. The variator changes the relationship of the switch lever and roller to the cam by moving the cam itself. This is accomplished by mounting the motor, cam, and gear train on a movable carriage.

HIGH-PRESSURE STEAM HEATING SYSTEMS

High-pressure steam heating systems operate at pressures above 15 psig (generally in the 25 to 150-psig range) and are usually found in large industrial buildings in which space heating or steam process equipment (e.g., water heaters and dryers) is used. An example of a typical two-pipe, high-pressure steam heating system is shown in Fig. 8-58. This system is also referred to as a *medium-pressure steam heating system* when the steam pressures are in the lower ranges.

One advantage of this system is that the high pressure of the steam permits the use of smaller pipe sizes. The high steam pressure also makes possible the elevation of the condensation return lines above the heating units because the return water can be lifted into the return mains. High-pressure condensation pumps and thermostatic traps are commonly used in these systems to handle the condensation and return it to the boiler.

High-pressure steam heating systems are more expensive to operate and maintain than low-pressure systems. In most cases, a licensed stationary engineer must be hired for the installation—a factor that tends to increase the operating cost.

STEAM BOILERS

The boiler is the source of heat for a steam heating system, and it will operate on a number of different fuels; however, regardless of the fuel used, the operating principle will be the same. Water is heated until it boils and changes to steam. The steam is then distributed to the heat-emitting units throughout the structure either by natural or mechanical means.

Low-pressure steam heating boilers are used in residences and small buildings. The design and construction of these boilers are very similar to the boilers used in hot-water heating systems in the same size. Both boilers are described in considerable detail in Chapter 15 (Boilers and Boiler Fittings).

CONTROL COMPONENTS

The controls used to regulate steam boilers and ensure their safe and efficient operation are similar in most respects to those used with hot-water boilers. A list of the principal controls used with steam boilers includes:

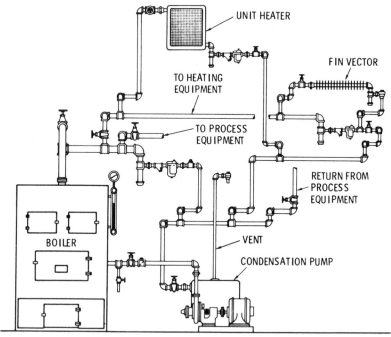

Fig. 8-58. A typical two-pipe, high-pressure, steam heating system.

1. ASME safety valve.
2. Steam pressure gauge.
3. Low-water cutoff.
4. Boiler water feeder.
5. High-limit pressure control.
6. Water gauge glass.
7. Water cocks.
8. Primary control (burner mounted).
9. Operating control (with tankless heater).

These and other boiler controls are described in considerable detail in Chapter 15 (Boilers and Boiler Fittings), Chapter 4 of Volume 2 (Thermostats and Humidistats), and Chapter 9 of Volume 2 (Valves and Valve Installation).

HARTFORD RETURN CONNECTION

A well-designed steam heating system will have a *Hartford return connection* (Fig. 8-59) in the condensation return line. The

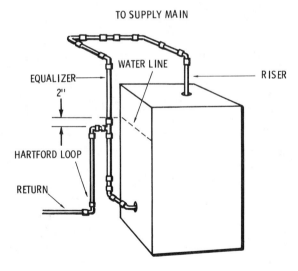

Fig. 8-59. The Hartford return connection.

purpose of the Hartford return connection (or *Hartford loop*, as it is sometimes called) is to prevent excessive loss of water for the boiler when a breakdown (such as a water leak) occurs in the return line.

As shown in Fig. 8-59, an equalizer connects the lower outlet to the steam outlet. The Hartford connection is taken off from the equalizer an inch or two below the normal water level. Evidently low pressure can only draw water out of the boiler connection. This gives a low water level, and it cannot recede further because steam will flow into the connection leg as the water recedes in the leg due to pressure difference.

More details about different boiler connections are found in Chapter 15 (Boilers and Boiler Fittings) and in Chapter 9 of Volume 2.

PIPES AND PIPING DETAILS

A steam heating system requires careful planning of the piping to ensure both an efficient and a safe operation. For example, the pipe material used (e.g., wrought iron, and Schedule-40 black steel) is important because the *capacity* of a particular weight pipe will depend on both its size and the material from which it is constructed.

The expansion of pipes when they become heated is another factor that must be considered when designing a steam heating system. Sufficient flexibility in the piping can be provided for by correctly designed offsets, slip joints or bellows, radiators and riser runouts, U-bends, or other expansion loops. The pitch of connections from risers must be sufficient to prevent the formation of water pockets when pipe expansion occurs.

Pipe materials, pipe sizes (and pipe-sizing methods), pipe expansion rates, pipe fittings, and piping details, such as wet and dry returns, drips, and connections to heat-emitting units, are covered in Chapter 8 of Volume 2 (Pipes, Pipe Fittings, and Piping Details).

STEAM TRAPS

A *steam trap* is an automatic device installed in a steam line to control the flow of steam, air, and condensation. In operation, it opens to expel air and condensation and closes to prevent the escape of steam. All steam traps operate on the principle that the pressure within the trap at the time of discharge will be slightly in excess of the pressure against which the trap must discharge.

The principal functions of steam traps in steam heating systems include: (1) *draining condensation* from the piping system, radiators, and steam processing equipment; (2) *returning condensation* to the boiler; (3) *lifting condensation* to a higher elevation in the heating system; and (4) *handling condensation* from one pressure to another.

Steam traps may be classified on the basis of their operating principles as follows:

1. Float traps.
2. Bucket traps.
3. Thermostatic traps.
4. Float and thermostatic traps.
5. Flash traps.
6. Impulse traps.
7. Lifting traps.
8. Boiler return traps.

PUMPS

Pumps are used in steam heating systems to dispose of the condensation or return it to the boiler and to discharge excess air and noncondensable gases to the atmosphere. The specific function (or functions) of a pump depends on the type of steamheating system in which it is used. The two basic steam heating pumps are:

1. Condensation return pumps.
2. Vacuum heating pumps

HEAT-EMITTING UNITS

A *heat-emitting unit* is a device that transmits heat to the interior of a room or space. The two heat-emitting units used in steam heating systems are radiators and convectors. Simply defined, a *radiator* is a heat-emitting unit that transmits heat from a direct heating surface principally by means of radiation. A *convector* may be defined as a heat-emitting unit that transmits heat from a heating surface principally by means of convection. The heating surface of a convector is usually of the extended finned tube construction.

A detailed description of the heat-emitting units used in steam heating systems is contained in Chapter 2 of Volume 3 (Radiators, Convectors, and Unit Heaters).

AIR SUPPLY AND VENTING

Two types of venting occur in heating systems. One deals with the passing to the outdoors of smoke and gases resulting from the burning of combustible fuels (e.g., coal, oil, and gas) and is described in the various chapters on furnaces and boilers. Another type of venting deals with the relief of pressure in steam heating systems by allowing a certain amount of air to escape (i.e., be vented) from the radiators. Venting radiators is described in Chapter 2 of Volume 3 (Radiators, Convectors, and Unit Heaters).

UNIT HEATERS

A *unit heater* is essentially a forced draft convector. A centrifugal fan or propeller is used to force the air over the heating surface and into the room or space through deflector vanes. The principal operating components of a steam unit heater are shown in Fig. 8-60. Unit heaters find their widest application in industrial and commercial buildings, gymnasiums, field houses, auditoriums, and other types of large buildings.

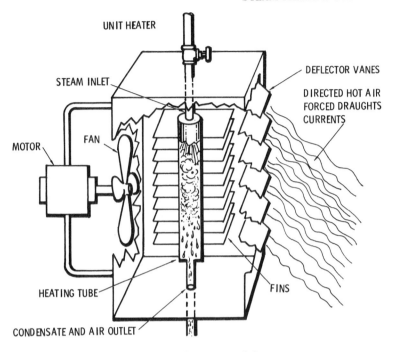

Fig. 8-60. Principal components of a steam unit heater.

All unit heaters are described in considerable detail in Chapter 2 of Volume 3 (Radiators, Convectors, and Unit Heaters).

AIR CONDITIONING

Central air conditioning can be added to a steam heated structure by installing a water chiller or a separate forced air cooling system (i.e., a split system).

The water chiller and steam boiler may be installed as separate units, or a complete package containing both units may be used. Water chiller installations are very rarely used in residential air conditioning. Split-system air conditioning dominates this field (i.e., with respect to steam heated residences).

CHAPTER 9

Electrical
Heating Systems

Different heating systems use electricity as a source of heat. Each of them employs one or more of the following methods of heat transfer:

1. Radiation.
2. Convection.
3. Forced air.

Radiation is the transmission of heat energy by means of electromagnetic waves. In other words, the heat is transferred directly from the surface of the heating element to the people, objects, and furnishings in the room *without* heating the air. The air will pick up a certain amount of heat from these surfaces (i.e., from the bodies of the people, furnishings, etc.), which then circulates by natural convection.

With *convection*, the temperature of the air is increased by contact with a heated surface. The air rises as it becomes warmer (and lighter), resulting in a circulation of the air in the space. Now note the difference between radiation and convection. In the radiation method of heat transmission, the heat energy passes through the air to the individual or object in much the same way that the sun transfers its heat. The air is *not* used as the medium of heat transfer. The surfaces are warmed, and the air picks up heat from these surfaces (convection), not directly from the heat energy passing through it.

In the *forced air* method of heat transmission, the heated or conditioned air is circulated by motor-driven fans integral to the heating unit or included elsewhere in the system. The heated air is forced directly into the room (as in the case of a wall heater) or through a duct system from a centrally located furnace (central forced warm-air heating systems).

CENTRAL HOT-WATER SYSTEMS

Some central hydronic systems (i.e., forced hot-water heating systems) use electric-fired boilers as the heat source. These boilers (Fig. 9-1) are compact units consisting of an insulated steel or cast-iron generator with replaceable immersion heating elements. The expansion tank, circulating pump, valves, and prewired controls are included within the boiler package, and the entire unit is assembled at the factory (Fig. 9-2).

An electric-fired boiler is compact, generally having a water capacity of only 2 or 3 gal. Some electric-fired boilers are small enough to be hung on a wall in the home. The small size is possible because water has a much higher heat-carrying capacity than air, and the electric elements are immersed directly in the water. This not only saves space but provides very rapid heating of the water because almost all the heat produced by the elements is transferred directly to the water. The heated water is piped directly to the room convectors where its stored heat is delivered to the air. Distribution of the water can be easily controlled by zone valves under the control of separate zone thermostats.

The controls for an electric boiler are similar to those found on boilers (e.g., oil- and gas-fired) in other central heating systems. The basic controls are:

1. A low-voltage thermostat.
2. Pressure and temperature limit controls.
3. Relay and sequence controls.
4. Circulating pump controls.

The circulating pump is activated by the low-voltage thermostat. Boiler overheating and pressure buildup is controlled by the boiler high-limit controls and the pressure-temperature relief valve.

A more detailed description of electric-fired boilers and their controls is found in other chapters of this book (see, for example, the section on electric-fired boilers in Chapter 15, Boilers and Boiler Fittings).

For information about the piping system used in electric hydronic heating, read the appropriate sections in Chapter 7

Fig. 9-1. Electric-fired boiler.

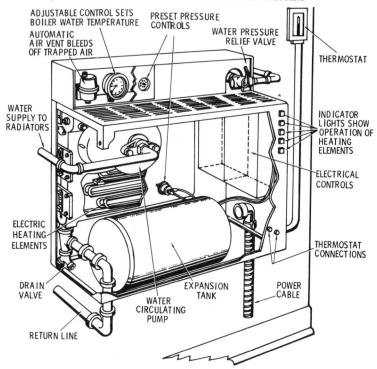

Fig. 9-2. Interior view of an electric boiler showing the basic components.

(Hot-Water Heating Systems), and in Chapters 8 and 9 of Volume 2.

Cooling for an electric hydronic system is generally provided by individual room air conditioners (i.e., through-the-wall units) or by chilled water pumped from a central refrigeration unit. The latter installation is usually found in multifamily structures, such as apartment buildings.

CENTRAL FORCED-WARM-AIR HEATING SYSTEMS

Central forced-warm-air heating systems that use electricity to heat or cool the air rely upon the following heat sources:

1. Electric-fired furnaces.
2. Duct heaters.
3. Heat pumps.

Regardless of the heat source, the air is circulated by fans through a duct system to the various rooms. Variations of these ducted air systems are shown in Fig. 9-3.

The electric-fired furnace is a complete unit designed for zero clearance and available in several models suitable for either horizontal or vertical installation. The heat is provided by fast-activating, coiled-resistance-wire heating elements. When there is more than one coil, the individual coils are energized at intervals in sequence to prevent the type of current overload possible if all were energized simultaneously. The heating coil sequencer is activated and controlled by a low-voltage thermostat mounted

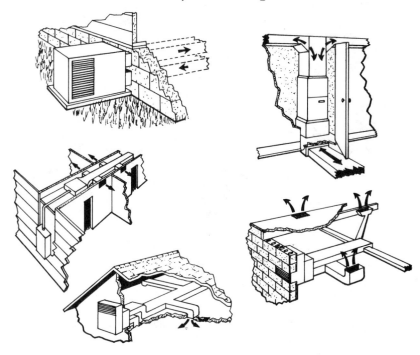

Fig. 9-3. Various duct systems for electric heating and cooling.

on a wall in some convenient location. Overheating of the furnace is prevented by the same type of high-limit control found in other types of furnaces. The warm air is forced through the duct system to outlets in the room by a blower. Evaporator coils can be added to the furnace for cooling. Additional modifications can provide humidity control (humidification and dehumidification), ventilation, and air filtration.

A more detailed description of electric-fired furnaces is found in Chapter 14 (Electric-Fired Furnaces). For further information on forced-warm-air heating systems, read the appropriate sections in Chapter 6 (Warm-Air Heating Systems). Controls for this heating system are described in the section "Electric Heating and Cooling Controls" in this chapter and in Chapter 4 of Volume 2 (Thermostats and Humidistats).

Duct heaters (Fig. 9-4) are factory-assembled units installed in the main (primary) or branching ducts leading from the blower

Courtesy Vulcan Radiator Co.

Fig. 9-4. Duct heaters.

unit. Heat is provided by parallel rows of resistance wire formed in the shape of spirals, which may be sheathed or left open. The blower is housed in a specially constructed and insulated cabinet to which the ducts are connected. This unit functions as the power source for circulating the air through the duct heater and the rest of the system. Cooling can be provided by adding a cooling unit to the blower. Humidity control is possible by first supercooling the air in the coil and then reheating it to the desired temperature and moisture content when it passes through the duct heater.

Some duct heaters are designed to be inserted into a portion of the duct through a hole cut in its side and are more or less permanent units. Other duct heaters are assembled at the factory in a portion of duct flanged for easy installation at the site.

Duct heaters may be installed in more than one duct to provide zoned heating. When this is the case, each duct heater is provided with means for independent temperature control. Another common method is to install a single duct heater in the main (primary) duct leading from the blower unit. Controls for either installation are similar to those used on electric-fired furnaces (see above).

The third type of heat source used in ducted, central forced, warm-air heating systems is the air-to-air or water-to-air heat pump. These are described briefly in another section of this chapter (see Electric Heat Pumps), and in greater detail in Chapter 12 of Volume 3 (Heat Pumps).

RADIANT HEATING SYSTEMS

Any electrical conductor that offers resistance to the flow of electricity will generate a certain amount of heat, the amount of heat generated being in direct relation to the degree of resistance. This method of generating heat is employed in *radiant heating systems*.

The conductor commonly used in radiant heating systems is an electric heating cable imbedded in the floors, walls, or ceilings (Fig. 9-5). Installation of the cables may be done at the site (as is often the case with new construction), or they may be obtained in the form of prewired, factory-assembled, panel-type units. The heat generated by the cables is transferred to the occupants and surfaces in the room by low-intensity radiation.

Site-installed heating cables or prewired and assembled panel units are used in the following radiant heating systems:

1. Radiant ceiling panel systems.
2. Radiant wall panel systems.
3. Radiant floor panel systems.

Radiant ceiling heating systems are by far the most commonly used type. The other two have certain disadvantages inherent in their construction. All three radiant heating systems are described in greater detail in Chapter 1 of Volume 3 (Radiant Heating).

The electric heating cables are activated and controlled by wall-mounted, low-voltage or line thermostats. Cooling can only be accomplished by adding a separate and independent system.

BASEBOARD HEATING SYSTEMS

Baseboard heating systems use electric baseboard units located at floor level around the perimeter of each room (Fig. 9-6). Each

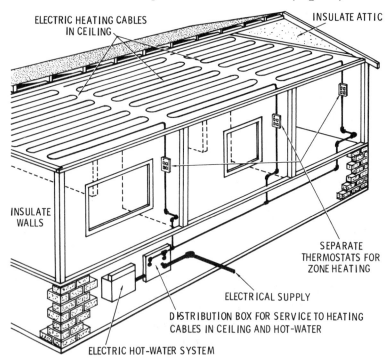

Fig. 9-5. Typical radiant heating system.

unit consists of a heating element enclosed in a thin metal housing. Zoning is possible within a single wall-mounted line or low-voltage thermostat located in each room. Heat is provided to the room *primarily* by convection (some radiation is involved) as the room air moves across the heated elements in the baseboard unit. These baseboard units are described more fully in Chapter 2 of Volume 3 (Radiators, Convectors, and Unit Heaters).

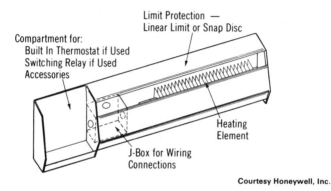

Limit Protection —
Linear Limit or Snap Disc

Compartment for:
Built In Thermostat if Used
Switching Relay if Used
Accessories

Heating
Element

J-Box for Wiring
Connections

Courtesy Honeywell, Inc.

Fig. 9-6. Typical electric baseboard heating unit.

As is the case with radiant heating systems, cooling can be provided by installing a separate cooling system (central or room air conditioners). The baseboard heating system represents the most widely used form of electric heating.

ELECTRIC UNIT VENTILATORS

Electric unit ventilators [similar in design to electric unit heaters (see Fig. 9-7)] are used to heat, ventilate, and cool large spaces that are by nature subject to periods in which there are high densities of occupancy. They are frequently found in offices, schools, auditoriums, and similar structures. The basic components of a typical unit ventilator are:

1. The housing.
2. Motor and fans.
3. Heating element.

4. Dampers.
5. Filters.
6. Grilles or diffusers.

Automatic controls activate the unit ventilator and vary the temperature of the air discharged into the room in accordance with room requirements. Outdoor air is drawn through louvers in the wall and into the unit ventilator before being discharged into the interior of the structure. These ventilators are usually floor- or ceiling-mounted, depending on the design of the room.

In addition to electricity, unit ventilators may also be gas-fired or use steam or hot water as the heat medium. Unit ventilators are described in greater detail in Chapter 2 of Volume 3 (Radiators, Convectors, and Unit Heaters).

ELECTRIC UNIT HEATERS

Electric unit heaters (Fig. 9-7) are used primarily for heating large spaces, such as offices, garages, warehouses, and similar commercial and industrial structures.

As is the case with unit ventilators, the heat conveying medium and combustion source may be other than electricity. For exam-

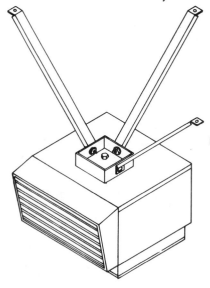

Fig. 9-7. Wall- or ceiling-mounted electric unit heater.

ple, the air may be heated by either gas-fired or oil-fired units. Steam or hot water can be substituted for air as the heat conveying medium. Selecting electricity as a medium will depend on such factors as:

1. The availability of cheap electrical power.
2. The need for supplementary housing.
3. The scarcity of other heat-conveying mediums.

The typical unit heater contains the following basic components:

1. Fan and motor.
2. Heating element.
3. Directional outlet.
4. Casing or housing.

Unit heaters may be floor-mounted or suspended from the ceiling, depending upon the design requirements of the structure.

ELECTRIC SPACE HEATERS

Smaller unit heaters designed and constructed for domestic heating purposes are commonly referred to as *space heaters*. Like the larger units, they are also designed to operate with other heat conveying mediums. However, the advantage of using electric space heaters over the fuel-burning types is that they do not have to be vented and require no flue.

Space heaters are available as either portable or permanent types. The latter can be fitted into ducts or mounted on walls or the ceiling.

For further information on space heaters see the appropriate sections in Chapter 2 of Volume 3 (Radiators, Convectors, and Unit Heaters).

HEAT PUMPS

A *heat pump* (Figs. 9-8 and 9-9) is an electrically powered, reversible-cycle refrigeration unit capable of both heating and cooling the interior of a structure.

Courtesy Electric Energy Association

Fig. 9-8. Outdoor section of a split-unit heat pump. It is connected to the inside section by refrigerant tubing and electrical wiring.

The heat source is commonly either outside air (in the air-to-air heat pump) or a closed loop of circulating water (in the water-to-air heat pump). The former is the most popular heat pump used for single-family dwellings. The water-to-air heat pump system is most often found in multifamily structures. These are essentially central heating systems, with the heat pump replacing the central furnace or boiler as the heat-generating unit.

Courtesy Electric Energy Association

Fig. 9-9. Outside-wall-mounted single-package heat pump.

Operating valves in the heat pump unit control the reversal of the refrigeration cycle. It removes heat from the interior of the structure, discharges it outside during hot weather, and supplies heat to the interior spaces during periods of cold weather.

The basic components of a heat pump installation are: (1) the compressor, (2) the condenser, (3) the evaporator, and (4) a low-voltage thermostat. These and other aspects of heat pumps are described in greater detail in Chapter 12 of Volume 3.

ELECTRIC HEATING AND COOLING CONTROLS

In an electric heating and cooling system, the best results are obtained by finding a comfortable thermostat setting and leaving it there. Constantly changing the thermostat setting results in consuming more energy (resulting in increased operating costs) and creates additional wear and tear on the equipment. Experts estimate that for each degree the thermostat is raised above the normal setting, there is a 3 percent increase in heating costs. A comfortable setting for the thermostat depends upon a number of variables, including (1) the type of system, (2) environmental conditions, and (3) your personal requirements. For example, 70°F is usually adequate for a radiant heating system (i.e., baseboard, panel-type heating installations, etc.) in a properly insulated structure. A comfortable indoor temperature setting for air conditioning is usually 76 to 78°F.

A variety of control methods are used in electric heating and cooling systems to maintain kilowatt demand at a level both economical in operation and suitable in performance. The basic controls used for these purposes are:

1. Thermostats.
2. Sequence switching devices.
3. Load-limiting controls.
4. Time-delay relays.

Thermostats can be located in each room or on each heating unit to provide decentralized control. A manual switch can be provided with each thermostat to shut off the current if the room is to be unoccupied for long periods of time.

Sequence switching devices are used to switch electrical current in sequence (rotation) from one room or circuit to the next. A *load-limiting control* is designed to shut off the current to one or more circuits when total electric demand exceeds a preset value. *Time-delays* are used to restore service after an interruption *nonsimultaneously* over a period of several minutes so that overloading is avoided. Each of these controls is examined in considerable detail in Chapter 14 (Electric-Fired Furnaces).

INSULATION FOR ELECTRICALLY HEATED AND COOLED STRUCTURES

Any structure electrically heated and cooled *must* be properly insulated or operation costs will be unacceptable. Converting an older structure to electric heat is therefore not recommended unless you have no objection to the additional expense of improving the insulation or you feel you can live with the higher operating costs. Consequently, electric heating is more often considered for new construction.

Insulating a structure that is to be electrically heated and cooled requires greater attention to construction details than do other types of systems; however, the results are well worth the efforts because reduced operating costs and other gains (e.g., the reduction of outside noise and quiet operating characteristics) are soon readily apparent. Because electric heating and cooling systems require an especially well-insulated structure for efficient operation, recommendations for the required minimum levels of insulation are included in this chapter. (These recommendations are based on information included in the Electric Energy Association's publication *Electric Space Conditioning in Residential Structures*.)

The recommendations found in the following paragraphs will contain references to *U*-factors and *R*-values. The *U-factor* is the overall coefficient of heat transfer and is expressed in Btu per hour per square foot of surface per degree Fahrenheit difference between air on the inside and air on the outside of a structural section. The *R-value* is the term used to express thermal resistance of the insulation. The *U*-factor is the reciprocal of the sum of the thermal resistance values (*R*-values) of each element of the structural section. Table 9-1 gives the insulation recommendations for electrically heated and cooled structures both in terms of *R*-values and overall coefficients of heat transfer in terms of *U*-factors.

Table 9-1. Maximum Winter Heat Loss for Electrically Heated Homes

Degree Days	Maximum Heat Loss Values	
	Watts/Sq. ft.	Btuh/Sq. ft.
Over 8000	10.0	34
7001 to 8000	9.4	32
6001 to 7000	8.8	30
4501 to 6000	8.2	28
3001 to 4500	7.6	26
2000 to 3000	7.0	24

Courtesy Electric Energy Association

Table 9-2 lists heat loss limits for electrically heated structures as recommended by the Electric Energy Association. The values listed are expressed in watts and Btuh (1 watt = 3.413 Btuh) per square foot of floor space measured to the exterior walls. The assumed infiltration rate on which this table is based is approximately three-quarters air change per hour.

The maximum summer-heat-gain limits for electrically air conditioned homes are indicated in Fig. 9-10. These figures are adapted from the FHA Minimum Property Standards for single- and multifamily structures.

Frame Walls

Frame walls (Fig. 9-11), with either wood or metal studs, are commonly insulated with rolls and batts of mineral, glass, or wood-fiber insulation material. More and more use of sprayed-on or foamed-in-place rigid insulation is also being made. Whatever the insulation decided upon, exterior opaque walls and walls

Table 9-2. Insulation Recommendations (*R*-Values) and Overall Coefficients of Heat Transfer (*U*-Factors) for Major Areas of Heat Loss and Heat Gain in Residential Structures

Type of Construction	Opaque Sections Adjacent to Unheated Spaces			Opaque Sections Adjacent to Separately Heated Dwelling Units		
	Walls	Floors	Ceilings	Walls	Floors	Ceilings
Frame	*R*-11 *U*:0.07	*R*-11 *U*:0.07	*R*-19 *U*:0.05	*R*-11 *U*:0.07	*R*-11 *U*:0.07	*R*-11 *U*:0.07
Masonry	*R*-7 *U*:0.11	*R*-7 *U*:0.11	*R*-11 *U*:0.07	*R*-7 *U*:0.11	*R*-7 *U*:0.11	*R*-11 *U*:0.07
Metal section	*R*-11 *U*:0.07	Not applicable	Not applicable	*R*-11 *U*:0.07	Not applicable	Not applicable
Sandwich	*R*-* *U*:0.07	*R*-* *U*:0.07	*R*-* *U*:0.05	*R*-* *U*:0.07	*R*-* *U*:07	*R*-* *U*:07
Heated basement or unvented crawl space	*R*-7 *U*:0.11	Insulation not required	Insulation not required	*Since the thermal resistance of sandwich construction depends upon its composition and thickness, the amount of additional insulation required to obtain the maximum *U*-factor must be calculated in each case.		
Unheated basement or vented crawl space	Insulation not required	See "Floors"	Insulation not required			

Note: Insulation *R*-values refer to the resistance of the insulation only.

Courtesy Electric Energy Association

between separately heated units should have a maximum *U*-factor of 0.007, which will require insulation rated at *R*-11 or greater. A thermal resistance of *R*-11 can be provided by 3½-in. fiber batts or other equally suitable materials (see Chapter 3, Insulation Principles).

Masonry

A masonry wall is a particularly poor thermal barrier. A typical masonry cavity wall (4-in facebrick on the exterior, an air space, 4-in. concrete block, an air space, and a layer of gypsum wall board) has a thermal resistance equal to approximately 1 in. of rigid insulation. Exterior opaque masonry walls should have maximum *U*-factor of 0.11 requiring insulation rated at *R*-7 or greater. This can be accomplished in the following ways:

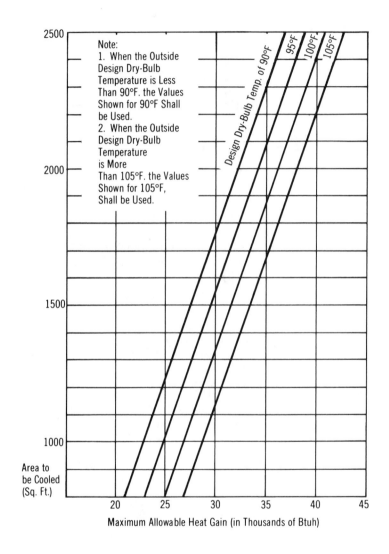

Fig. 9-10. Maximum summer heat-gain limits for air conditioned homes.

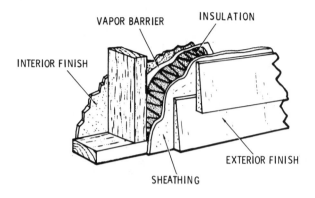

VAPOR BARRIER　　INSULATION

INTERIOR FINISH

EXTERIOR FINISH

SHEATHING

Fig. 9-11. Frame wall construction.

1. Place a layer of preformed rigid insulation between the exterior facebrick (or other finishing material) and the concrete block during construction (Fig. 9-12). Another method is to leave the space empty and then fill it with a foam insulation before capping the wall.

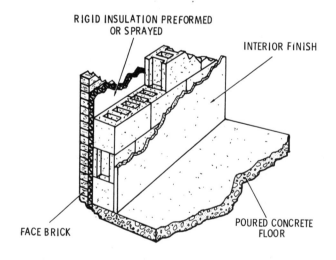

RIGID INSULATION PREFORMED
OR SPRAYED

INTERIOR FINISH

FACE BRICK

POURED CONCRETE
FLOOR

Fig. 9-12. Masonry wall construction with insulation installed on the outside portion.

2. If no air space exists between the exterior facebrick or other exterior finish and the concrete block, it will be necessary to add a layer of suitable insulating material (rated R-7 or better) on the inside surface of the block (Fig. 9-13). The insulating material can be rigid insulation attached to furring strips or preformed rigid insulation secured in place with adhesives. An interior wall finish is then applied over the insulating material.

If the wall is constructed entirely of concrete, apply preformed rigid insulation to either side by adhesive bonding and cover the layer of insulation with a wall-finish material (Fig. 9-14). The insulation must be rated R-7 or better.

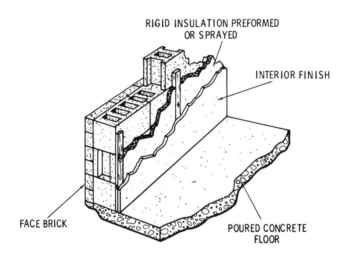

RIGID INSULATION PREFORMED OR SPRAYED

INTERIOR FINISH

FACE BRICK

POURED CONCRETE FLOOR

Fig. 9-13. Masonry wall construction with insulation installed on the inside portion.

Metal

Exterior opaque metal walls (Fig. 9-15) should have a maximum U-factor of 0.07. This will require insulation having a thermal resistance rated at R-11 or better, which can be provided by a layer of rigid insulation either preformed or foamed in place.

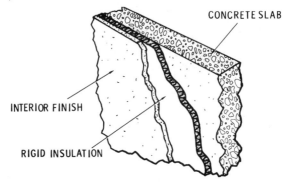

Fig. 9-14. Concrete slab construction.

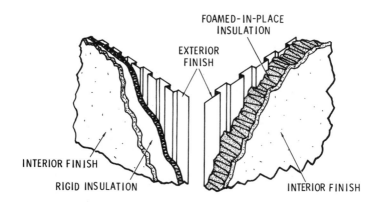

Fig. 9-15. Metal sections used for load-bearing walls and certain walls in residential construction.

Sandwich Construction

Sandwich or layered construction will vary in composition and thickness, but will typically consist of three layers: two facings and a core. Exterior opaque walls of a sandwich construction should have a maximum U-factor of 0.007. The thermal resistance (R-value) must be calculated on the basis of its composition and thickness.

Basement Walls

There is no need to insulate the exterior walls of unheated basements because there is no need for reducing heat loss. The exterior walls of *heated* basements (or interior walls separating heated from unheated areas) are a different matter because heat loss is a concern here. An insulating layer having a thermal resistance rated at R-7 or greater should be applied. Fig. 9-16 illustrates two methods for insulating these walls.

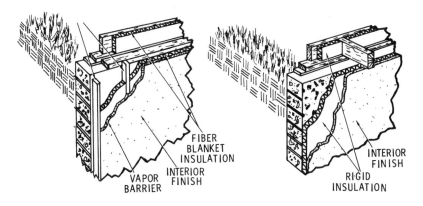

FIBER
BLANKET
INSULATION
VAPOR BARRIER
INTERIOR FINISH
INTERIOR FINISH
RIGID INSULATION

Fig. 9-16. Exterior walls of heated basements.

Crawl-Space Exterior Walls

The exterior walls of *unvented* crawl spaces should be insulated with a material providing a thermal resistance of at least R-7 or better. A method for doing this is illustrated in Fig. 9-17. No insulation is required for the exterior walls of *vented* crawl spaces.

Walls Between Separately Heated Dwelling Units

Where walls separating adjacent but independently heated dwelling units occur, use insulation with a thermal resistance of at least R-11. The insulation may be placed in either wall or divided proportionally between them (Fig. 9-18).

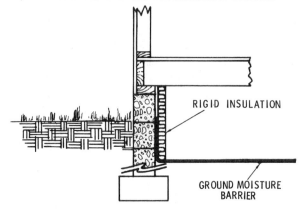

RIGID INSULATION

GROUND MOISTURE BARRIER

Fig. 9-17. Exterior walls of unvented crawl space.

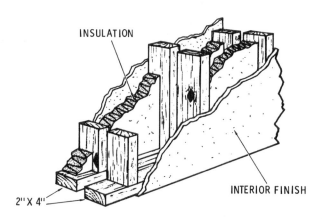

INSULATION

INTERIOR FINISH

2" X 4"

Fig. 9-18. Walls between adjacent separately heated dwelling units.

Wood or Metal Joist Frame Floors

Wood or metal joist floors over unheated basements or vented crawl spaces should have a maximum U-factor of 0.07 (requiring insulation rated at R-11 or greater). Three methods for providing insulation are illustrated in Figs. 9-19 to 9-21.

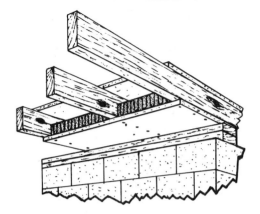

Fig. 9-19. Frame floor construction with insulation between the joist.

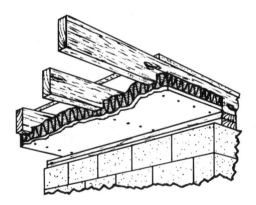

Fig. 9-20. Frame floor construction with insulation covering the joist.

Concrete Floors

Concrete floors over unheated spaces should have a maximum U-factor of 0.11 requiring insulation rated at R-7 or greater. The following three methods are recommended for applying a layer of insulation:

1. Spray rigid insulation in a suitable thickness on the *underside* of the floor.

2. Bond preformed rigid insulation to the top or bottom of the surface.

3. Form concrete around a core of rigid insulation.

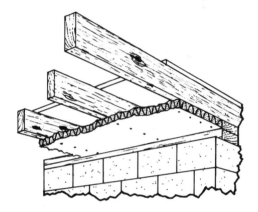

Fig. 9-21. Frame floor construction with rigid insulation panels attached to the bottom of the joist.

Slab-on-Grade Floors

Insulation rated R-5 is required for all slab-on-grade floors. Two methods for insulating slab-on-grade floors are illustrated in Fig. 9-22. Table 9-3 lists the maximum heat losses in terms of watts and Btuh per linear foot of exposed slab edge as recommended by the Electric Energy Association. More information can be obtained from the FHA Minimum Property Standards.

Floors of Sandwich Construction

Floors of sandwich construction should be insulated in the same manner as sandwich walls (see above).

Frame Ceilings and Roofs

Any ceiling below an unheated space should be insulated with materials having a rating of R-19 or better. The ceilings should have a maximum U-factor of 0.05. Among the various methods for providing such insulation are:

1. Staple blankets of batts to the joists before the finished ceiling is applied (Fig. 9-23).
2. Blow or manually apply loose insulation between ceiling joists from above.
3. Spray on insulation between the joists from above.

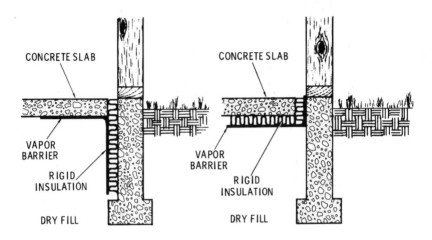

Fig. 9-22. Slab-on-grade construction.

Table 9-3. Slab Edge Heat Loss

Outdoor Design Temperature, °F	Total Width of Insulation, in. Rated R-5	Heat Loss per Foot of Exposed Slab Edge	
		Watts	Btuh
−30 and Colder	24	10.0	34
−25 to −29	24	9.4	32
−20 to −24	24	8.8	30
−15 to −19	24	8.2	28
−19 to −14	24	7.9	27
−15 to −9	24	7.3	25
0 to −4	24	7.0	24
+5 to +1	24	6.4	22
+10 to +6	18	6.2	21
+15 to +11	12	6.2	21
+20 to +16	Edge only	6.2	21
+21 and Warmer	None	—	—

Courtesy Electric Energy Association

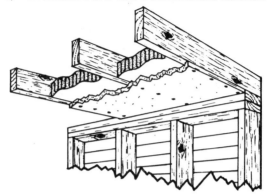

Fig. 9-23. Ceiling and roof frame construction.

Concrete Ceilings

Concrete ceilings below unheated spaces should have a maximum U-factor of 0.07, requiring insulation rated R-11 or greater. The following methods are suggested for insulating concrete ceilings:

1. Attach preformed rigid insulation to either surface with adhesives.
2. Apply sprayed-on rigid insulation to either surface.
3. Make a sandwich construction with preformed rigid insulation formed around a concrete core.
4. Apply rigid insulation to the *top surface* where the ceiling also serves as the roof of the structure.

Sandwich Ceilings

Ceilings of sandwich construction below unheated spaces should have a maximum U-factor of 0.05, requiring insulation rated at R-19 or better. Sandwich ceilings that also serve as the roof of a structure meet the same requirements.

DOORS AND WINDOWS

Two major heat loss areas for all types of construction are doors and windows. It is particularly important to reduce the rate of heat loss at these points to a minimal level when an electric heating system is used.

An exterior door constructed with a core of rigid insulation and faced with metal or acrylic film will have a very low U-factor (as low as 0.074). Infiltration heat loss around the edges of the door can be effectively reduced with suitable weather stripping.

A typical 1-in. solid wood door will have a U-factor of approximately 0.64 (compare this to the U-factor for the door described in the preceding paragraph). Doubling the thickness of the door will reduce the U-factor to about 0.43. The addition of metal of glass storm doors will further reduce the U-factor (0.39 and 0.29, respectively).

Windows with double glazing will have a U-factor of 0.58, as compared to 1.13 for single glazing. Double glazing also reduces noise transmission.

FURTHER INFORMATION

Further and more detailed information about insulation, the problems of heat loss and gain, and methods for calculating heat loss can be found by reading the appropriate sections of Chapter 2 (Heating Fundamentals), Chapter 3 (Insulation Principles), and Chapter 4 (Heating Calculations).

ADVANTAGES OF ELECTRIC HEATING AND COOLING

Among the principal advantages of using electric heating and cooling are:

1. Greater safety.
2. Quiet operation.
3. Economy of space.
4. Reduction of drafts.
5. Reduction of outside noise.
6. Uniform temperature.
7. Structural design flexibility.

Although the chances of an explosion in gas- or oil-fired heating systems are very rare because of the safety features built into these systems, they simply *cannot* occur in total electric heating.

Electric heating units are very compact and therefore utilize very little space. In baseboard systems, there is no need for ducts or pipes to carry a heat medium from its source to the space being heated. These factors offer a great degree of structural design flexibility because duct and pipe arrangements do not have to be taken into consideration. In addition, no chimney or flue arrangement is required.

In a structure properly insulated for electric heating and cooling, there is a marked reduction of drafts and the degree of outside noise penetration. Moreover, uniform temperatures will prevail.

Finally, an electric heating system is generally quieter than other types because fewer mechanical parts are involved. This quietness of operation is particularly characteristic of baseboard-type installations.

DISADVANTAGES OF ELECTRIC HEATING AND COOLING

An electrically heated and cooled structure offers certain inherent disadvantages when compared with other types of heating and cooling systems. Of course, this is true of *any* system regardless of the energy source. When choosing an energy source for a heating and cooling system, it is always necessary to carefully weigh the advantages and disadvantages of the system and its energy source against the requirements you demand from them.

The principal disadvantages of electric heating and cooling are:

1. Generally higher operating costs.
2. Lack of ventilation.
3. Higher installation costs.

CHAPTER 10

Furnace Fundamentals

The American Society of Heating, Refrigerating, and Air-Conditioning Engineers (ASHRAE) defines a furnace as being *a complete heating unit for transferring heat from fuel being burned to the air supplied to a heating system.* The *Standard Handbook for Mechanical Engineers* (Baumeister and Marks, seventh edition, 12-104) provides a definition that differs only slightly from the one offered by the ASHRAE. It defines a furnace as being *a self-enclosed, fuel-burning unit for heating air by transfer of combustion through metal directly to the air.* Contained within these closely similar definitions are the two basic operating principles of a furnace; that is, some sort of fuel is used to produce combustion, and the heat resulting from this combustion is transferred to the air within the structure. Note that *air,* not steam, water, or some other form of liquid, is used as the heat conveying medium. This feature distinguishes warm-air heating systems from the other types (see Chapter 6, Warm-Air Heating Systems).

Most modern furnaces are used in warm-air heating systems in which the furnace is a centralized unit, and the heat produced in the furnace is forced or rises by means of gravity through a system of ducts or pipes to the various rooms in the structure. This is what one commonly refers to as a *central heating system.* In other words, the furnace is generally in a centralized location within the heating system in order to obtain the most economical and efficient distribution of heat (although this is not an absolute necessity when a forced-warm-air furnace is used).

Ductless or pipeless furnaces are also used in some heating applications but are limited in the size of the area which they can effectively heat. They are installed in the room or area to be heated but are provided with no means for distributing the heat beyond the immediate adjacent area. This is a far less efficient and economical method of heating than the central heating system, but it is found to be adequate for a room, an addition to an existing structure, or a small house or building.

CLASSIFYING FURNACES

There are a number of different ways in which furnaces can be classified. One of the more popular methods is based on the fuel used to fire the furnace. Using this method, the following four types of furnaces are recognized:

1. Gas-fired furnaces.
2. Oil-fired furnaces.
3. Coal-fired furnaces.
4. Electric-fired furnaces.

The first three categories of furnaces use a fossil fuel to produce the combustion necessary for heat transfer. The last one, electric-fired furnaces, uses electricity. Whether or not electricity can be justifiably called a fuel is not of great importance here because in this particular instance it functions in the *same* manner as the three fossil fuels: It heats the air being distributed. Furnaces can also be classified by the means with which the heated air is distributed to the room (or rooms) in the structure. This method of classifying furnaces establishes two broad classes: (1) gravity warm-air furnaces and (2) forced-warm-air furnaces. The gravity warm-air furnaces rely *primarily* upon the principle of gravity for circulating the heated air. Because warm air is lighter than cold air, it will rise and pass through ducts or directly into the rooms to be heated. After passing off its heat, the cooler and heavier air descends through returns to the furnace, where it is reheated. Gravity-type furnaces represent the earliest designs used in warm-air heating systems. They were sometimes equipped with fans (integral or booster) to increase the rate of air flow. They have been largely replaced in popularity by forced-warm-air furnaces, which are equipped with integral fans.

Forced-warm-air furnaces are often divided into three principal classes, based primarily upon the location on the furnace of the warm-air discharge outlet and the return-air inlet. Furthermore, additional design considerations dependent upon the planned location of the furnace also enter into the classification of warm-air furnaces, resulting in three types, or classes, of warm-air furnaces:

1. Upflow furnaces:
   a. Upflow highboy furnaces.
   b. Upflow lowboy furnaces.
2. Downflow furnaces.
3. Horizontal furnaces.

Upflow Highboy Furnace

A typical *upflow highboy furnace* (also referred to as an *upflow furnace* or a *highboy furnace*) is shown in Fig. 10-1. These are compact heating units that stand no higher than 5 or 6 ft. and occupy a floorspace of approximately 4 to 6 sq. ft. (2 ft. × 2 ft. or 2 ft. × 3 ft.).

The heated air is discharged through the top of the upflow furnace (hence the name), and the return air enters the furnace through air intakes in the bottom or sides. Cooling coils can be easily added to the top of the furnace or in the duct system.

Upflow Lowboy Furnace

The *upflow lowboy furnace* (also referred to as an *upflow furnace* or *lowboy furnace*) (Fig. 10-2) is designed for low clearances and stands only about 4 to 4½ ft. high. Although shorter than either the upflow highboy or downflow types (see below), it is longer from front to back.

Both the return-air inlet and the warm-air discharge outlet are usually located on the top of the lowboy furnace. The lowboy furnace is found in heating installations where the ductwork is located above the furnace.

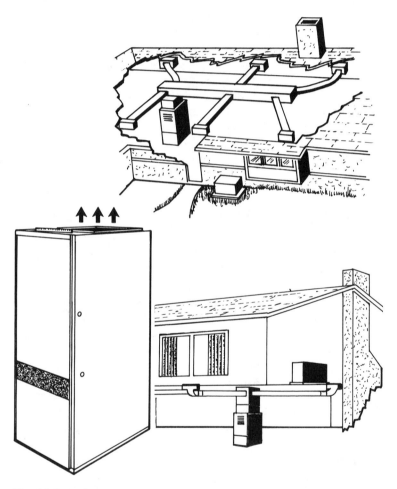

Fig. 10-1. Upflow (highboy) furnace.

AIR FLOW

Fig. 10-2. Upflow (lowboy) furnace.

Downflow Furnace

Fig. 10-3 shows an example of a *downflow furnace* (also referred to as a *counterflow furnace* or a *downdraft furnace*). It is very similar in size and shape to the upflow furnace, but it discharges its warm air at the bottom rather than the top. The return-air intake is located at the top.

A *downflow furnace* is used primarily in heating installations where the duct system is either imbedded in a poured concrete slab or suspended beneath the floor in a crawl space.

Horizontal Furnace

Horizontal furnaces (Fig. 10-4) are designed for installation in low, cramped spaces. They are often installed in attics (and referred to as *attic furnaces*) where they are positioned in such a way that a minimum of ductwork is used. This type of furnace is also frequently installed in crawl spaces.

303

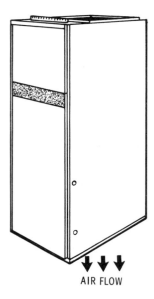

AIR FLOW

Fig. 10-3. Downflow furnace.

Although dimensions will vary slightly among the various manufacturers, the typical horizontal furnace is about 2 ft. wide by 2 ft. high and 4½ to 5 ft. long.

The terms used in this classification system (*upflow furnace, downflow furnace*, etc.) are commonly employed by furnace manufacturers in the advertising literature describing their products. This is equally true of their installation and operation manuals. Because of the widespread usage of these terms, they will be employed in the more detailed description of furnaces in the following chapters.

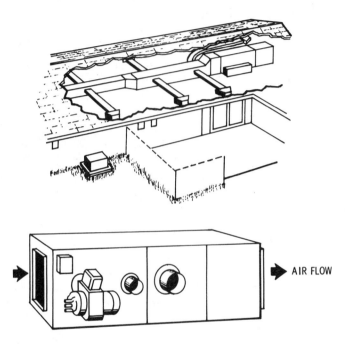

AIR FLOW

Fig. 10-4. Horizontal furnace.

GRAVITY WARM-AIR FURNACES

Gravity warm-air furnaces rely upon the fact that warm air is lighter than cold air. As a result, the warmer air rises through the ducts or pipes in the structure, gives off its heat to the rooms, and descends to the furnace as it becomes cooler and heavier. It is then reheated and rises once again to the rooms. A continuous circulation path is thus established through the heating and cooling of the air. Sometimes a fan is added to increase the rate of flow, but the primary emphasis is still the effect of gravity on the differing weights of air.

Depending upon the design, a gravity warm-air furnace will fall into one of the following categories:

1. A gravity warm-air furnace without a fan.
2. A gravity warm-air furnace with an *integral* fan.
3. A gravity warm-air furnace with a *booster* fan.

Each of these three categories represents different types of warm-air furnaces used in *central* heating systems.

Any gravity warm-air furnace not equipped with a fan relies entirely upon gravity for air circulation. The flowrate is very slow, and extreme care must be taken in the design and placement of the ducts of pipes. Sometimes an integral fan is added to reduce the internal resistance to airflow and thereby speed up air circulation. A booster fan provides the same function but is designed not to interfere with air circulation when it is not use.

The round-cased, gravity warm-air furnace illustrated in Fig. 10-5 is a coal-fired unit that can be converted to gas or oil (see Chapter 16, Boiler and Furnace Conversions). Depending upon the model, these furnaces are capable of developing up to 108,266 Btu at register and up to 144,319 Btu at bonnet.

Floor, wall, pipeless furnaces, and some unit (space) heaters also operate on the principle of the gravity warm-air furnace. They are distinguished by the fact that the warm air is discharged directly into the room (or rooms) without the use of ducts or pipes.

SELECTING A FURNACE FOR A NEW HOUSE

One of the many decisions involved in building a new house is the selection of the heating, ventilating, and air conditioning system. This is probably one of the most important decisions because the cost of operating and maintaining the system and its efficiency will be a daily fact of life for as long as you live there. It is therefore in your best interest to select the most efficient heating system that you possibly can.

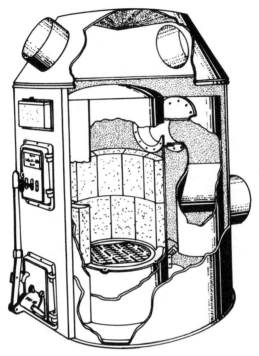

Courtesy Oneida Heater Co., Inc.

Fig. 10-5. Coal-fired, gravity warm-air furnace.

The type of heating system selected (forced-warm-air, hydronic, steam, electric baseboard, etc.) will depend upon such factors as the design of the house, the climate where the house is located, fuel costs, and personal preference. Comments on the advantages and disadvantages of the various types of heating systems are found in Chapters 6 to 9 (e.g., see Chapter 6, Warm-Air Heating Systems).

Assuming that, like the majority of home owners, you have decided to install a *forced*-warm-air heating system, you should pay particular attention to selecting a suitable furnace. Should this decision be left entirely to your building contractor? Certainly not. For convenience, friendship, or other reasons, he may prefer to subcontract the work to a heating firm with which he customarily does business. As a result, you may not be getting the best furnace for your needs. For example, John Doe Heating, Ventilating, and Air Conditionzing Company may deal in furnace Brand X (or Brands X, Y, and Z). It may be that none of these furnaces is as efficient and economical to operate as the one you might wish to use.

There are many manufacturers of forced-warm-air furnaces doing business in the country. The furnaces produced by these manufacturers differ from one another in a variety of ways, including:

1. Installation cost.
2. Design factors.
3. Type of fuel used.
4. Cost of furnace.
5. Furnace gross output (Btuh).
6. Furnace net output (Btuh).
7. Furnace efficiency.

The *installation cost* depends upon the area in which you are living and can vary widely for the same furnace model from the same manufacturer. It is not something you can easily control because it depends upon local labor costs, delivery charges, and availability of equipment.

Design factors involve developments in the heating equipment that improve (or detract from) their performance. For example, some electric-fired furnaces use a direct-drive blower motor instead of a belt-driven one. The direct-drive motor results in a smaller, more compact furnace. The belt-driven blower motor has the advantage of being more easily maintained (it is simply a matter of replacing the standard motor used in the unit). Other design developments include improvement in flame retention (e.g., in oil-fired furnaces), more efficient combustion chambers, or improvements in other components. It may take a little research, but a number of professional journals in the heating field carry articles reviewing new products. Consumer magazines also provide this service. Do not take the manufacturer's (or the sales representative's) claim at face value. Go to your local library and read what the experts have to say.

The type of fuel used will be an important factor in determining future operating costs. What is the cost of the various fuels in your area? Remember, the cheapest may not necessarily be the best for your purposes. Operating costs depend on the total amount of fuel necessary to heat your home on a month-to-month basis. As a general rule, electricity is more expensive than the fossil fuels, and gas slightly more expensive than oil. Read Chapter 5 (Heating Fuels) for additional information.

In considering the *cost of a furnace*, remember that the most expensive furnace is not necessarily the best one, and the cheapest is not always that bargain you expect it to be. Your *first* consideration should be furnace performance. Although the initial cost of a suitable furnace may be high relative to others, it will soon pay for itself with lower operating costs.

Furnace efficiency is expressed as a percentage and is found by subtracting *furnace net output* (the actual amount of heat delivered to the rooms) from *furnace gross output*. For example, an oil-fired furnace with a gross output of 150,000 Btuh and a net output of 120,000 Btuh will have an approximate efficiency of 80 percent.

By way of review, the three basic steps involved in selecting a suitable furnace for a new house are:

1. Determine the *type* of heating system you want for the house.
2. Estimate the amount of heat loss from the structure (see Chapter 4, Heating Calculations).
3. Select a furnace with a heating capacity capable of replacing the lost heat.

SELECTING
A FURNACE FOR AN OLDER HOUSE

Sometimes an older house or building will have a furnace that needs replacing. If you are planning to install a newer model of the same kind of furnace, you should have no serious difficulties. However, be sure that the newer model develops a similar Btu rating at both register and bonnet. In a forced-warm-air heating system, the ducts are sized in accordance with the overall requirements of the system and a furnace is selected with a capacity to meet these requirements.

Switching from coal to gas or oil also should not present any great difficulties. Conversion burners have been designed and manufactured for just this purpose (see Chapter 16, Boiler and Furnace Conversions).

Installing an electric-fired furnace in an older structure is not generally recommended unless the construction is particularly tight and well insulated.

FURNACE COMPONENTS AND CONTROLS

Most forced-warm-air furnaces have the following basic components and controls:

1. Heat exchanger.
2. Blower.
3. Air filter.
4. Primary control.
5. Fan and limit controls.
6. Thermostat control.

The *heat exchanger* (Fig. 10-6 and 10-7) is a metal surface (0.05- to 0.06-in. steel) located between the burning fuel and the circulating air in the furnace. The metal becomes hot and transfers its heat to the air above, which is then circulated through the ducts by the *blower*.

Most forced-warm-air furnaces are equipped with either a disposable or permanent (and washable) *air filter* to clean the circulating air. An air filter is not recommend for a gravity warm-air furnace because it tends to restrict the airflow.

Most warm-air furnaces are thermostatically controlled. In addition, there will be a primary control as well as fan and limit controls. Among the functions of the primary control is the regulation or stoppage of the flow of fuel to the combustion chamber when the fire is out or the thermostat indicates that no further fuel is needed. The fan and limit controls are also actuated by the thermostat and are designed to start or stop the fan when the

BLOWER

Courtesy Meyer Furnace Co.

Fig. 10-6. Heat exchanger for a horizontal gas-fired furnace.

temperature in the furnace bonnet reaches predetermined and preset temperatures. In warm-air furnaces, it is suggested that the limit-control switch be placed in the warm-air plenum, with a recommended setting of 200°F for a forced-warm-air furnace and 300°F for a gravity warm-air furnace.

These and other controls are offered by furnace manufacturers in a variety of different combinations. Their primary function is to provide safe, smooth, and automatic operation of the warm-air furnace. More detailed descriptions of the controls used in furnaces and heating systems are found in Chapter 4 (Thermostats and

311

FILTER

BLOWER

FAN AND LIMIT
SWITCH

Courtesy Meyer Furnace Co.

Fig. 10-7. Heat exchanger for an upflow gas-fired furnace.

Humidistats), Chapter 5 (Gas and Oil Controls), Chapter 6 (Other Automatic Controls), and Chapter 9 of Volume 2 (Valves and Valve Installation). (Other chapters also include sections on automatic controls. Check the Index) .

Most warm-air furnaces (except hand-fired coal furnaces and electric-fired furnaces) are provided with an apparatus designed to automatically prepare the fuel for combustion or to feed it directly to the fire. These apparatuses (oil burners, gas burners, and automatic coal stokers) are described in the appropriate chapters. Electric-fired, warm-air furnaces utilize electric resistance heaters.

Cooling coils, electronic air cleaners, and humidifiers are examples of optional equipment that can be added to most forced-warm-air furnaces to provide total environmental control. It is best to include this optional equipment when the furnace is installed rather than add them at a later date. By doing so, you avoid complications that might arise in the overall design of your heating system. For example, it is not simply a matter of adding cooling coils to obtain air conditioning. Duct sizes must also be considered, and they are not necessarily the same as those used for heating. This optional equipment is considered in detail in Chapters 7 and 8 of Volume 3.

PIPELESS FLOOR AND WALL FURNACES

A *pipeless furnace* (Fig. 10-8) is commonly a gravity warm-air furnace installed in a central location beneath the floor. A single grille for the warm air and a return is provided for air circulation. This type of pipeless furnace is sometimes referred to as a *floor furnace*, although the latter is actually a permanently installed room heater and should be distinguished as such. *Wall furnaces* also belong to this category.

Both gas- and oil-fired furnaces are manufactured for installation in recesses cut from the floor. They are available in a wide range of Btu ratings for different types of installations. Some are thermostatically equipped for automatic temperature control and with blowers for forced-air circulation (although those that operate on the gravity principle of air circulation are more common). There is also a choice between electric ignition or pilot flame. All gas- or oil-fired floor furnaces *must* be vented.

Another type of pipeless furnace is the gas- or oil-fired vertical furnace installed in a closet or a wall recess. The counterflow types discharge the warm air from grilles located at the bottom of the furnace, as shown in Fig. 10-9.

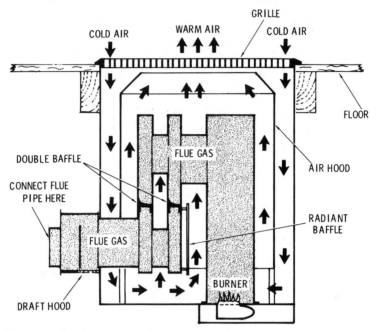

GRILLE

COLD AIR   WARM AIR   COLD AIR

FLOOR

DOUBLE BAFFLE

FLUE GAS

AIR HOOD

CONNECT FLUE
PIPE HERE

RADIANT
BAFFLE

FLUE GAS

DRAFT HOOD

BURNER

Fig. 10-8. Gas-fired, gravity floor furnace.

DUCT FURNACES

A *duct furnace* (Fig. 10-10) is a unit heater designed for installation in a duct system where a blower (or blowers) is used to circulate the air. It is commonly designed to operate on natural or propane gas, although electric duct heaters are also available (see Chapter 7 of Volume 2).

FURNACE INSTALLATION

No attempt should be made to install a warm-air furnace until you have consulted the local codes and standards. The American Gas Association and the National Fire Protection Association have also established codes for installing warm-air furnaces; these are available to the public through their publications (see,

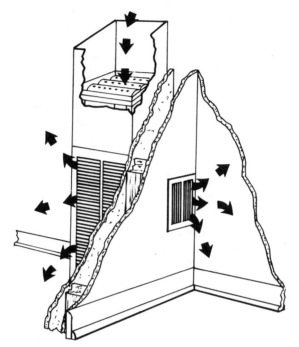

Fig. 10-9. Counterflow vertical furnace.

for example, NFPA No. 31, "Installation of Oil Burning Equipment 1972"; NFPA No. 90B, "Residence Warm Air Heating 1971"; and NFPA No. 204, "Smoke and Heat Venting 1968"). These publications can be obtained free or at a modest price from these organizations. Their addresses appear in Appendix A (Professional and Trade Associations) of this book.

As soon as the heating equipment is shipped to your building site or existing structure, check it for missing or damaged parts. The manufacturer should be notified immediately of any discrepancies so that replacements can be made.

Follow the manufacturer's instructions for installing the heating equipment, but give precedence to local codes and regulations should any conflict arise. Most manufacturers base their installation instructions on existing codes formulated by the

315

American Gas Association and the National Fire Protection Association.

Locate the furnace so that it has the shortest possible flue run containing as few elbows (turns in the run) as possible. Clearances around the furnace should be kept to the required minimum but within the regulations established by local codes and standards. Air ventilation must be adequate for efficient operation.

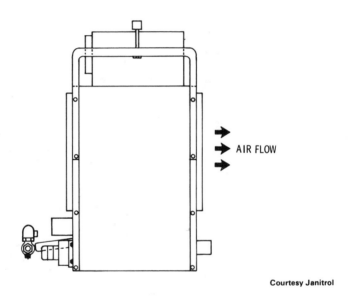

AIR FLOW

Fig. 10-10. Gas-fired duct furnace.

After the furnace has been installed, check all gas or oil lines for possible leaks. Make the necessary repairs if any leaks are found.

More detailed instructions for installing furnaces are found in Chapter 11 (Gas-Fired Furnaces), Chapter 12 (Oil-Fired Furnaces), Chapter 13 (Coal-Fired Furnaces), and Chapter 14 (Electric-Fired Furnaces).

FURNACE MAINTENANCE

Furnace maintenance is a very important part of the efficient operation of a warm-air heating system and should never be neglected.

The manufacturer of the heating equipment will provide recommendations for proper maintenance, and these recommendations should be carefully followed. By doing so, you will extend the life of the equipment, improve its efficiency, and reduce operating costs.

Maintenance recommendations specific to gas-fired, oil-fired, coal-fired, and electric-fired furnaces are included in the appropriate chapters (see, for example, Chapter 11, Gas-Fired Furnaces). Certain maintenance recommendations are general in nature and will apply to any type of furnace regardless of the fuel used. These general maintenance recommendations are as follows:

1. *Always* read and closely follow the manufacturer's instructions (if they are available).
2. Check gas and oil pipes for leaks.
3. Check and clean (or replace) air filters.
4. Clean the furnace once a year (preferably before the heating season starts).
5. Oil the motor of a gas or oil burner two to three times each year.

GENERAL TROUBLESHOOTING HINTS

Each type of equipment described in this book contains a troubleshooting section in which problems specific to that equipment are discussed. A complaint common to all heating systems is that the system will not deliver heat. Because this is a troubleshooting problem that is general in nature, recommended corrective procedures will be included in this section.

317

The first thing that should be done if the system fails to deliver heat is to check the fuse box. Many times the whole problem can be traced to a blown fuse. By simply replacing it, you will save the cost of a service call. By the way, do not look at the manufacturer's name on the room thermostat and call the nearest factory or representative. Thermostat manufacturers do not make furnaces! If the fuse box does not contain the problem, then check the following:

1. Make certain the power is on at the main switch.
2. Make certain the burner motor fuse is not blown.
3. Check the burner on-off switch (if the switch is off, make sure the combustion chamber is free of fuel or fuel vapors, then turn the switch to *on* position).
4. Check the fuel supply.
5. Make certain all manual valves are open.
6. Make certain the limit switches are closed.
7. Reset the safety switch, and set the thermostat to call for heat.

CHAPTER 11

Gas-Fired Furnaces

Gas-fired furnaces are available in upflow, downflow (counterflow), or horizontal-flow models for installation in attics, basements, closet spaces, or at floor level (Figs. 11-1, 11-2, and 11-3). These furnaces are also available in a wide range of heating capacities for different-size structures.

This chapter is primarily concerned with a description of the procedures recommended for the installation, operation, servicing, and repair of gas-fired, forced-warm-air furnaces. Additional information about furnaces is contained in Chapter 10 (Furnace Fundamentals).

Only gas-fired furnaces approved by the American Gas Association should be used in a heating installation. Not only does AGA certification ensure a certain standard of quality, it is often required by many local codes and regulations. The furnace should be installed in accordance with procedures outlined by the National Environmental Systems Contractors Association.

PLANNING SUGGESTIONS

The first step in planning a heating system is to calculate the maximum heat loss for the structure. This should be done in accordance with procedures described in the manuals of the National Warm Air Heating and Air Conditioning Association or

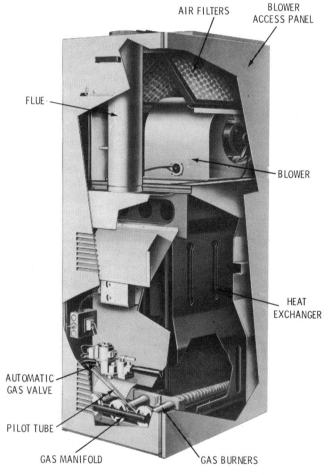

Courtesy Coleman Co., Inc.

Fig. 11-1. Gas-fired downflow furnace.

by a comparable method. *Correctly* calculating the maximum heat loss for the structure is very important because the data will be used to determine the size (capacity) of the furnace selected for the installation.

If the heating or heating/cooling installation is to be approved by either the FHA or VA, heat loss and heat gain calculations should be made in accordance with the procedure described in *National Environmental Systems Contractors Manual J.*

Locations and Clearances

The proper location of a gas-fired furnace depends upon a careful consideration of the following three factors:

1. Length of heat runs.
2. Chimney location.
3. Clearances.

A gas-fired, forced-warm-air furnace should be located as near as possible to the center of the heat-distribution system and

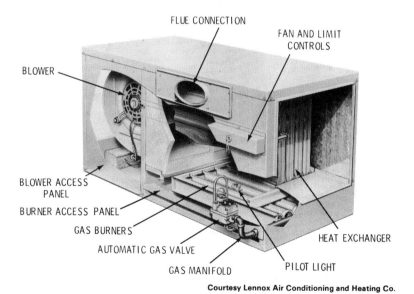

FLUE CONNECTION

FAN AND LIMIT CONTROLS

BLOWER

BLOWER ACCESS PANEL

BURNER ACCESS PANEL

GAS BURNERS

AUTOMATIC GAS VALVE

GAS MANIFOLD

HEAT EXCHANGER

PILOT LIGHT

Courtesy Lennox Air Conditioning and Heating Co.

Fig. 11-2. Horizontal gas-fired furnace.

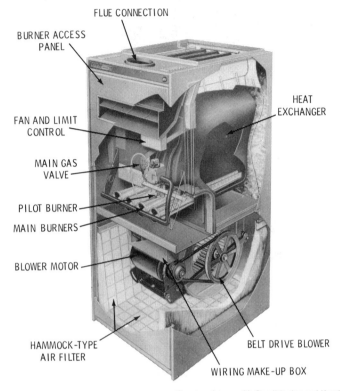

FLUE CONNECTION

BURNER ACCESS PANEL

HEAT EXCHANGER

FAN AND LIMIT CONTROL

MAIN GAS VALVE

PILOT BURNER

MAIN BURNERS

BLOWER MOTOR

HAMMOCK-TYPE AIR FILTER

BELT DRIVE BLOWER

WIRING MAKE-UP BOX

Courtesy Lennox Air Conditioning and Heating Co.

Fig. 11-3. Gas-fired upflow furnace.

chimney. Centralizing the furnace reduces the need for one or more long supply ducts, which tend to lose a certain amount of heat. The number of elbows should also be kept to a minimum for the same reason. Installing the furnace near the chimney will reduce the length of the horizontal run of flue pipe. The flue pipe should always be kept as short as possible.

Sufficient clearance should be provided for access to the draft hood and flue pipe. It is particularly important to locate a gas-fired furnace so that the draft hood is at least 6 in. from any combustible material.

Allowing access for lighting the furnace and servicing it is also an important consideration. Most furnace manufacturers recommend a clearance of 24 to 30 in. in front of the unit for access to the burner and controls.

Clearances between the furnace and any combustible materials are usually provided by the manufacturer in the furnace specifications.

Installation Recommendations

New furnaces for residential installations are shipped preassembled from the factory with all internal wiring completed. In order to install the new furnace, the gas piping from the supply main, the electrical service from the line voltage main, and the low-voltage thermostat must be connected. Directions for making these connections are to be found in the furnace manufacturer's installation instructions.

Make certain you have familiarized yourself with all local codes and regulations that govern the installation of a gas-fired furnace. Local codes and regulations take precedence over national standards. Your furnace installation *must* comply with the local codes and regulations.

Always check a new furnace for damages as soon as it arrives from the factory. If shipping damages are found, the carrier (*not the factory*) should be notified and a claim filed immediately. Heating equipment is shipped FOB from the factory, and it is the responsibility of the carrier to see that it arrives undamaged.

Always place the furnace on a solid level base. This will reduce vibrations from the equipment and keep operating noise to a minimum. In a crawl-space installation, the furnace is either supported on a slab or on concrete blocks, or it is suspended from floor joists with ⅜-in. hanger rods.

A furnace installed in an attic should be placed on a fiberboard sheeting base to absorb vibrations. Many furnaces designed for attic installations are certified for installation directly on combustible material such as ceiling joists or attic floors. In an attic installation, the furnace should be placed over a load-bearing partition for additional support.

Duct Connections

The warm-air plenum, warm-air duct, and return-air duct should be installed in accordance with the recommendations of the National Environmental Systems Contractors Association.

Detailed information about the installation of an air-duct system is also contained in the following two publications of the National Fire Protection Association:

1. *Residence Type Warm Air Heating and Air Conditioning System* (NFPA No. 90B).
2. *Installation of Air Conditioning and Ventilating Systems of Other than Residence Type* (NFPA No. 90A).

Additional information about duct connections can be found in Chapter 7 of Volume 2 (Ducts and Duct Systems)

It is important for you to remember the following facts about furnace duct connections and the air distribution ducts:

1. Duct lengths should be kept to a minimum by centralizing the furnace location as much as possible.
2. The duct system must be properly sized. Undersizing will cause a higher external static pressure than the one for which the furnace was designed. This may result in a noisy blower or insufficient air distribution.
3. Duct sizing should include an allowance for the future installation of air conditioning equipment.
4. The furnace should be leveled before any duct connections are made.
5. Seal around the base of the furnace with a calking compound to prevent air leakage if a bottom air return is used.
6. Ducts should be connected to the furnace plenums with tight fittings.
7. Warm-air-plenum and return-air-plenum connections should be the same size as the openings on the furnace.
8. Use tapered fittings or starter collars between the ducts and the furnace plenum.
9. Ductwork located in unconditioned spaces (e.g., unheated attics, basements, and crawl spaces) should be insulated if air conditioning is planned for some future date.
10. Install canvas connectors between air plenums and the casing of the furnace to ensure quiet operation. Check local codes and regulations to make certain canvas connectors are in compliance.
11. Line the first 10 ft. or so of the supply and return ducts with

acoustical material when extremely quiet operation is necessary.

12. Install locking-type dampers in each warm-air run to facilitate balancing the system.

Electrical Wiring

All of the electrical wiring *inside* the furnace is completed and inspected at the factory before the unit is shipped. Wiring instructions will accompany each furnace, and the internal (factory installed) wiring will be clearly marked. The wiring to controls and to the electrical power supply must be completed by the installer, and the wiring should be done in accordance with the diagram supplied by the manufacturer. This wiring is usually indicated on wiring diagrams by broken lines.

All wiring connected to the furnace must comply with the National Electric Code and any local codes and regulations. Local codes and regulations will always take precedence over national ones when there is any conflict. In order to validate the furnace warranty, many manufacturers require that a local electrical authority approve all electrical service and connections.

The two types of electrical connections required to field wire a furnace are: (1) line-voltage field wiring, and (2) control-voltage field wiring. Both are shown in the wiring diagram for a Carrier heating/cooling unit (Fig.11-4).

The line-voltage wiring connects the furnace to the building power supply. It runs directly from the building power panel to a fused disconnect switch. From there, the wiring runs to terminals L_1 and L_2 on the power-supply terminal block (i.e., the furnace junction box).

The unit must be properly grounded either by attaching a ground conduit for the supply conductors or by connecting a separate wire from the furnace ground lug to a suitable ground. The furnace must be electrically grounded in accordance with the National Electric Code (ANSI-C1-1971).

Furnaces equipped with motors in excess of ¾-hp or 12-amp ratings, must be wired to a separate 240-volt service in accordance with the National Electric Code and local codes and regulations. A 240-volt transformer should be installed whenever 240-

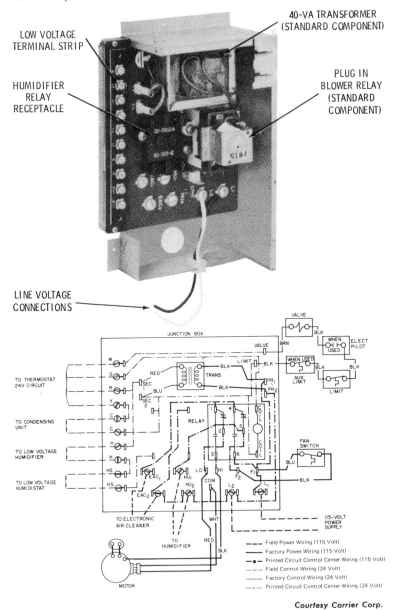

Courtesy Carrier Corp.

Fig. 11-4. Wiring diagram for a Carrier heating and cooling unit.

volt service is required and/or a motor larger than ¾ hp is used. The external control-voltage circuitry consists of the wiring between the thermostat and the low-voltage terminal block located in the control-voltage section of the furnace. Instructions for the control-voltage wiring are generally shipped with the thermostat.

Ventilation and Combustion Air

A gas-fired furnace requires an unobstructed supply of air for combustion. This combustion air supply is generally supplied through natural air infiltration (cracks in the walls and around windows and doors) or through ventilating ducts.

If a gas-fired furnace is located in an open area (basement or utility room) and the ventilation is relatively unrestricted, there should be a sufficient supply of air for combustion and draft hood dilution. On the other hand, if the heating unit is located in an enclosed furnace room or if normal air infiltration is effectively reduced by storm windows or doors, then certain provisions must be made to correct this situation. Fig. 11-5 illustrates one type of modification that can be made to provide an adequate supply of combustion and ventilation air to a furnace room. As shown, two permanent grilles are installed in the walls of the furnace room, each of a size equal to 1 sq. in. of free area per 1000 Btu per hour of burner output. One grille should be located approximately 6 in. from the ceiling, and the other one near the floor. A possible alternative is to provide openings in the door, ceiling, or floor as shown in Fig. 11-6.

If the furnace is located in a tightly constructed building, it should be directly connected to an outside source of air. A permanently open grille sized for at least 1 sq. in. of free area per 5000 Btu per hour of burner output should also be provided. Connection to an outside source of air is also recommended if the building contains a large exhaust fan.

Provisions for ventilation and combustion air are described in greater detail in *Installation of Gas Appliances and Gas Piping* (ASA Z21.30—NFPA No. 54). These standards were adopted and approved by the National Fire Protection Association and the National Board of Fire Underwriters.

Venting

Provisions must be made for venting the products of combustion to the outside in order to avoid contamination of the air in the living or working spaces of the structure. The four basic methods of venting the products of combustion to the outside are:

1. Masonry chimneys.
2. Low-heat Type A prefabricated chimneys.
3. Type B gas vents.
4. Type C vents.
5. Wall venting.

Masonry and prefabricated chimneys are described below (see "Chimneys" and "Troubleshooting Chimneys").

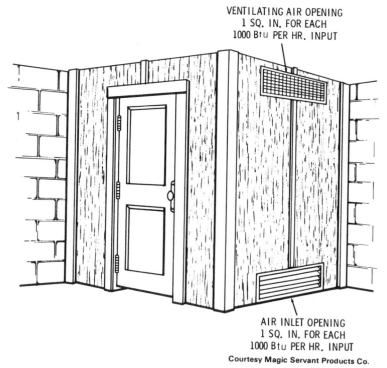

VENTILATING AIR OPENING
1 SQ. IN. FOR EACH
1000 Btu PER HR. INPUT

AIR INLET OPENING
1 SQ. IN. FOR EACH
1000 Btu PER HR. INPUT

Courtesy Magic Servant Products Co.

Fig. 11-5. Recommended air openings in furnace room wall.

Low-heat Type A flues are metal, prefabricated chimneys that have been tested and approved by the Underwriters' Laboratories. These chimneys are easier to install than masonry chimneys and are generally safer under abnormal firing conditions.

Type B gas vents (Fig. 11-7) are listed by the Underwriters' Laboratories and are recommended for venting all standard gas-fired furnaces. They are not recommended for incinerators, combination gas-oil appliances, or appliances convertible to solid fuel. Check the furnace specifications for the type of vent to use. It will usually be AGA certified for use with a Type B gas vent. A Type C vent is used to vent gas-fired furnaces in attic installations (Fig. 11-8).

Wall venting involves gas-fired appliances which have their combustion process, combustion air supply, and combustion

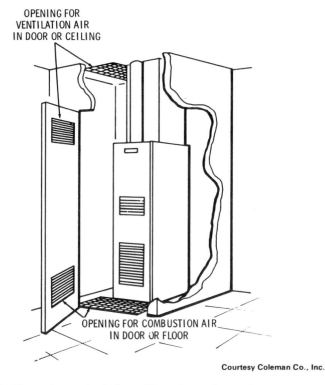

OPENING FOR
VENTILATION AIR
IN DOOR OR CEILING

OPENING FOR COMBUSTION AIR
IN DOOR OR FLOOR

Fig. 11-6. Alternative method of providing air supply openings.

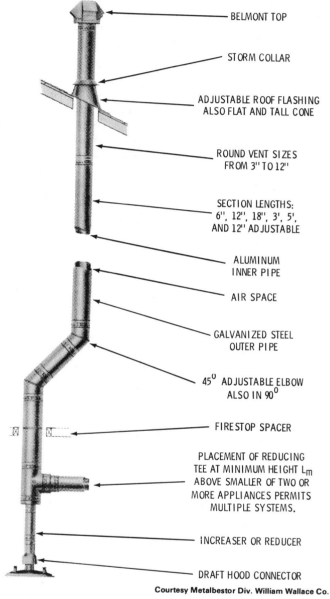

BELMONT TOP

STORM COLLAR

ADJUSTABLE ROOF FLASHING
ALSO FLAT AND TALL CONE

ROUND VENT SIZES
FROM 3" TO 12"

SECTION LENGTHS:
6", 12", 18", 3', 5',
AND 12" ADJUSTABLE

ALUMINUM
INNER PIPE

AIR SPACE

GALVANIZED STEEL
OUTER PIPE

45° ADJUSTABLE ELBOW
ALSO IN 90°

FIRESTOP SPACER

PLACEMENT OF REDUCING
TEE AT MINIMUM HEIGHT L_m
ABOVE SMALLER OF TWO OR
MORE APPLIANCES PERMITS
MULTIPLE SYSTEMS.

INCREASER OR REDUCER

DRAFT HOOD CONNECTOR

Courtesy Metalbestor Div. William Wallace Co.

Fig. 11-7. Double-wall, gas-vent pipe and fittings for single or multiple Type B gas-vent system.

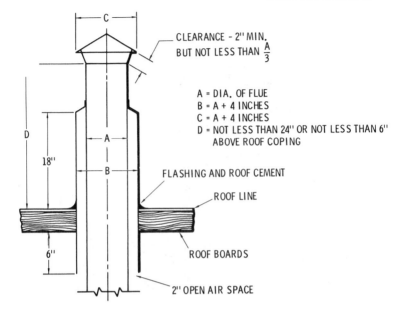

CLEARANCE - 2" MIN.
BUT NOT LESS THAN $\frac{A}{3}$

A = DIA. OF FLUE
B = A + 4 INCHES
C = A + 4 INCHES
D = NOT LESS THAN 24" OR NOT LESS THAN 6"
ABOVE ROOF COPING

FLASHING AND ROOF CEMENT

ROOF LINE

ROOF BOARDS

2" OPEN AIR SPACE

Fig. 11-8. Type C vent.

products isolated from the space which is being heated. These appliances are generally vented through the wall (Figs. 11-9 and 11-10).

Flue Pipe

The *flue* is the general term used to describe the passage through which the flue gases pass from the combustion chamber of the furnace (or boiler) to the outside. A flue is also referred to as a *flue pipe, vent pipe*, or *vent connector*.

The term *appliance flue* refers to the flue passages inside a furnace or boiler. A *chimney flue* is the vertical flue passage running up through the chimney. A *vent connector* is specifically the flue passage between the heating unit and the chimney. It is also variously referred to as a *chimney connector* or *smoke pipe*. A *flue outlet*, or *vent*, is the opening in a furnace or boiler through which the flue gases pass.

The flue pipe (vent pipe) connects the smoke outlet of the furnace with the chimney. It should never extend beyond the

331

Fig. 11-9. Single-package heating and cooling unit with through-the-wall venting.

inner liner of the chimney and should never be connected to the flue of an open fireplace. Furthermore, flue connections from two or more sources should never enter the chimney at the same level from opposing sides.

The horizontal run of flue pipe should be pitched toward the chimney with a rise of at least ¼ in. per running foot. Some furnace manufacturers recommend as much as a 1-in. minimum rise per running foot. Do not allow the horizontal run to exceed, in length, 75 percent of the vertical run. Vent pipe crossovers in an attic must not extend at an angle of less than 60° from the vertical.

At intervals, fasten the horizontal run of flue pipe securely with sheet-metal screws and support the pipe with straps or hangers to prevent sagging. Sagging can cause cracks to develop at the joints, which may result in releasing toxic flue gases into the living and working spaces of the structure.

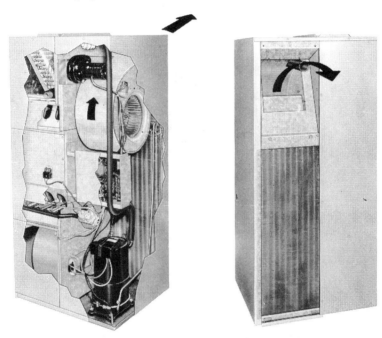

Courtesy Climatrol

Fig. 11-10. Cutaway of a gas-fired heating/cooling unit with through-the-wall venting design.

The point at which the flue pipe enters the chimney should be at least 2 ft. above the cleanout opening of the chimney (Fig 11-11). The flue pipe must be the same size as the outlet of the flue collector (furnace flue collar). Never install a damper in a flue pipe or reduce the size of the flue pipe. Run the flue pipe from the draft hood to the chimney in as short a distance as possible. All joints in the flue pipe should be made with the length nearest the furnace overlapping the other.

If excessive condensation is encountered, install a drip tee in place of an elbow. Never use dampers or other types of restrictors in the vent pipe. A minimum distance should be maintained between the vent pipe and the nearest combustible material.

When more than one appliance is vented into a common flue, the area of the common flue should be equal to the area of the largest flue plus 50 percent of the area of the additional flue.

CHIMNEYS

Most chimneys are constructed of brick or metal. If a brick chimney is used, it should be lined with a protective material to prevent damage from water vapor. The most commonly used liner is smooth faced tile. Prefabricated factory-built chimneys are also used, but only those listed by Underwriters' Laboratories are suitable for use with fuel-burning equipment.

The standard chimney must be at least 3 ft. higher than the roof or 2 ft. higher than any portion of the structure within 10 ft. of the chimney flashing in order to avoid downdrafts (Fig. 11-12).

An existing chimney should always be checked to make sure it is smoketight and clean. Any dirt or debris must be cleaned out before the furnace or boiler is used.

TROUBLESHOOTING THE CHIMNEY

The chimney must give sufficient draft for combustion or the furnace will not operate efficiently. It must also provide a means for venting the products of combustion to the outside. Although a chimney-produced draft is not as important for the combustion process in gas-fired appliances as it is in other types of fuel-burning equipment, the venting capacity of the chimney is extremely important. The chimney must be of a suitable area and height to vent all the products of the combustion process.

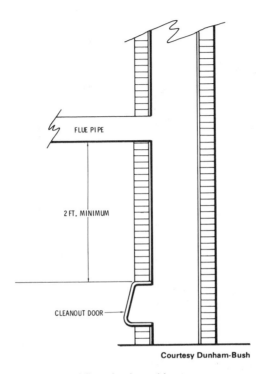

Courtesy Dunham-Bush

Fig. 11-11. Entry point of flue pipe into chimney.

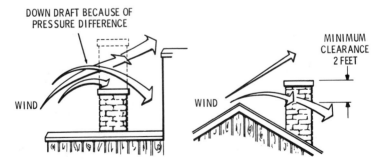

Fig. 11-12. Examples of correct and incorrect chimney designs.

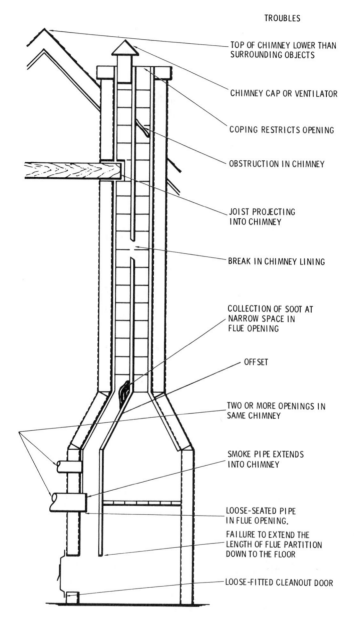

TROUBLES

TOP OF CHIMNEY LOWER THAN
SURROUNDING OBJECTS

CHIMNEY CAP OR VENTILATOR

COPING RESTRICTS OPENING

OBSTRUCTION IN CHIMNEY

JOIST PROJECTING
INTO CHIMNEY

BREAK IN CHIMNEY LINING

COLLECTION OF SOOT AT
NARROW SPACE IN
FLUE OPENING

OFFSET

TWO OR MORE OPENINGS IN
SAME CHIMNEY

SMOKE PIPE EXTENDS
INTO CHIMNEY

LOOSE-SEATED PIPE
IN FLUE OPENING.

FAILURE TO EXTEND THE
LENGTH OF FLUE PARTITION
DOWN TO THE FLOOR

LOOSE-FITTED CLEANOUT DOOR

Fig. 11-13. Common chimney problems and their corrections.

Fig. 11-13 illustrates some of the common chimney problems that can cause insufficient draft and improper venting. Some of these problems are detectable by observation; others require the use of a draft gauge.

Problem	Remedy
Top of chimney is lower than surrounding objects.	Extend chimney above all objects within 20 ft.
Chimney cap or ventilator.	Remove.
Coping restricts opening.	Make opening as large as inside of chimney.
Obstruction in chimney.	Use rod or weight on string or wire to break and dislodge.
Joist projecting into chimney.	Change support for joist so that chimney will be clear. Should be handled by a competent brick contractor.
Break in chimney lining.	Rebuild chimney with a course of brick between flue tiles.
Collection of soot at narrow space in flue opening.	Clean out with weighted brush or bag of loose gravel on end of line. May be necessary to open chimney.
Offset.	Change to straight or long offset.
Two or more openings in same chimney.	Least important opening must be closed using some other chimney flue.
Smoke pipe extends into chimney.	Shorten pipe so that end is flush with inside of tile.
Loose-fitted cleanout door.	Leaks should be eliminated by cementing all pipe openings.
Failure to extend the length of flue partition down to the floor.	Extend partition to floor level.
Loosely fitted cleanout door.	Close all leaks with cement.

Sometimes a chimney will be either too small or too large for the installation. When this is the case, the chimney should be rebuilt with the necessary corrections made in its design.

DRAFT HOOD

A gas-fired furnace should be equipped with a draft hood attached to the flue outlet of the appliance. The draft hood used on the appliance should be certified by the American Gas Association. Only gas conversion furnaces equipped with power-type burners and conversion burner installations in large steel boilers with inputs in excess of 400,000 Btuh are not required to have draft hoods.

A draft hood is a device used to ensure the maintenance of constant low draft conditions in the combustion chamber. By this action, it contributes to the stability of the air supply for the combustion process. A draft hood will also prevent excessive chimney draft and downdrafts that tend to extinguish the gas burner flame. Because of this last function, a draft hood is often referred to as a *draft diverter*.

Draft hoods may be either internally or externally mounted, depending upon the design of the furnace. Never use an external draft hood with a furnace already equipped with an *internal* draft hood. Either vertical or horizontal discharge from the draft hood is possible (Fig. 11-14). Locations of draft hoods in conversion burner installations are illustrated in Fig. 11-15.

NEUTRAL PRESSURE-POINT ADJUSTER

In some installations, a neutral pressure-point adjuster is installed in the flue pipe between the furnace and the draft hood. The procedure for making a neutral pressure-point adjuster is illustrated in Fig. 11-16. *Always* leave the neutral pressure-point adjuster in wide open position until after the burner rating has been established.

BASIC COMPONENTS

The basic components of a gas-fired, forced-warm-air furnace are:

338

1. Automatic controls.
2. Heat exchanger.
3. Gas burner.
4. Gas pilot assembly.
5. Blower and motor.
6. Air filter(s).

Each of these components is described in the sections that follow. Additional information is contained in Chapter 10 (Fur-

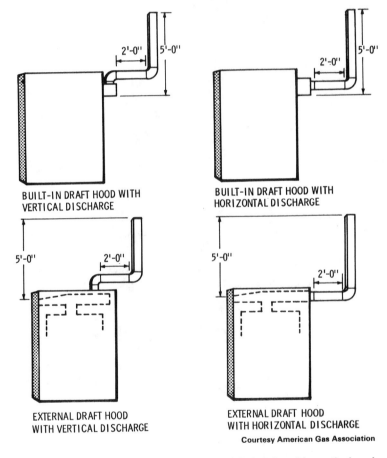

BUILT-IN DRAFT HOOD WITH
VERTICAL DISCHARGE

BUILT-IN DRAFT HOOD WITH
HORIZONTAL DISCHARGE

EXTERNAL DRAFT HOOD
WITH VERTICAL DISCHARGE

EXTERNAL DRAFT HOOD
WITH HORIZONTAL DISCHARGE

Courtesy American Gas Association

Fig. 11-14. External and internal (built-in) draft hoods with vertical and horizontal discharge.

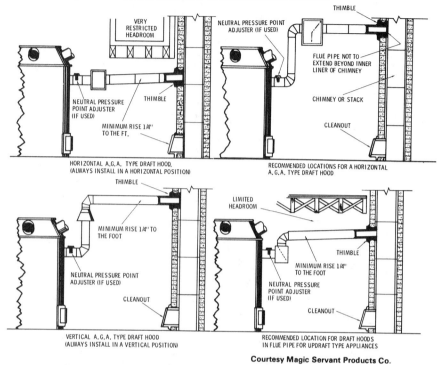

Fig. 11-15. Location of draft hoods in conversion burner installations.

nace Fundamentals) and the various chapters in which furnace controls and fuel-burning equipment are described.

AUTOMATIC CONTROLS

The automatic controls used in a heating system are designed to ensure its safe and efficient operation. Detailed descriptions of these controls are found in Chapter 4 Thermostats and Humidistats), Chapter 5 (Gas and Oil Controls); and Chapter 6 of Volume 2 (Other Automatic Controls). Additional information can be found in the appropriate sections of Chapter 16 (Boiler and Furnace Conversions) and in Chapter 2 of Volume 2 (Gas Burners). This section is primarily concerned with outlining the

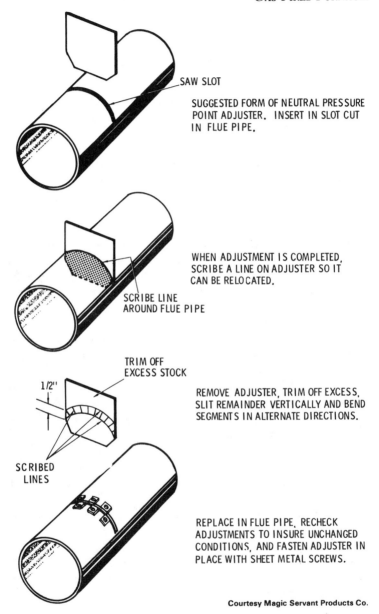

SAW SLOT

SUGGESTED FORM OF NEUTRAL PRESSURE
POINT ADJUSTER. INSERT IN SLOT CUT
IN FLUE PIPE.

WHEN ADJUSTMENT IS COMPLETED,
SCRIBE A LINE ON ADJUSTER SO IT
CAN BE RELOCATED.

SCRIBE LINE
AROUND FLUE PIPE

TRIM OFF
EXCESS STOCK

1/2"

REMOVE ADJUSTER, TRIM OFF EXCESS,
SLIT REMAINDER VERTICALLY AND BEND
SEGMENTS IN ALTERNATE DIRECTIONS.

SCRIBED
LINES

REPLACE IN FLUE PIPE, RECHECK
ADJUSTMENTS TO INSURE UNCHANGED
CONDITIONS, AND FASTEN ADJUSTER IN
PLACE WITH SHEET METAL SCREWS.

Courtesy Magic Servant Products Co.

Fig. 11-16. Suggested construction of a neutral pressure-point adjuster.

operating principles of the automatic controls used with a gas-fired furnace. These controls include:

1. Room thermostat.
2. Fan and limit controls.
3. Pilot safety valve.
4. Automatic main gas valve.
5. Gas pressure regulator.

The operation of the furnace is controlled by the room thermostat. The thermostat senses air-temperature changes in the space or spaces being heated and sends an electrical signal to open or close the automatic main gas valve. The thermostat is generally wired in series with the pilot safety valve, the automatic main gas valve, and the limit control. [See Chapter 4 of Volume 2 (Thermostats and Humidistats) for a detailed description].

The fan and limit control is a safety device that operates on a thermostatic principle. This combined control is designed to govern the operation of the furnace within a specified temperature range (usually 80 to 150°F), and is located in the furnace plenum where it responds to outlet air temperatures.

One of the functions of this control is to prevent the furnace from overheating by shutting off the gas supply when the plenum air temperature exceeds the upper temperature setting (180°F) on the control. The limit-control switch (and pilot safety relay) responds to excessive plenum temperatures by closing the automatic main gas valve and thereby shutting off the flow of gas to the burner or burner assembly.

Another function of the fan and limit control is to stop the fan when the gas burner or burner assembly has been turned off. When the plenum air drops below the lower setting (80°F) on the control, the fan is automatically shut off. For additional information about fan and limit controls, read the appropriate sections in Chapter 6 of Volume 2 (Other Automatic Controls).

The function of the automatic pilot safety valve is to shut off the main gas supply when the pilot flame goes out. There are several pilot safety devices on the market, but the most commonly used ones employ a thermocouple lead in conjunction with a valve switch.

The automatic main gas valve controls the flow of gas to the burner manifold (in the case of a burner assembly) or to the gas burner. It is located between the burner (or burners) and the gas pressure regulator. Chapter 6 of Volume 2 (Other Automatic Controls) contains additional information about both pilot safety and main gas valves.

The gas in the supply main is frequently subjected to pressure variations. A gas pressure regulator installed in the piping between the supply riser and the automatic main gas valves ensures a constant gas pressure in the burner manifold.

HEAT EXCHANGER

Each gas-fired furnace has a burner compartment containing the gas burner and pilot. The gas and air necessary for combustion are mixed in the venturi or mixing tube prior to ignition. This mixture is then ignited by the pilot flame as it leaves the burner. The burner flame is encased in a combustion chamber or fire box located directly below the furnace heat exchanger. The heat of the combustion process is transferred to the metal walls of the heat exchanger and then to the air rising through it (Fig. 11-17).

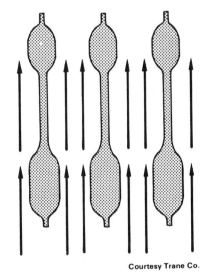

Fig. 11-17. Uniform air flow across all sections of the heat exchanger.

Courtesy Trane Co.

The air flowing through a heat exchanger is subject to temperatures which frequently approach the 700 to 800°F range. Because of these high temperatures, the airflow must be uniform across all sections of the heat exchanger and must move at a sufficient rate of speed to keep these temperatures from building any higher. A low volume of air flowing across one section of the heat exchanger will result in overheating and the failure·of the entire heat exchanger assembly (Fig 11-18).

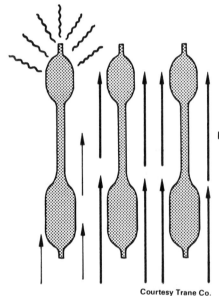

Fig. 11-18. Low air flow across one heat exchanger section may result in premature failure due to overheating.

Courtesy Trane Co.

Overheating can also be caused by a portion of the gas flame impinging on the inner surface of the heat exchanger (Fig. 11-19). This results in the generation of extremely high temperatures. Under normal operating conditions, the gas flame should be directed upward through the center of each section.

Another cause of heat exchanger failure is overfiring. This condition is usually traced to an excessively high manifold pressure or an orifice that is too large. The result is a firing rate that exceeds the design limits.

If the heat exchanger becomes so covered with soot, carbon

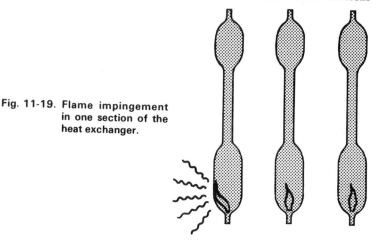

Fig. 11-19. Flame impingement in one section of the heat exchanger.

deposits, and other contaminants that it reduces the operating efficiency of the furnace, it should be thoroughly cleaned. Gaining access to the heat exchanger is not easy because it requires removing the burners and manifold assembly, vent connector, draft hood (diverter), and flue baffles. For this reason, you are advised to call a local service repairman if you want to clean the heat exchanger.

GAS BURNERS

Gas burners and gas burner assemblies are designed to provide the proper mixture of gas and air for combustion purposes. The gas burners used in residential heating installations operate primarily on the same principle as the Bunsen burner. [See Chapter 2, of Volume 2 (Gas Burners)].

Gas-fired furnaces require both primary and secondary air for combustion process. The gas passes through the small orifice in the mixer head, which is shaped to produce a straight-flowing jet moving at high velocity (Fig. 11-20). As the gas stream enters the throat of the mixing tube, it tends to spread and induce air in through the opening of the adjustable shutter. This is the *primary air*, which mixes with the gas *before* the air-gas mixture is forced through the burner ports (Fig. 11-20). The energy in the gas stream forces the mixture through the mixing tube into the burner manifold casting, from which it issues through ports where additional air must be added to the flame to complete combustion. This *secondary air* supply flows into the heat exchanger and around the burner. Its function is to mix with unburned gas in the heat exchanger in order to complete the combustion process.

The primary air is admitted at a ration of about 5 parts primary air to 1 part gas for manufactured gas, and at a 10 to 1 ratio for natural gas. These ratios are generally used as theoretical values of air for purposes of complete combustion. Most burners used in gas-fired furnaces operate efficiently on 40 to 60 percent of the theoretical value. Primary air is regulated by means of an adjustable shutter. When burning natural gas, the air adjustment is generally made to secure as blue a flame as possible. Burner manufacturers provide a number of different ways to field adjust primary air. For example, shutters for primary air control are part of the main burners. Factory adjusted, angled orifices are also used on some burners.

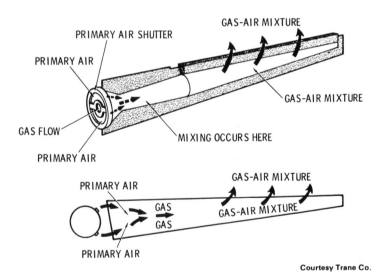

Courtesy Trane Co.

Fig. 11-20. Schematic of air movements in a typical gas burner.

The secondary air is drawn into the burner by natural draft. No provision is made by manufacturers to field adjust secondary air.

346

Secondary air is controlled primarily by the design of the flue restrictors and draft diverter. The general configuration of the heat exchanger and burners are also important factors in secondary air control. Excess secondary air constitutes a loss and should be reduced to a proper minimum (usually not less than 25 to 35 percent).

As shown in Fig. 11-21, a typical gas flame consists of a well-defined inner cone surrounded by an outer envelope. Combustion of the primary air and gas mixture occurs along the *outer* surface of the inner cone. Any unburned gases combine with the

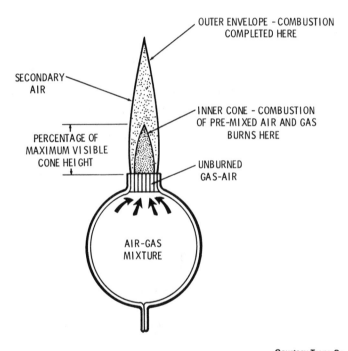

OUTER ENVELOPE - COMBUSTION COMPLETED HERE

SECONDARY AIR

INNER CONE - COMBUSTION OF PRE-MIXED AIR AND GAS BURNS HERE

PERCENTAGE OF MAXIMUM VISIBLE CONE HEIGHT

UNBURNED GAS-AIR

AIR-GAS MIXTURE

Fig. 11-21. Typical gas flame.

secondary air and complete the combustion process in the outer envelope of the flame.

As Fig. 11-21 illustrates, the major portion of the inner cone consists of an unburned air-gas mixture. It is along the outer surface of the inner cone that combustion takes place. The height of the inner cone can be reduced by increasing the amount of primary air contained in the air-gas mixture. The extra oxygen increases the burning rate, which causes the inner cone to decrease in size.

A burner is generally considered to be properly adjusted when the height of the inner cone is approximately 70 percent of the maximum visible cone height. The gas should flow out of the burner ports fast enough so that the flame cannot travel or flash back into the burner head. The velocity must not be so high that it blows the flame away from the port(s).

CROSSOVER BURNER

Most modern gas-fired furnaces consist of several individual heat exchanger sections with a gas burner inserted in each section. In this arrangement, all burners must fire almost simultaneously or there will be the danger of unburned gas flowing into the combustion chamber. Simultaneous ignition can be provided by installing a crossover burner across the top of the main burners near the port area. Fig. 11-22 and 11-23 show the types of crossovers in Trane gas-fired furnaces.

GAS PILOT ASSEMBLY

The gas pilot assembly consists of a pilot burner and a pilot safety device (Fig. 11-24). The most commonly used pilot safety is the thermocouple lead operating in conjunction with a companion valve or switch. The thermocouple itself is actually a miniature generator that converts heat (from the pilot flame) into a small electrical current. The upper half of the thermocouple is located in the pilot flame. As long as the pilot flame is operating, an electrical circuit is maintained between the pilot burner and the gas valve. If the pilot flame goes out, the circuit between the thermocouple and the gas valve is broken. As a result, the gas valve closes and shuts off the gas flow to the pilot burner.

CROSSOVER

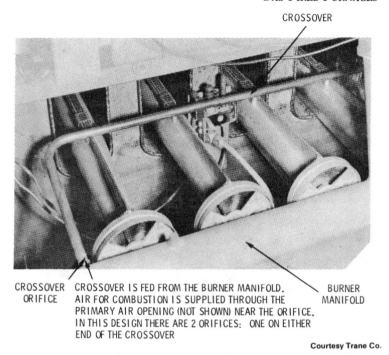

CROSSOVER   CROSSOVER IS FED FROM THE BURNER MANIFOLD.   BURNER
ORIFICE   AIR FOR COMBUSTION IS SUPPLIED THROUGH THE   MANIFOLD
            PRIMARY AIR OPENING (NOT SHOWN) NEAR THE ORIFICE.
            IN THIS DESIGN THERE ARE 2 ORIFICES: ONE ON EITHER
            END OF THE CROSSOVER

Courtesy Trane Co.

Fig. 11-22. Crossover located perpendicular to and across the top of the main burners near the port area.

Other pilot safety devices used with gas-fired furnaces are described in Chapter 5 of Volume 2 (Gas and Oil Controls).

PILOT BURNER ADJUSTMENT

It is sometimes necessary to adjust the pilot flame. On some combination gas controls a pilot adjustment screw is provided for this purpose. On units using natural gas, the screw should be adjusted until the pilot flame completely envelopes the thermocouple. If propane gas is used, the screw should be wide open.

Sometimes a pilot gas regulator is used on a furnace to control pressures in areas where gas pressure variation is great. The pilot valve is placed in wide open position, and pilot flame adjustments are made by adjusting the gas pressure regulator.

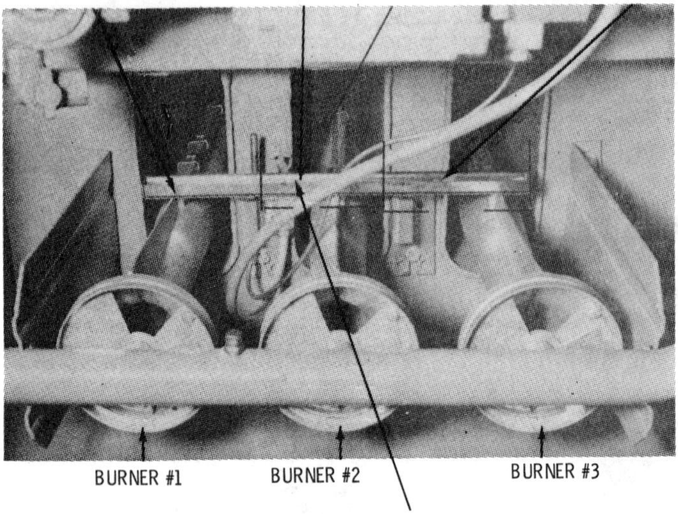

MUST FIT SNUGLY
AGAINST THE BURNER

CROSSOVER

GAS-AIR MIXTURE FED TO THIS SECTION
OF CROSSOVER BY BURNER #3

BURNER #1   BURNER #2   BURNER #3

BURNER #2 FEEDS THIS SECTION OF CROSSOVER.
SCOOP IN BURNER DIRECTS FLOW UP INTO CROSSOVER

Courtesy Trane Co.

Fig. 11-23. Crossover arrangement in which each burner feeds a portion of the gas-air mixture into the crossover where it is ignited by the pilot flame as it leaves the crossover part.

GAS INPUT ADJUSTMENT

The gas input of a furnace must not be greater than that specified on the rating plate. However, it is permissible to reduce the gas input to a *minimum* of 80 percent of the rated input in order to balance the load.

The gas input can be checked by timing the flow through the meter. The procedure for clocking the meter is as follows:

1. Turn off all gas units connected through the meter except the furnace. The main burners of the furnace must be on for the timing check.
2. Time one complete revolution of the hand (or disk) on the meter. The hand should be the one with the smallest cubic

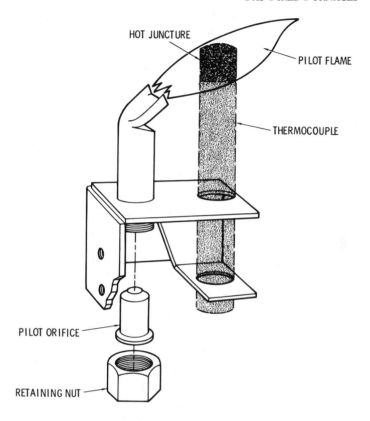

HOT JUNCTURE

PILOT FLAME

THERMOCOUPLE

PILOT ORIFICE

RETAINING NUT

Fig. 11-24. Pilot burner assembly.

footage per revolution (usually ½ or 1 cu. ft. per revolution on small meters and 5 cu. ft. on large meters).

3. Using the size of the test meter dial (½, 1, 2, or 5 cu. ft.) and the number of seconds per revolution, determine the cubic feet per hour from the appropriate column in Table 11-1.

4. Determine the actual Btuh input of the burner by multiplying the cubic feet per hour rate (obtained in Step 3) by the Btu per cubic feet of gas (obtained from the local gas company).

By the way of example, let's assume that you have contacted the local gas company and they have given you a value of 1040

Table 11-1. Gas Rate in Cubic Feet per Hour

Seconds for One Revolu- tion	Size of Test Meter Dial				Seconds for One Revolu- tion	Size of Test Meter Dial			
	½ Cu. Ft.	1 Cu. Ft.	2 Cu. Ft.	5 Cu. Ft.		½ Cu. Ft.	1 Cu. Ft.	2 Cu. Ft.	5 Cu. Ft.
	Cubic Feet per Hour					Cubic Feet per Hour			
10	180	360	720	1800	50	36	72	144	360
11	164	327	655	1636	51	35	`71	141	353
12	150	300	600	1500	52	35	69	138	346
13	138	277	555	1385	53	34	68	136	340
14	129	257	514	1286	54	33	67	133	333
15	120	240	480	1200	55	33	65	131	327
16	112	225	450	1125	56	32	64	129	321
17	106	212	424	1059	57	32	63	126	316
18	100	200	400	1000	58	31	62	124	310
19	95	189	379	947	59	30	61	122	305
20	90	180	360	900	60	30	60	120	300
21	86	171	343	857	62	29	58	116	290
22	82	164	327	818	64	29	56	112	281
23	78	157	313	783	66	29	54	109	273
24	75	150	300	750	68	28	53	106	265
25	72	144	288	720	70	26	51	103	257
26	69	138	277	692	72	25	50	100	250
27	67	133	267	667	74	24	48	97	243
28	64	129	257	643	76	24	47	95	237
29	62	124	248	621	78	23	46	92	231
30	60	120	240	600	80	22	45	90	225
31	58	116	232	581	82	22	44	88	220
32	56	113	225	562	84	21	43	86	214
33	55	109	218	545	86	21	42	84	209
34	53	106	212	529	88	20	41	82	205
35	51	103	206	514	90	20	40	80	200
36	50	100	200	500	94	19	38	76	192
37	49	97	105	486	98	18	37	74	184
38	47	95	189	474	100	18	36	72	180
39	46	92	185	462	104	17	35	69	173
40	45	90	180	450	108	17	33	67	167
41	44	88	176	440	112	16	32	64	161
42	43	86	172	430	116	15	31	62	155
43	42	84	167	420	120	15	30	60	150
44	41	82	164	410	130	14	28	55	138
45	40	80	160	400	140	13	26	51	129
46	39	78	157	391	150	12	24	48	120
47	38	77	153	383	160	11	22	45	113
48	37	75	150	375	170	11	21	42	106
49	37	73	147	367	180	10	20	40	100

Courtesy Dunham-Bush, Inc.

Btu per cu. ft. When you clock the meter, you find that it takes exactly 18 seconds for the hand to make one complete revolution on the ½ cu. ft. meter dial. With this information, turn to Table 11-1 and find 18 seconds in the extreme left-hand column. Opposite the value 18 seconds under the ½ cu. ft. column heading read 100 cu. ft. The rest is simple multiplication:

100 cu. ft. per hour × 1040 Btu per cu. ft.

= 104,000 Btuh

If the Btuh rate is not within 5 percent of the desired value, change the burner orifices or adjust the pressure regulator. Pressure regulator setting changes are *only* used for *minor adjustments* in Btu input. Major adjustments are made by changing the burner orifices.

When making adjustments, remember the following: *Btu input may never be less than the minimum or more than the maximum on the rating plate.*

There is no meter to clock on furnaces fired with propane. The burner orifices on these units are sized to give the proper rate at 11 in. W.C. (water column) manifold pressure.

After the correct Btu input rate has been determined, adjust the burner for the most efficient flame characteristics (see "Combustion Air Adjustments").

HIGH-ALTITUDE ADJUSTMENT

For elevations *above* 2000 ft., the Btu input should be reduced (derated) 4 percent for each 1000 ft. above sea level. This adjustment is required because of the increase in air volume at higher elevations. Air expands at approximately 4 percent per 1000 ft. of elevation. At elevations above 2000 ft., the volume of air required to supply enough oxygen for complete combustion exceeds the maximum air-inducing ability of the burners. Because the primary air volume cannot be increased, the gas flow (input) must be decreased (i.e., derated).

CHANGING BURNER ORIFICES

A gas-fired furnace is supplied with standard burner orifices for the gas shown on the rating plate. The orifice size will depend on the calorific value of the gas (Btu per cubic foot) and the manifold pressure (3.5 in. W.C. for natural gas and 11 W.C. for

LP gas). The burner orifice *size* partially determines the firing rate by allowing only a predetermined volume of gas to pass. An orifice that is too large will allow too much gas to pass through into the burner. This condition sometimes results in overheating the heat exchanger or restricting the air-inducing ability of the burner.

The gas burner orifices supplied with a furnace are suitable for *average* calorific values of a gas at the listed manifold pressures. Table 11-2 lists the various orifice capacities for different drill sizes. The procedure for determining the suitability of a particular drill size is as follows:

1. Divide the total Btu input of the furnace by the *total* number of burner orifices in the unit.
2. From Table 11-2, select the burner orifice size closest to the calculated value (plus or minus 5 percent) for the gas used.

This procedure can be illustrated by determining the most suitable drill size for a 150,000-Btu-input, gas-fired furnace. This furnace has 4 burners, with 2 orifices per burner, and uses natural gas.

$$\text{capacity of each burner orifice} = \frac{150,000}{4 \times 2}$$

$$= \frac{150,000}{8}$$

$$= 18,750$$

In Table 11-2, the figure closest to 18,750 in the left-hand column (natural gas) is 19,000. The drill size opposite this figure is #46.

MANIFOLD PRESSURE ADJUSTMENT

A suitable manifold pressure is important to efficient furnace operation. If the gas pressure is too low, it will cause rough ignition, incomplete and inefficient combustion, and incorrect fan control response. An excessively high manifold pressure may

Table 11-2. Orifice Capacity Table

Drill Size	Approximate Orifice Capacity (Btuh)*	
	Natural Gas @ 3½" W.C.	Propane Gas @ 11" W.C.
60	4650	13,000
59	4900	13,750
58	5150	14,450
57	5400	15,150
56	6300	17,600
55	7900	22,000
54	8800	24,750
53	10,400	28,800
52	11,750	33,000
51	13,000	36,700
50	14,250	40,000
49	15,500	43,500
48	16,750	47,200
47	17,900	50,200
46	19,000	53,500
45	19,500	55,000
44	21,500	60,050
43	23,000	
42	25,250	
41	26,750	
40	27,800	
39	28,600	
38	29,900	
37	31,100	
36	33,000	
35	35,000	

*Values determined using the following data: propane at 1.52 specific gravity/2500 Btu per cu. ft.; and natural gas at 0.62 specific gravity/1000 Btu per cu. ft.

Courtesy The Trane Company

cause the burners to overfire the heat exchanger. Overfiring the heat exchanger not only reduces the life of this component, but it also may result in repeated cycling of the burner on the high limit control.

Manifold pressure should be set at 3.5 in. W.C. for natural gas. Control manufacturers normally preset the pressure regulator in natural-gas valves at the time of manufacture so that natural-gas units are fired at this 3.5 in. W.C. rate. Only small variations in the gas flow should be made by adjusting the pressure regulator. This adjustment should never exceed plus or minus 0.3 in. W.C. Major changes in the gas flow should be made by changing the size of the burner orifice (see below).

The manifold gas pressure can be tested by using a U-tube water manometer (Fig. 11-25). The manometer is connected to the manifold through an opening covered by a ⅛-in. plug cap. The pressure test *must* be run while the unit is operating. On Dunham-Bush, gas-fired furnaces, the test is run through an opening in the gas valve (Fig. 11-26). The furnace manufacturer's installation literature should contain instructions for testing the manifold pressure.

Once you have determined the manifold pressure, you may find it necessary to adjust it for better operating characteristics. Manifold pressure can be increased by turning the adjusting screw clockwise, and decreased by turning it counterclockwise.

LP gas units are fired at 11 in. W.C. manifold pressure. LP gas (propane and butane) is heavier than natural gas and has a higher heating value. As a result, LP gas needs more primary air for

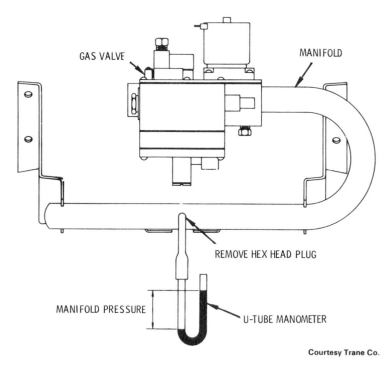

Courtesy Trane Co.

Fig. 11-25. Applying a V-tube manometer to gas manifold.

356

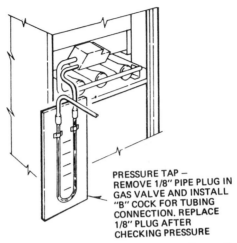

PRESSURE TAP —
REMOVE 1/8" PIPE PLUG IN
GAS VALVE AND INSTALL
"B" COCK FOR TUBING
CONNECTION. REPLACE
1/8" PLUG AFTER
CHECKING PRESSURE

Courtesy Dunham-Bush, Inc.

Fig. 11-26. Applying a V-tube manometer to gas valve.

combustion. Thus, a higher manifold pressure is required to induce the greater volume of primary air into the burner than is the case in natural gas units.

An LP furnace does not have a pressure regulator in the gas valve. Pressure adjustments are made by means of the regulator at the supply tank. These adjustments *must* be made by the installer and regularly checked by the serviceman for possible malfunctions.

COMBUSTION AIR ADJUSTMENT

Primary air shutters are provided on all furnaces to enable adjustment of the primary air supply. The purpose of this adjustment is to obtain the most suitable flame characteristics. Experience has shown that the most efficient burner flame for natural gas has a soft blue cone without a yellow tip. For propane gas, there should be a little yellow showing in the tips of the flame.

The procedure for adjusting the primary air on a Coleman gas-fired furnace may be summarized as follows:

1. Light the pilot burner.
2. Turn up the setting on the room thermostat until the main burners come on.
3. Allow the main burners 10 minutes to warm up.
4. Loosen locknut on adjusting screw (Fig. 11-27).
5. Turn adjusting screw *in* (clockwise) until the yellow tip appears in the flame.
6. Turn adjusting screw *out* (counterclockwise) until the yellow tip *just* disappears.
7. Hold adjusting screw and tighten locknut.
8. Repeat Steps 4 to 7 on each of the other burners.

On a Fedders gas-fired furnace, a locking screw is located above the primary air opening on each gas burner (Fig. 11-28). The primary air adjustment procedure is as follows:

1. Loosen the locking screw at the base of the burner.
2. Adjust the air-shutter opening to a position that gives a slight yellow tip on the end of the flame.

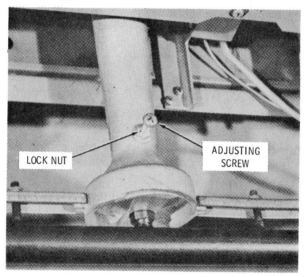

LOCK NUT

ADJUSTING SCREW

Courtesy Coleman Co., Inc.

Fig. 11-27. Position of the locknut and adjusting screw on a Coleman gas-fired furnace.

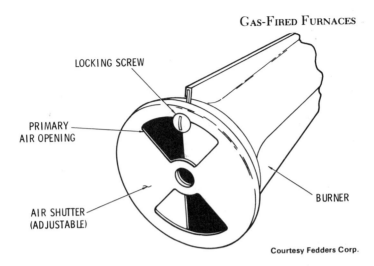

LOCKING SCREW

PRIMARY
AIR OPENING

AIR SHUTTER
(ADJUSTABLE)

BURNER

Courtesy Fedders Corp.

Fig. 11-28. Primary air adjustment on a Fedders gas-fired furnace.

3. Open the air shutter until the yellow tip just disappears.
4. Tighten the locking screw.
5. Repeat Steps 1 to 4 for each of the burners.

A primary air-shutter assembly is provided on all Carrier gas-fired furnaces in order to simplify the adjustment procedure. As shown in Fig. 11-29, adjustment of the end burner results in a simultaneous adjustment of all burners.

After the primary air adjustments have been made, check to see that the burners are level. The burner flames should be uniform and centered in the heat exchanger (see "Heat Exchanger" in this chapter).

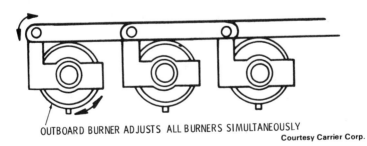

OUTBOARD BURNER ADJUSTS ALL BURNERS SIMULTANEOUSLY

Courtesy Carrier Corp.

Fig. 11-29. Primary air-shutter assembly on Carrier furnace.

Before allowing the furnace to continue operating, you must check to see that it has the proper draft. This can be done by passing a match along the draft-hood opening. If the vent is drawing properly, the match flame will be drawn inward (i.e., into the draft hood). If the furnace is not receiving proper draft, the products of combustion escaping the draft hood will extinguish the flame. The draft must be corrected before operating the furnace.

GAS SUPPLY PIPING

The *gas supply piping* (also referred to as the *gas service piping*) must be sized and installed in accordance with the recommendations contained in the latest edition of *American National Standards for Installation of Gas Appliances and Gas Piping* (ANSI Z21.30-1964) and *Piping for Gas Appliances* (NFPA No. 54-1964). The location of the main shutoff valve will usually be specified by local codes or regulations. Consult the local codes or regulations for recommended sizes of pipe for the required gas volumes. Local codes or regulations always take precedence over national standards when there is a conflict between the two.

Generally, 1-in. supply pipe is adequate for furnace inputs up to 125,000 Btuh. A 1¼-in. pipe is recommended for higher inputs. The inlet gas supply pipe size should never be smaller than ½ in. The gas line from the supply should serve only a single unit.

A *drip leg* (also called a *dirt leg* or a *dirt trap*) should be installed at the bottom of the gas supply riser to collect moisture or impurities carried by the gas. A manual main shutoff valve should be installed either on the gas supply riser or on the horizontal pipe between the riser and ground union joint Figs. 11-30 and 11-31). The location of the main shutoff valve will usually be specified by local codes and regulations.

The *ground union joint* should be installed between the manual main shutoff valve and the gas control valve on the furnace or boiler (Fig. 11-32). The ground union joint provides easy access to the gas controls on the unit for servicing or repair.

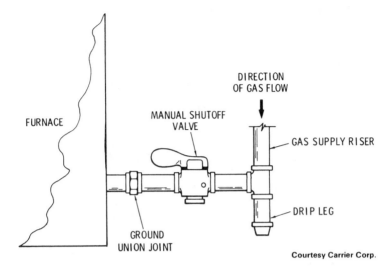

Courtesy Carrier Corp.

Fig. 11-30. Location of manual valve on horizontal gas supply line.

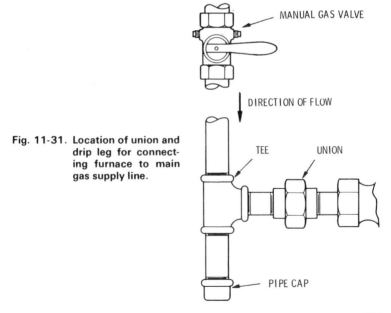

Fig. 11-31. Location of union and drip leg for connecting furnace to main gas supply line.

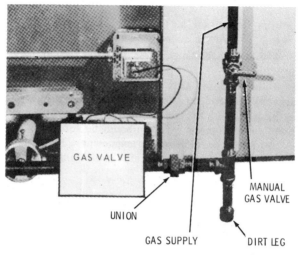

Courtesy Coleman Co.

Fig. 11-32. Location of ground union joint.

SIZING GAS PIPING

The gas supply pipe should be sized per volume of gas used and allowable pressure drop. The pipe must also be of adequate size to prevent undue pressure drop. The diameter of the supply pipe must *at least* equal that of the manual shutoff valve in the supply riser. Minimum acceptable pipe sizes are listed in Table 11-3.

Table 11-3. Capacity of Pipe (cu. ft./hr.)
(Pressure Drop 0.3—Specific Gravity 0.60)

Length of Pipe (feet)	Diameter of Pipe (inches)			
	$^3/_4$	1	$1^1/_4$	$1^1/_2$
15	172	345	750	1220
30	120	241	535	850
45	99	199	435	700
60	86	173	380	610
75	77	155	345	545
90	70	141	310	490
105	65	131	285	450
120		120	270	420

Courtesy Janitrol

The volume rate of gas in cubic feet per hour is determined by dividing the Btuh required by the furnace by the Btu per cubic foot of gas being used. The cubic feet per hour value can then be used to determine pipe size for gas of a specific gravity other than 0.60. The data used for these calculations are listed in Tables 11-3 and 11-4. The procedure is as follows:

1. Multiply the cubic feet per hour required by the multiplier (Table 11-4).
2. Find recommended pipe size for length of run in Table 11-3.

Tables 11-3 and 11-4 apply to the average piping installation where four or five fittings are used in the gas piping to the fur nace. Existing pipe should be converted to the proper size of pipe where necessary. In no case should pipe less than ¾-in. diameter be used. Any extensions to existing piping should conform to Tables 11-3 and 11-4.

Table 11-4. Multiplier Table

Specific Gravity	Multiplier	Specific Gravity	Multiplier
0.40	0.813	1.30	1.47
0.50	0.910	1.40	1.53
0.60	1.00	1.50	1.58
0.70	1.08	1.60	1.63
0.80	1.15	1.70	1.68
0.90	1.22	1.80	1.73
1.00	1.29	1.90	1.77
1.10	1.35	2.00	1.83
1.20	1.42		

Courtesy Janitrol

INSTALLING GAS PIPING

The principal recommendations to be followed when installing gas piping are:

1. A single gas line must be provided from the main gas supply to the furnace.

2. The gas supply pipe must be sized according to the volume of gas used and the allowable pressure drop.

3. The supply riser and drop leg must be installed adjacent to the furnace on either side. Never install it in front of the furnace where it might block access to the unit.

4. A manual main shutoff valve or plugcock should be installed on the gas supply piping (usually on the riser) outside the unit in accordance with local codes and regulations.

5. Install a tee for the drop leg in the pipe at the same level as the gas supply (inlet) connection on the furnace.

6. Extend the drip leg and cap it.

7. Install a ground union joint between the tee and the furnace controls.

8. Use a pipe compound resistant to LP gas on all threaded pipe connections in an LP gas installation.

9. Test for pressure and leaks. Use a solution of soap or detergent and water to detect any leaks. *Never* check for leaks with a flame.

STARTUP INSTRUCTIONS

The startup procedure for most gas-fired furnaces is very similar, particularly insofar as use of the thermostat is concerned. The major difference is in the lighting of the pilot burner, and this will depend upon the gas valve used.

The startup procedure for a gas-fired furnace may be summarized as follows:

1. Open all warm air registers.

2. Shut off main gas valve and pilot gas cock, and wait at least 5 minutes to allow gas that may have accumulated in the burner compartment to escape.

3. Set the room thermostat at the lowest setting.

4. Open the water supply valve if a humidifier is installed in the system.

5. Turn off the line voltage switch if one is provided in the furnace circuit.

6. Light the pilot (see below).

7. Turn electrical power on.
8. Replace all access doors.
9. Light main burners by setting the room thermostat *above* the room temperature.
10. Set the room thermostat at the desired temperature after the main burners are lighted.

Lighting the Pilot

The pilot lighting instructions should be included with the furnace and prominently displayed. Lighting instructions for the 24-volt Honeywell V8280, V8243, and Robertshaw (Unitrol) 7000 combination gas valves basically summarize the pilot-lighting procedure (Fig. 11-33). The procedure is as follows:

1. Depress gas cock dial and turn it counterclockwise to "pilot" position.
2. Keep gas cock dial depressed and light pilot with a match.
3. Keep dial depressed until pilot remains lighted when released.
4. Depress dial and turn counterclockwise to *on* position.

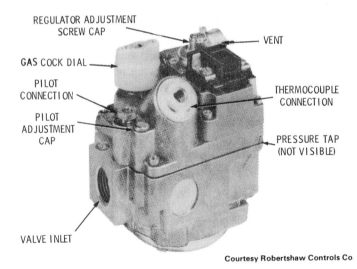

Courtesy Robertshaw Controls Co

Fig. 11-33. Robertshaw Unitrol 7000 series combination gas valve.

BLOWERS AND MOTORS

A forced-warm-air furnace should be equipped with a variable-speed electric motor designed for continuous duty. The motor should be provided with overcurrent protection in accordance with the National Electrical Code (ANSI C1-1971).

Blower motors are available as either direct-drive or belt-driven units. On direct-drive units, the blower wheel is mounted directly on the motor shaft and runs at motor speed. The blower wheel of a belt-driven unit is operated by a V-belt connecting a fixed pulley mounted on the blower shaft with a variable-pitch drive pulley mounted on the motor shaft.

The blower motors used in residential furnaces are single-phase electric motors. Depending upon the method used for starting, they may be classified as:

1. Shaded pole.
2. Permanent split capacitor.
3. Split-phase.
4. Capacitor-start motors.

A direct-drive blower generally uses a shaded pole (SHP) or permanent capacitor (PSC) motor. Belt-driven blower motors are usually split-phase (SPH) types up to the $\frac{1}{3}$-hp size. Larger motors are capacitor-start, split-phase motors.

Blower motors are fractional horsepower motors commonly available in the following sizes: $\frac{1}{12}$, $\frac{1}{10}$, $\frac{1}{6}$, $\frac{1}{4}$, $\frac{1}{3}$, $\frac{1}{2}$, and $\frac{3}{4}$ hp. A fractional horsepower motor should be protected with temperature- or current-sensitive devices to prevent motor winding temperatures from exceeding those allowed in the specifications.

Provisions should be made for the periodic lubrication of the blower and motor. Use only the type and grade of lubricant specified in the operating instructions. Do not overlubricate. Too much oil can be just as bad as too little. A direct-drive blower should be capable of operating within the range of air temperature at the static pressures for which the furnace is designed and at the voltage specified on the motor nameplate.

Motors using belt drives must be supplied with adjustable pulleys. The only exception is a condenser motor used on a forced-warm-air furnace with a cooling unit. *Always* provide the means for adjusting a belt-driven motor in an easily accessible location.

Most forced-warm-air furnaces are shipped with the blower motors factory adjusted to drive the blower wheel at a correct speed for heating. This is a *slower* speed than the one used for air conditioning, but it is relatively easy to alter the air delivery to meet the requirements of the installation (see "Air Conditioning" in this chapter).

Air Delivery and Blower Adjustment

The furnace blower is used to force air through the heat exchanger in order to remove the heat produced by the burners. This removal of heat by the blower serves a twofold purpose: (1) It distributes the heat through the supply ducts to the various spaces in the structure; and (2) it protects the surfaces of the heat exchanger from overheating.

Most blowers are factory set to operate with a specific air temperature rise. For example, Janitrol Series 37 gas-fired furnaces are certified by the American Gas Association to operate with-in a range of 70 to 100°F air temperature heat rise. In other words, the temperature of the air within the supply ducts may be 70 to 100°F higher than the air temperature in the return duct. A temperature rise of 85 to 90°F with approximately 0.1-in. external pressure in the area beyond the unit and filter is generally considered adequate for most residential applications.

The furnace air delivery rate should be adjusted at the time of installation to obtain a temperature rise within the range specified on the furnace rating plate. Many furnace manufacturers recommend that the temperature rise be set *below* the maximum to ensure that temperature rise will not exceed the maximum requirement should the filter become excessively dirty.

Although the blower and motor are factory-set to operate within a specific temperature-rise range, the unit can be field adjusted to deliver more or less air as required. The air delivery rate is adjusted by changing the blower speed. On direct-drive motors, this is accomplished by moving the line lead to the desired terminal. Belt-driven motors are adjusted by changing the effective diameter of the driver pulley (see "Direct-Drive Blower Adjustment" and "Belt-Drive Blower Adjustment").

Direct-Drive Blower Adjustment

The blower (fan) of a direct-drive unit is operated by a drive shaft connected to the motor. Three common direct-drive motors used with blowers are:

1. Two-speed, direct-drive shaded pole motors.
2. Three-speed, direct-drive PSC motors.
3. Four-speed, direct-drive PSC motors.

These blower motors are shipped from the factory set for low-speed heating operation. If a higher operating speed is desired, the red blower motor wire must be moved from the low-speed red terminal to a higher speed terminal (Fig. 11-34).

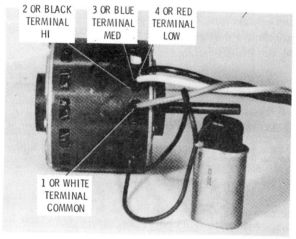

Courtesy Coleman Co., Inc.

Fig. 11-34. Blower motor speed terminals.

Belt-Drive Blower Adjustment

Air-temperature rise can be decreased by increasing the blower speed. A belt-drive unit can be adjusted for a higher speed as follows:

1. Loosen the setscrew in the movable outer flange of the motor pulley.

2. Reduce the distance between the fixed flange and the movable flange of the pulley.

3. Tighten the setscrew.

The procedure for decreasing the blower speed (and thereby increasing the air-temperature rise) is similar except that in Step 2 above the distance between the fixed flange and the movable flange of the pulley are *increased*.

The motor and blower pulley should be aligned to minimize belt wear. The belt should be parallel to the blower scroll, as shown in Fig. 11-35. Adjustments can be made by moving the motor or adjusting the pulley on the shaft.

Correct belt tension is also important. Belt slippage occurs

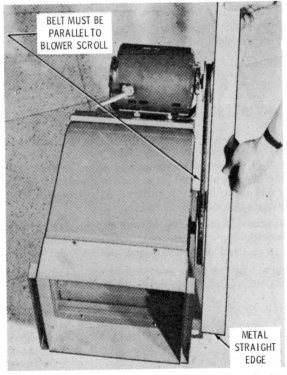

Courtesy Coleman Co., Inc.

Fig. 11-35. Checking belt alignment.

when there is too little tension. Too much tension will cause motor overload and bearing wear. Correct belt tension will usually be indicated by the amount of depression in the belt midway between the blower and motor pulleys. Usually this will be a 1-in. depression at the center of the belt. On the Coleman belt driven unit shown in Fig. 11-36 it is only ½ in.

Courtesy Coleman Co., Inc.

Fig. 11-36. Checking belt tension.

AIR FILTERS

A forced-warm-air, gas-fired furnace is supplied with either a disposable air filter or a permanent (washable) one. The type of filter will be indicated on a label attached to the filter. *Never* use a filter with a gravity warm-air furnace because it will obstruct the airflow.

Proper maintenance of the air filter is very important to the operating efficiency of the furnace. A dirty, clogged air filter reduces the airflow through the furnace and causes air temperatures inside the unit to rise. This increase in air temperature results in a reduced life for the heat exchanger and a lower operating efficiency for the furnace.

If a disposable air filter is used, it should be inspected on a monthly basis. If it is dirty, it should be replaced. This is usually done at the beginning and middle of the heating season. The same schedule should be followed during the summer months when air conditioning is used.

If a permanent filter is provided, it must be removed and cleaned from time to time. Cleaning the filter involves the following procedure:

1. Shake it to remove dust and dirt particles.
2. Vacuum it.
3. Wash it in a solution of soap or detergent and water.
4. Allow time for it to dry.
5. Replace it in the furnace.

Read the label on the filter for any special instructions that may apply to cleaning it. The location of filters in upflow and counterflow furnaces is shown in Figs. 11-37 and 11-38. Note that in each case the air filter is placed directly in the path of the air entering the furnace. An external filter unit for a horizontal, gas-fired furnace is shown in Fig. 11-39.

A filter should be installed in the return plenum or duct when air conditioning is added to the system. Because air conditioning normally requires a greater volume of air than heating, the filter should be sized for air conditioning rather than heating.

Access to internally installed filters is generally provided through a service panel on the furnace. Instructions for cleaning or replacing the filters should be placed on the service panel. It would be a good idea to include the dimensions of the replacement filter(s) in the information.

An adequate filter should provide at least 50 sq. in. of area per cfm of air circulated. For an air velocity of 1000 cfm, this would mean a filler area of 400 sq. in. Filters sized on 50 sq. in. for each 100 cfm meet the requirements of air conditioning—a greater volume of air than heating.

Never use a smaller-size filter than the one required in the specifications. To do so would create excessive pressure drop at the filter, which would reduce the operating efficiency of the furnace.

Additional information about air filters can be found in Chapter 14 of Volume 3 (Air Cleaners and Filters).

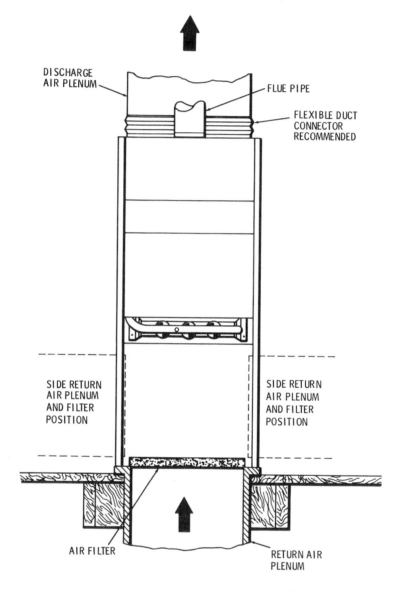

DISCHARGE AIR PLENUM

FLUE PIPE

FLEXIBLE DUCT CONNECTOR RECOMMENDED

SIDE RETURN AIR PLENUM AND FILTER POSITION

SIDE RETURN AIR PLENUM AND FILTER POSITION

AIR FILTER

RETURN AIR PLENUM

Fig. 11-37. Location of air filter in upflow furnace.

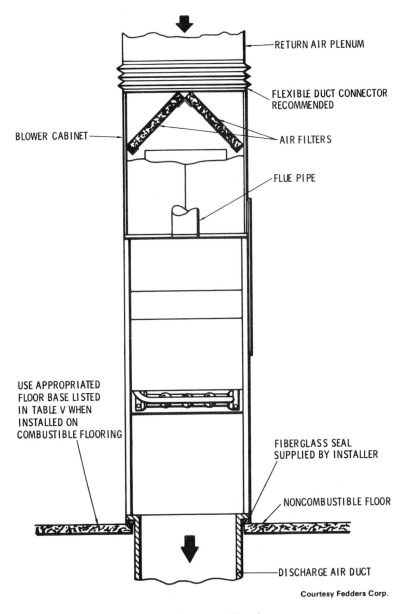

RETURN AIR PLENUM

FLEXIBLE DUCT CONNECTOR
RECOMMENDED

BLOWER CABINET

AIR FILTERS

FLUE PIPE

USE APPROPRIATED
FLOOR BASE LISTED
IN TABLE V WHEN
INSTALLED ON
COMBUSTIBLE FLOORING

FIBERGLASS SEAL
SUPPLIED BY INSTALLER

NONCOMBUSTIBLE FLOOR

DISCHARGE AIR DUCT

Courtesy Fedders Corp.

Fig. 11-38. Location of air filter in counterflow furnace.

AIR CONDITIONING

If you intend to install air conditioning equipment at some future date, the duct sizing should include an allowance for it. Air conditioning involves a greater volume of air than heating. Quite often, the blowers and motors of the furnace will be sized for the addition of cooling with matching evaporator coil and condensing unit. Check the specifications to be sure.

All ductwork located in unconditioned areas (e.g., attics, crawl spaces) *downstream* from the furnace should be insulated. A furnace used in conjunction with a cooling unit should be installed in parallel or on the upstream side of the evaporator coil to avoid condensation on the heating element. In a parallel installation, dampers or comparable means should be provided to prevent chilled air from entering the furnace.

INSTALLATION AND MAINTENANCE CHECKLIST

A properly installed and maintained furnace will operate efficiently and economically. The following installation checklist is offered as a guide for the installer.

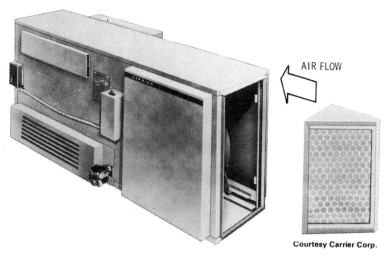

AIR FLOW

Courtesy Carrier Corp.

Fig. 11-39. External air filter unit in a Carrier horizontal gas-fired furnace.

1. Adjust primary air shutter.
2. Provide sufficient space for service accessibility.
3. Check all field wiring.
4. Supply line fuse or circuit breaker must be of proper size and type for furnace protection.
5. Line voltage must meet specifications while furnace is operating.
6. Airflow must be sufficient.
7. Ventilation and combustion air must be sufficient.
8. Ductwork must be checked for proper balance, velocity, and quietness.
9. Measure the manifold pressure.
10. Check all gas piping connections for gas leaks (*use soap and water, not a flame*).
11. Cycle the burners.
12. Check limit switch.
13. Check fan switch.
14. Adjust blower motor for desired speed.
15. Make sure air filter is properly secured.
16. Make sure all access panels have been secured.
17. Pitch air conditioning equipment condensation lines toward a drain.
18. Check thermostat heat anticipatory setting.
19. Check thermostat for normal operation.
20. Clear and clean the area around the furnace.

A general checklist for both furnace installation and maintenance is shown in Fig. 11-40. Other maintenance recommendations are contained in various sections of this chapter.

TROUBLESHOOTING
A GAS-FIRED FURNACE

The list that follows contains the most common operating problems associated with gas-fired furnaces. Each problem is given in the form of a symptom, the possible cause, and a suggested remedy. This list is intended to provide the operator with a quick reference to the cause and correction of a specific problem.

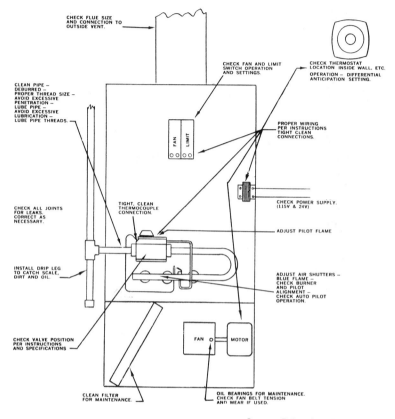

Courtesy Robertshaw Controls Co.

Fig. 11-40. Installation and maintenance checklist.

| Symptom and Possible Cause | Possible Remedy |

Flame Too Large

(a) Pressure regulator set too high.

(a) Reset using manometer.

(b) Defective regulator.

(b) Replace.

(c) Burner orifice too large.

(c) Replace with correct size.

Symptom and Possible Cause   *Possible Remedy*

Noisy Flame

(a) Too much primary air.
(b) Noisy pilot.
(c) Burr in orifice.

(a) Adjust air shutters.
(b) Reduce pilot gas.
(c) Remove burr or replace orifice.

Yellow Tip flame

(a) Too little primary air.
(b) Clogged burner ports.
(c) Misaligned orifices.
(d) Clogged draft hood.

(a) Adjust air shutters.
(b) Clean ports.
(c) Realign.
(d) Clean.

Floating flame

(a) Blocked venting.
(b) Insufficient primary air.

(a) Clean.
(b) Increase primary air supply.

Delayed Ignition

(a) Improper pilot location.
(b) Pilot flame too small.

(c) Burner ports clogged near.
(d) Low pressure.

(a) Reposition pilot.
(b) Check orifice; clean, increase pilot gas.
(c) Clean ports.

(d) Adjust pressure regulator.

Failure to Ignite

(a) Main gas supply off.
(b) Burned out fuse.
(c) Limit switch defective.
(d) Poor electrical connections.
(e) Defective gas valve.
(f) Defective thermostat.

(a) Open manual valve.
(b) Replace.
(c) Replace.
(d) Check, clean, and tighten.

(e) Replace.
(f) Replace.

377

Symptom and Possible Cause   *Possible Remedy*

Burner Will Not Turn Off

(a) Poor thermostat location.   (a) Relocate.
(b) Defective thermostat.   (b) Check calibration; check
                                     switch and contacts;
                                     replace.
(c) Limit switch maladjusted.   (c) Replace.
(d) Short circuit.   (d) Check operation at valve;
                                     check for short and
                                     correct.
(e) Defective or sticking   (e) Clean or replace.
    automatic valve.

Rapid Burner Cycling

(a) Clogged filters.   (a) Clean or replace.
(b) Excessive anticipation.   (b) Adjust thermostat antici-
                                     patory for longer cycles.
(c) Limit setting too low.   (c) Readjust or replace limit.
(d) Poor thermostat location.   (d) Relocate.

Rapid Fan Cycling

(a) Fan switch differential   (a) Readjust or replace.
    too low.
(b) Blower speed too high.   (b) Readjust to lower speed.

Blower Will Not Stop

(a) Manual fan on.   (a) Switch to automatic.
(b) Fan switch defective.   (b) Replace.
(c) Shorts.   (c) Check wiring and correct.

Noisy Blower and Motor

(a) Fan blades loose.   (a) Replace or tighten.
(b) Belt tension improper.   (b) Readjust (usually allow 1
                                     in. slack).
(c) Pulleys out of alignment.   (c) Realign.

378

Symptom and Possible Cause	*Possible Remedy*
(d) Bearings dry.	(d) Lubricate.
(e) Defective belt.	(e) Replace.
(f) Belt rubbing.	(f) Reposition.

Burner Will Not Turn On

(a) Pilot flame too large or too small.	(a) Readjust.
(b) Dirt in pilot orifice.	(b) Clean.
(c) Too much draft.	(c) Shield pilot.
(d) Defective automatic.	(d) Replace.
(e) Defective thermocouple.	(e) Replace.
(f) Improper thermocouple.	(f) Properly position thermopile.
(g) Defective wiring.	(g) Check connections; tighten and repair shorts.
(h) Defective thermostat.	(h) Check for switch closure and repair or replace.
(i) Defective automatic valve.	(i) Replace.

Blower Will Not Run

(a) Power not on.	(a) Check power switch; check fuses and replace if necessary.
(b) Fan control adjustment.	(b) Readjust or replace.
(c) Loose wiring.	(c) Check and tighten.
(d) Defective motor overload, protector, or motor.	(d) Replace motor.

Not Enough Heat

(a) Thermostat set too low.	(a) Raise setting.
(b) Lamp or some other heat source too close to thermostat.	(b) Move heat source away from thermostat.
(c) Thermostat improperly located.	(c) Relocate thermostat.

Symptom and Possible Cause	Possible Remedy
(d) Dirty air filter.	(d) Clean or replace.
(e) Thermostat out of calibration.	(e) Recalibrate or replace.
(f) Limit set too low.	(f) Reset or replace.
(g) Fan speed too low.	(g) Check motor and fan belt and tighten if too loose.

Too Much Heat

(a) Thermostat set too high.	(a) Lower setting.
(b) Thermostat out of calibration.	(b) Recalibrate or replace.
(c) Short in wiring.	(c) Locate and correct.
(d) Valve sticks open.	(d) Replace valve.
(e) Thermostat in draft or on cold wall.	(e) Relocate thermostat to sense average temperature.
(f) Bypass open.	(f) Close bypass.

CHAPTER 12

Oil-Fired Furnaces

Oil-fired furnaces are available in upflow, downflow (counter-flow), or horizontal flow models and in a wide range of heating capacities for installations in attics, basements, closet spaces, or at floor level (Figs. 12-1 and 12-2). The furnace, oil burner, and controls are generally available from the manufacturer as a complete package for residential installations. All internal wiring is done at the factory.

Oil-fired furnaces should be listed by the Underwriters' Laboratories for construction and operating safety. Furnaces approved by the agency are marked *UL Approved*. Furnaces not made under UL standards should not be installed.

Another source of information concerning standards for oil-fired furnaces is the publication *Installation of Oil Burning Equipment 1972* (NFPA No. 31) from the National Fire Protection Association. See also *American Standard Performance Requirements for Oil-Powered Central Furnaces* (ANSI Z91.1-

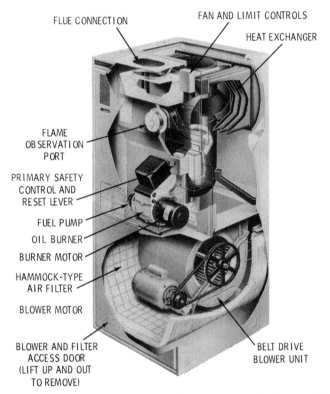

FLUE CONNECTION

FAN AND LIMIT CONTROLS

HEAT EXCHANGER

FLAME
OBSERVATION
PORT

PRIMARY SAFETY
CONTROL AND
RESET LEVER

FUEL PUMP

OIL BURNER

BURNER MOTOR

HAMMOCK-TYPE
AIR FILTER

BLOWER MOTOR

BLOWER AND FILTER
ACCESS DOOR
(LIFT UP AND OUT
TO REMOVE)

BELT DRIVE
BLOWER UNIT

Courtesy Lennox Air Conditioning and Heating Co.

Fig. 12-1. Lennox upflow oil-fired furnace.

1972), a publication of the American National Standards Institute.

PLANNING SUGGESTIONS

The first step in planning a heating system is to calculate the maximum heat loss for the structure. This should be done in accordance with procedures described in the manuals of the National Warm Air Heating and Air Conditioning Association or by a comparable method. This is very important because the

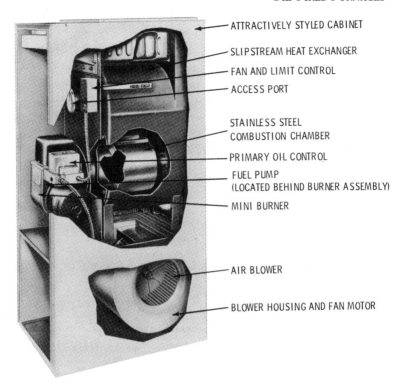

ATTRACTIVELY STYLED CABINET

SLIPSTREAM HEAT EXCHANGER

FAN AND LIMIT CONTROL

ACCESS PORT

STAINLESS STEEL
COMBUSTION CHAMBER

PRIMARY OIL CONTROL

FUEL PUMP
(LOCATED BEHIND BURNER ASSEMBLY)

MINI BURNER

AIR BLOWER

BLOWER HOUSING AND FAN MOTOR

Courtesy Carrier Corp.

Fig. 12-2. Cutaway of a Carrier upflow oil furnace.

data will be used to determine the size (capacity) of the furnace selected for the installation.

Location and Clearance

An oil-fired, warm-air furnace should be located as near as possible to the center of the heat-distribution system and chimney. A centralized location for the furnace usually results in the best operating characteristics because long supply ducts and the heat loss associated with them are eliminated. The furnace should be installed as close as possible to the chimney to reduce the horizontal run of flue pipe.

383

An oil-fired furnace should be located a safe distance from any combustible material. Observe the minimum clearances suggested by the manufacturer for the furnace except where they may come into conflict with local codes and regulations (Table 12-1). Local standards *always* take precedence over the manufacturer's specifications or national standards. Where there are no local standards specifying clearances, use those listed in *Installation of Oil Burning Equipment 1972* (NFPA No. 31-1972). Contents of the 1972 edition have been approved by the American National Standards Institute (ANSI).

Allow access for servicing the furnace and oil burner. Sufficient clearance should also be provided for access to the barometric draft-control assembly.

Installation Recommendations

The installation of an oil-fired furnace should comply with all local codes and regulations and *Installation of Oil Burning Equipment 1972.*

Table 12-1. Minimum Clearances Specified for Carrier Oil-Fired Upflow (58HV), Loboy Upflow (58HL), Downflow (58HC), and Horizontal (58HH) Furnace Models

From	Furnace				
	58HV	58HL	58HC	58HH	
			Clearance (in.)		
Top of furnace or plenum	2	2	2	6	2*
Bottom of furnace	0	0	—	6	2*
Front of furnace	15	24	15	24	48*
Rear of furnace	1	6	2	6	2*
Sides of furnace	1	6	1	6	4*
Top of horizontal warm-air duct within 6 ft. of furnace	2	—	4	6	3*
Flue pipe: Horizontally or below pipe	9	—	9	—	—
Vertically	18	18	15	18	9*
Any side of supply plenum and warm-air duct within 6 ft. of furnace	2	—	6	—	—

*Alcove installation.   Courtesy Carrier Corp.

New furnaces for residential installations are shipped preassembled from the factory with all internal wiring completed. The fuel tank must be installed and the fuel line connected to the furnace by the installer. The electrical service from the line-voltage main and the low-voltage thermostat will also have to be connected. Directions for making these connections are included in the furnace manufacturer's installation literature.

Make certain you have familiarized yourself with all local codes and regulations that govern the installation of an oil-fired furnace. Local codes and regulations *always* take precedence over national standards. Read the section "Installation Recommendations" in Chapter 11 (Gas-Fired Furnaces) for additional information.

Dust Connections

Detailed information concerning the installation of an air-duct system is contained in the following two publications of the National Fire Protection Association:

1. *Installation of Air Conditioning and Ventilating Systems of Other Than Residence Type* (NFPA No. 90A).
2. *Residence Type Warm Air Heating and Air Conditioning Systems* (NFPA No. 90B).

Additional information about duct connections can be found in Chapter 7 of Volume 2 (Duct and Duct Systems). The specific recommendations concerning furnace duct connections and the air-distribution ducts contained in Chapter 11 (Gas-Fired Furnaces) also apply to gas-fired furnaces.

Electric Wiring

Always follow all national and local codes when wiring a unit. The local code should take precedence when there is a conflict between the two. If there is no local electrical code in force, do all wiring in accordance with the National Electric Code.

Closely follow the wiring diagrams provided by the furnace manufacturer and those who manufacture the automatic controls. Failure to do this may result in losing the factory warranty.

A typical wiring diagram for an oil-fired furnace is shown in Fig. 12-3. The comments made in Chapter 11 (Gas-Fired Furna-

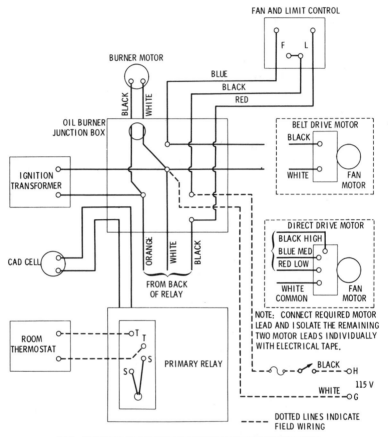

FAN AND LIMIT CONTROL

BURNER MOTOR

BLUE
BLACK
RED

OIL BURNER
JUNCTION BOX

BELT DRIVE MOTOR
BLACK

WHITE   FAN
               MOTOR

IGNITION
TRANSFORMER

DIRECT DRIVE MOTOR
BLACK HIGH
BLUE MED
RED LOW

CAD CELL

WHITE   FAN
COMMON   MOTOR

FROM BACK
OF RELAY

NOTE: CONNECT REQUIRED MOTOR
LEAD AND ISOLATE THE REMAINING
TWO MOTOR LEADS INDIVIDUALLY
WITH ELECTRICAL TAPE.

ROOM
THERMOSTAT

PRIMARY RELAY

BLACK   H
                   115 V
WHITE   G

DOTTED LINES INDICATE
FIELD WIRING

NOTE: MAKE FIELD CONNECTIONS TO WIRES CAPPED WITH RED WIRE NUTS.

Courtesy Trane Co.

Fig. 12-3. Wiring diagram for a Trane oil-fired furnace.

ces) concerning line-voltage and control-voltage wiring connections also pertain to the wiring of oil-fired units.

Ventilation and Combustion Air

An oil-fired furnace should be located where a sufficient
supply of air is available for combustion, proper venting, and the

maintenance of suitable ambient temperature. In buildings of exceptionally tight construction, an outside air supply should be introduced for ventilation and combustion purposes.

If the furnace is located in a confined space such as a furnace room, closet, or utility room, the enclosure should be provided with two permanent openings for the passage of the air supply. One opening should be located approximately 6 in. from the top of the enclosure, and the other opening approximately 6 in. from the bottom. Each opening should have free area of at least 1 sq. in. per 1000 Btuh (minimum size 100 sq. in.) of the total input rating, or 1 sq. ft. per gallon of oil per hour.

The size of the openings must correspond to the bonnet capacity of the furnace. Furnace manufacturers generally recommend suitable sizes for air openings in their equipment specifications (Table 12-2).

Table 12-2. Recommended Sizes of Ventilation Openings for Trane Counterflow Oil-Fired Furnace

Bonnet Capacity, 1,000 Btuh	Length, in.	Height, in.
85	24	12
100	24	12
125	26	13

Courtesy The Trane Co.

Air from an attic or a crawl space can also serve as a ventilation and combustion air supply to a confined furnace. Suitable openings near (or on) the floor and near (or on) the ceiling must be provided for the passage of the air.

When an oil furnace is installed in an unconfined area, such as a full basement, the full amount of air necessary for combustion is generally supplied by infiltration. Only when the construction is exceptionally tight is it necessary to provide openings to the outdoors.

Combustion Draft

An oil-fired furnace requires sufficient draft for the proper burning of oil. The average combustion draft is usually 0.02 in. wg over the fire (the so-called *overfire draft*), and between 0.04

387

and 0.06 in. wg draft in the stack (the *flue draft*) for units having inputs up to 2.0 gph. For higher inputs, 0.06 to 0.08 in. wg will be the rule. These figures are *average* ones and do not apply to all installations. For example, a twentieth-century Zeph-Air Low Boy furnace only requires 0.01 in. wg draft over the fire and a flue draft of 0.02 to 0.03 in. wg. *Always* consult the furnace manufacturer's operating manual first for recommended drafts.

Venting

Oil-fired appliances use the same methods for venting the products of combustion to the outside air as do gas-fired units, except when the venting method is specifically limited to a particular fuel (e.g., Type B gas vents). Additional information about venting methods is contained in Chapter 11 (Gas-Fired Furnaces). The type of venting to be used with a furnace is generally recommended by the manufacturer in the installation literature.

Flue Pipe

Use 24-gauge galvanized flue pipe (or its equivalent) for the flue connection between the furnace and the chimney. The horizontal run of flue pipe should slope upward at least ¼ in. per running foot. The pitch should not exceed 75 percent of the vertical vent. Locate the unit so that the furnace flue outlet is no more than 10 flue pipe diameters from the chimney connection. If the distance is greater, use the next-size-larger flue pipe.

Insulate the flue pipe when it passes near combustible material. The Underwriters' Laboratories require that any uninsulated flue pipe be installed with a *minimum* clearance of 6 in. from any combustible materials.

Draft Regulator

A *draft regulator* (or *barometric draft regulator*) is a device designed to maintain a proper draft in an oil-fired furnace by automatically reducing the chimney draft (i.e., the flue draft) to the desired value (Fig. 12-4).

Draft regulators are recommended on all installations for each

388

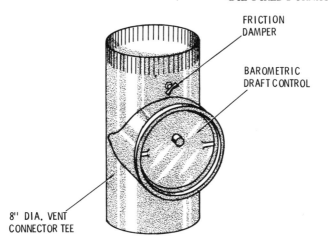

FRICTION
DAMPER

BAROMETRIC
DRAFT CONTROL

8" DIA. VENT
CONNECTOR TEE

Fig. 12-4. Barmetric draft-control assembly.

oil-fired appliance connected to a chimney *unless* the appliance is listed for use without one.

Do not use a small draft regulator because they will not do a satisfactory job. The draft regulator should be installed in the horizontal flue pipe as close as possible to the chimney. When this flue pipe is too short for a satisfactory installation, the draft regulator should be installed in the chimney either above or below the flue pipe entry into the chimney. Some authorities oppose installing the draft regulator in the chimney, but experience has shown that this can be a very satisfactory installation.

CHIMNEYS

The chimney used with an oil-fired furnace should be of sufficient height and large enough in its cross-sectional area to meet the requirements of the furnace. The chimney area should be at least 20 percent greater than the area of the glue outlet. To ensure the adequate removal of flue gases, the height and size of the chimney must be sufficient to create 0.02 to 0.04 in. wg draft over the fire for inputs up to 2.0 gph. For higher inputs, 0.06 to 0.08 in. wg is recommended. If the stack height is limited, it may be necessary to apply mechanically induced draft to the installation.

Ideally, *only* the furnace should be connected to the chimney. Read the section "Chimneys" in Chapter 11 (Gas-Fired Furnaces) for additional information.

Chimney Troubleshooting

Read the section "Chimney Troubleshooting" in Chapter 11 (Gas-Fired Furnaces) for a description of common chimney problems and suggested remedies. The chimney used with oil-fired furnaces are essentially identical to those used with gas-fired furnaces.

BASIC COMPONENTS

The basic components of an oil-fired, forced-warm-air furnace are:

1. Automatic controls.
2. Combustion chamber.
3. Heat exchanger.
4. Oil burner.
5. Fuel pump.
6. Blower and motor.
7. Air filter(s).

Each of these components is described in the sections that follow. Additional information is contained in Chapter 10 (Furnace Fundamentals) and the various chapters in which furnace controls and fuel-burning equipment are described.

AUTOMATIC CONTROLS

The automatic controls for an oil-fired furnace can be divided into the following three principal categories:

1. Room thermostat.
2. Fan and limit control.
3. Primary control.

The thermostat and limit controls are essentially identical to those used with gas-fired furnaces. The thermostat controls the furnace operation by signaling the primary control to turn the burner on and off in response to temperature changes in the space or spaces being heated. The fan and limit control is designed to govern the operation of the furnace within a specified temperature range.

Until comparatively recently, the primary control consisted of the oil-burner relay and a stack switch to shut off the oil flow to the burner if the oil failed to ignite or the flame went out. Stack switches have largely been replaced by burner-mounted cadmium detection cells.

A typical automatic-control system for an oil-fired furnace is shown in Fig. 12-5. Additional and more detailed information about the various types of controls can be found in Chapter 4 (Thermostats and Humidistats), Chapter 5 (Gas and Oil Controls), and Chapter 6 of Volume 2 (Other Automatic Controls).

COMBUSTION CHAMBER

The combustion chamber is the portion of the furnace that surrounds the oil-burner and flame and within which the combustion process takes place.

The typical combustion chamber is made of stainless steel or some sort of refractory material such as firebrick (Fig. 12-6 and 12-7). Round combustion chambers are probably the most common, although square, rectangular, and other shapes do occur.

HEAT EXCHANGER

The heat exchanger is a device used to transfer the heat of the combustion process to the air being circulated through the ducts of the heating system. The heat exchanger is generally con-

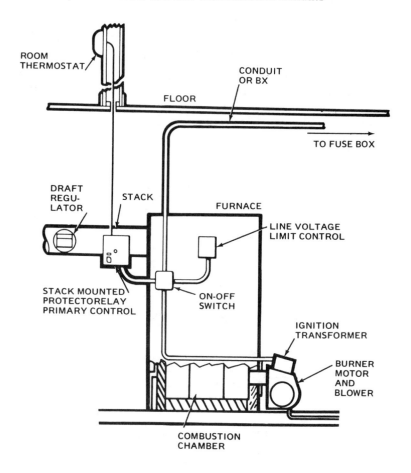

Fig. 12-5. Typical automatic control system for an oil-fired furnace.

structed of heavy gauge steel (14–16 gauge). The wrap-around, radiator-type heat exchanger is one of the most commonly used designs. It consists of an upper and lower chamber, each with an extension, or pouch, that must be aligned when assembled.

Courtesy Bard Mfg. Co.

Fig. 12-6. Stainless steel combustion chamber and heat exchanger on a Bard oil-fired furnace.

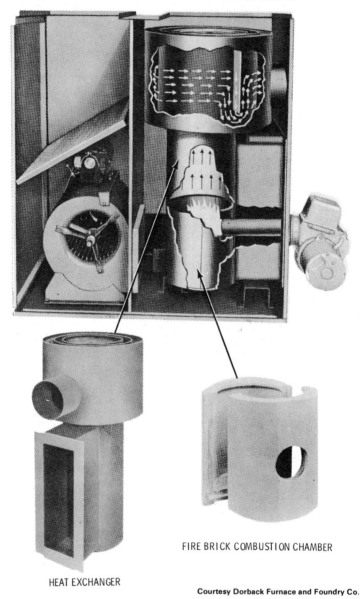

FIRE BRICK COMBUSTION CHAMBER

HEAT EXCHANGER

Courtesy Dorback Furnace and Foundry Co.

Fig. 12-7. Heat exchanger and combustion chamber on a Dorback oil-fired furnace.

OIL BURNER

An oil burner is a mechanical device used to combine fuel oil with the proper amount of air for combustion and deliver it to the point of ignition. Oil burners are described in considerable detail in Chapter 1 of Volume 2 (Oil Burners).

When installing the burner nozzle, make sure that the proper-size nozzle for the installation has been selected. Electrodes are adjusted at the time of manufacture, but they should be checked when the unit is installed to make sure they are set to the burner manufacturer's specifications.

FUEL PUMPS AND FUEL UNITS

Almost all oil burners used in residential heating installations are equipped with fuel units. It is very important to distinguish between a fuel unit and a fuel pump because the two are not identical. A fuel unit is an oil-pumping device that consists of a pump, a pressure regulating valve, and a filter. A pump is *only* an oil-pumping device. A pressure-regulating valve and a filter can be attached to the mountings on the pump if so desired. Fuel pumps and fuel units are described in Chapter 1 of Volume 2 (Oil Burners).

Manufacturers produce single-stage and two-stage pumps. The single-stage pumps are recommended for installations where the fuel tank is located at a level *above* the burner (Fig. 12-8). Two-stage pumps are used in installations with underground tanks where the combination of lift, horizontal run, fittings, and filters does not exceed the manufacturer's rating in inches of vacuum.

The capacity of the integral pump on an oil burner should be sufficient to handle the total vacuum in the system. The vacuum is expressed in inches and can be determined by the following procedure:

1. 1 in. of vacuum for each foot of lift.
2. 1 in. of vacuum for each 90° elbow in either the suction or return lines.

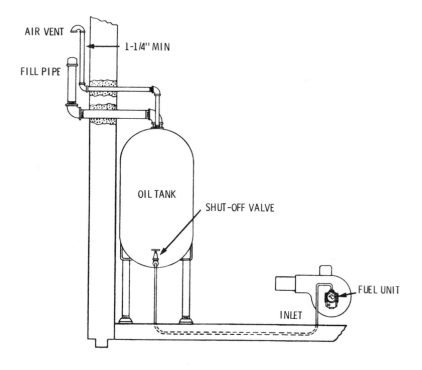

Fig. 12-8. Fuel tank located above burner.

3. 1 in. of vacuum for each 10 ft. of horizontal run ($\frac{3}{8}$-in. OD line).

4. 1 in. of vacuum for each 20 ft. of horizontal run ($\frac{1}{2}$-OD line).

After you have calculated the total vacuum, you can use these data to select the most suitable pump for the burner. Table 12-3 lists various vacuums and suggests appropriate pump capacities.

396

Table 12-3. Types of Pumps Recommended for Different Vacuums

Total Vacuum	Type of Pump
Up to 3 in. vacuum	Single-stage pump
4–13 in. vacuum	Two-stage pump
14 in. vacuum or more	Single-stage pump for the burner and a separate lift pump with a reservoir

TROUBLESHOOTING FUEL PUMPS AND FUEL UNITS

The following troubleshooting list contains the most common operating problems associated with fuel pumps and fuel units. Each problem is given in the form of a symptom, the possible cause, and a suggested remedy. The purpose of this list is to provide the operator with a quick reference to the cause and correction of a specific problem. Much of the information used in this troubleshooting list was obtained from the *Sundstrand Technical Fuel Unit Manual*, courtesy the Sundstrand Corporation.

Symptom and Possible Cause   *Possible Remedy*

No Oil Flow at Nozzle

(a) Clogged strainer or filter.

(b) Air binding in two-pipe system.

(c) Frozen pump shaft.

(a) Remove and clean strainer (Fig. 12-9); repack filter element.

(b) Check and insert bypass plug.

(c) Remove unit and return to factory for repair.

Oil Leak

(a) Loose plugs or fittings.

(a) Dope with good-quality thread sealer.

397

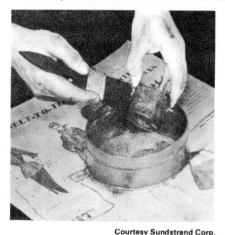

12-9. Cleaning the strainer with clean fuel or kerosene.

Courtesy Sundstrand Corp.

Symptom and Possible Cause	*Possible Remedy*
(b) Leak at pressure-adjusting end cap nut.	(b) Fiber washer may have been left out after adjustment of valve spring; replace washer.
(c) Blown seal.	(c) Replace fuel unit.
(d) Seal leaking.	(d) Replace fuel unit.

Noisy Operation

(a) Air inlet line.	(a) Tighten all connections and fittings in the intake line and unused intake port plugs (Fig. 12-10).
(b) Bad coupling alignment.	(b) Loosen mounting screws and shift fuel unit to a position where noise is eliminated. Retighten mounting screws.
(c) Pump noise.	(c) Work in gears by continued running or replace.

Fig. 12-10. Tightening connections and fittings.

Courtesy Sundstrand Corp.

Symptom and Possible Cause   *Possible Remedy*

Pulsating Pressure

(a) Air leak in intake line.

(b) Air leaking around
strainer cover.
(c) Partially clogged strainer.

(d) Partially clogged filter.

(a) Tighten all fittings and
valve packing in intake
line.
(b) Tighten strainer cover
screws.
(c) Remove and clean
stainer.
(d) Replace filter element.

Low Oil Pressure

(a) Nozzle capacity is greater
than fuel-unit capacity.
(b) Defective gauge.

(a) Replace fuel unit with
unit of correct capacity.
(b) Check against another
and replace if necessary.

Improper Nozzle Cutoff

(a) Filter leaks.

(b) Partially clogged nozzle
strainer.

(a) Check face of filter cover
and gasket for damage.
(b) Clean strainer or change
nozzle.

399

Symptom and Possible Cause   *Suggested Remedies*

(c) Air leak in intake line.   (c) Tighten intake fittings and packing nut on shut-off valve. Tighten unused intake port plug.

(d) Strainer cover loose.   (d) Tighten screws.

(e) Air pockets between cutoff valve and nozzle.   (e) Start and stop burner until smoke and after-fire disappears.

The cause of improper cutoff can be determined by inserting a pressure gauge in the nozzle port of the fuel unit (Fig. 12-11). Allow the fuel unit to operate for a few minutes and then shut off the burner. If the pressure drops to 0 psi on the gauge, the fuel unit should be replaced. A gauge reading above 0 psi indicates that the fuel unit is operating properly. The probable cause of improper cutoff in those cases where a gauge reading is obtained is usually air in the system.

Courtesy Sundstrand Corp.

Fig. 12-11. Checking fuel pressure with a gauge.

FUEL TANK AND LINE

The fuel tank and line must be installed in accordance with the requirements of the Underwriters, Laboratories and any local codes and regulations. Local codes and regulations take precedence over national standards. Additional information can be obtained from *Installation of Burning Equipment 1972* (NFPA

No. 31), a publication of the National Fire Protection Association. Use only an approved tank and line for the installation.

A one-pipe system is recommended for an oil-tank installation when the following conditions are present:

1. Tank is installed in the basement or similar location and/or below the pump inlet port.
2. Vacuum at pump does not exceed 2 in.

A two-pipe (return line) system is recommended for a tank installation when the following conditions exist:

1. Tank is located outside and/or below the oil-pump inlet port.
2. Vacuum at pump does not exceed 10 in.

Use a two-stage pump for installations having especially long lines or high lifts *if* the vacuum at the inlet port does not exceed 20 in.

A shutoff valve should be installed in the suction line close to the oil burner. This valve enables the operator to disconnect the line without draining it. Do *not* place a shutoff valve in the return line. Closing a shutoff valve on the return line while the oil burner is operating could damage the pump. All joints in the fuel line should be sealed with a suitable oil-piping compound.

OIL FILTER

An oil filter should be included in the fuel line inside the building and as close to the fuel tank as possible (Fig. 12-12). One filter is adequate on oil burners using more than 1.00 gph.

Change the filter cartridge at least once a year. The filter body should be thoroughly cleaned before installing a new cartridge.

PRIMING THE FUEL PUMP

On occasion, the oil burner may fail to pump oil. When this occurs, check the oil supply line to the furnace for leaks. If there are no leaks, it may be necessary to prime the fuel pump. Pumps

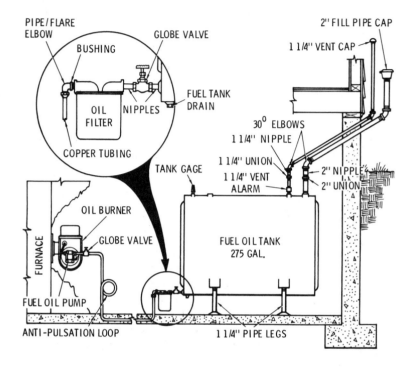

Fig. 12-12. Typical location of oil filter.

are self-priming for single-stage, two-pipe systems and for two-stage pumps. A single-stage pump (one-pipe system) should be primed as follows:

1. Turn off the electrical power supply to the unit.
2. Read and follow the priming instructions provided by the manufacturer.
3. Prime the pump until the oil is free from bubbles.

When a *new* pump fails to prime, it may be due to dry pump conditions, which can be corrected by removing the vent plug and filling the pressure cavity slowly so that the fuel oil "wets" the gears (Fig. 12-13).

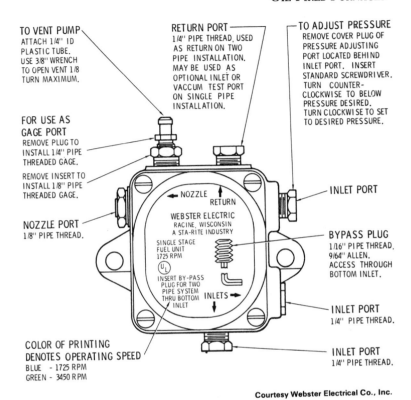

TO VENT PUMP
ATTACH 1/4" ID
PLASTIC TUBE.
USE 3/8" WRENCH
TO OPEN VENT 1/8
TURN MAXIMUM.

FOR USE AS
GAGE PORT
REMOVE PLUG TO
INSTALL 1/4" PIPE
THREADED GAGE.

REMOVE INSERT TO
INSTALL 1/8" PIPE
THREADED GAGE.

NOZZLE PORT
1/8" PIPE THREAD.

COLOR OF PRINTING
DENOTES OPERATING SPEED
BLUE   - 1725 RPM
GREEN - 3450 RPM

RETURN PORT
1/4" PIPE THREAD, USED
AS RETURN ON TWO
PIPE INSTALLATION.
MAY BE USED AS
OPTIONAL INLET OR
VACCUM TEST PORT
ON SINGLE PIPE
INSTALLATION.

TO ADJUST PRESSURE
REMOVE COVER PLUG OF
PRESSURE ADJUSTING
PORT LOCATED BEHIND
INLET PORT. INSERT
STANDARD SCREWDRIVER.
TURN COUNTER-
CLOCKWISE TO BELOW
PRESSURE DESIRED.
TURN CLOCKWISE TO SET
TO DESIRED PRESSURE.

← NOZZLE
RETURN
WEBSTER ELECTRIC
RACINE, WISCONSIN
A STA-RITE INDUSTRY
SINGLE STAGE
FUEL UNIT
1725 RPM
UL
INSERT BY-PASS
PLUG FOR TWO
PIPE SYSTEM
THRU BOTTOM
INLET
INLETS →

INLET PORT

BYPASS PLUG
1/16" PIPE THREAD.
9/64" ALLEN.
ACCESS THROUGH
BOTTOM INLET.

INLET PORT
1/4" PIPE THREAD.

INLET PORT
1/4" PIPE THREAD.

Courtesy Webster Electrical Co., Inc.

Fig. 12-13. Webster model M series fuel pump.

ADJUSTING OIL-PUMP PRESSURE

The oil-pressure regulator on the fuel pump is generally factory set to give nozzle oil pressures of 100 psig. The firing rate is indicated on the nameplate and can be obtained with standard nozzles by adjusting the pump pressures as follows (Fig. 12-13):

1. Turn the adjusting screw clockwise to increase pressure.
2. Turn the adjusting screw counterclockwise to decrease pressure.
3. Never exceed the pressures indicated in Table 12-4.

403

STARTING THE BURNER

The procedure for starting an oil burner may be summarized as follows:

1. Open all warm-air registers.
2. Check to be sure all return air grilles are unobstructed.
3. Open the valve on the oil supply line.
4. Reset the burner primary relay.
5. Set the thermostat above the room temperature.
6. Turn on the electric supply to the unit by setting the main electrical switch to the *on* position.
7. Change the room thermostat setting to the desired temperature.

The oil burner should start after the electric power has been switched on (Step 6). There is no pilot to light as is the case with gas-fired appliances. The spark for ignition is provided automatically on demand from the room thermostat.

Allow the burner to operate at least 10 minutes before making any final adjustments. Whenever possible, use instruments to adjust the fire.

COMBUSTION TESTING
AND ADJUSTMENTS

The instruments recommended for combustion testing and adjustments are:

1. Draft gauge.
2. Smoke tester.
3. Carbon dioxide tester.
4. 200/1000°F stack thermometer.
5. 0/150-psig pressure gauge.
6. 0/30-in. mercury vacuum gauge.

Smoky combustion indicates poor burner performance. The amount of smoke in the flue gas can be measured with a smoke tester (Fig. 12-14). The tube of the smoke tester is inserted through a ⅜-in. hole drilled in the flue pipe, and the test is run as

Table 12-4. Maximum Recommended Pressures

Bonnet Capacity (1000 Btuh)	Firing Rate (gph)	Standard Nozzle Size	Pump Pressure (psig)
85	0.76	0.75	103
100	0.90	0.90	100
125	1.12	1.10	104
150	1.35	1.35	100
200	1.80	1.75	112
250	2.25	2.25	100
335	3.00	1.50	100

Courtesy Carrier Corp.

shown in Fig. 12-15. Any smoke in the air drawn into the smoke tester will register on a filter paper inserted in the device. The results are interpreted according to the smoke scale in Table 12-5.

One of the most common causes of smoky combustion is soot formation on the heating surfaces. This is easily corrected by cleaning. Other possible causes of smoky combustion include:

1. Insufficient draft.
2. Poor fuel supply.
3. Fuel pump malfunctioning.

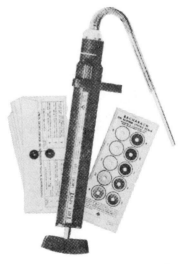

Fig. 12-14. Bacharach true-spot smoke tester.

Courtesy Bacharach Instrument Co.

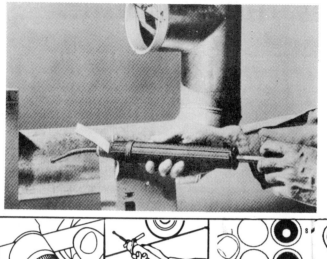

Insert filter test paper into
TRUE-SPOT

Withdraw gas sample from
flue pipe by 10 pump strokes

Grade soot spot on test paper
by comparison with shadings on scale

Courtesy Bacharach Instrument Co.

Fig. 12-15. A smoke tester in use.

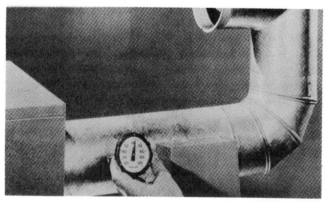

Courtesy Bacharach Instrument Co.

Fig. 12-16. Running a stack-gas temperature test with a stack thermometer.

Table 12-5. Smoke Scale

BACHARACH SMOKE SCALE NO.	RATING	SOOTING PRODUCED
1	EXCELLENT	EXTREMELY LIGHT, IF AT ALL
2	GOOD	SLIGHT SOOTING WHICH WILL NOT INCREASE STACK TEMPERATURE APPRECIABLY
3	FAIR	MAY BE SOME SOOTING BUT WILL RARELY REQUIRE CLEANING MORE THAN ONCE A YEAR
4	POOR	BORDERLINE CONDITION SOME UNITS WILL REQUIRE CLEANING MORE THAN ONCE A YEAR
5	VERY POOR	SOOT RAPIDLY AND HEAVILY

Courtesy Bacharach Instrument Co.

4. Defective firebox.
5. Incorrectly adjusted draft regulator.
6. Defective oil-burner nozzle.
7. Wrong size oil-burner nozzle.
8. Improper fan delivery.
9. Excessive air leaks in boiler or furnace.
10. Unsuitable fuel-air ratio.

Net stack temperatures in excess of 700°F for conversation units and 500°F for packaged units are considered abnormally high. The *net* stack temperature is the difference between the temperature of the flue gases inside the pipe and the room air temperature outside. For example, if the flue-gas temperature is 600°F and the room temperature is 75°F, then the net stack temperature is 525°F (600°F − 75°F = 525°F).

A 200/1000°F stack thermometer is used to measure the flue gas temperature. The thermometer stem is inserted through a hole drilled in the flue pipe (Fig. 12-16). A high stack temperature may be caused by any of the following:

1. Undersized furnace.
2. Defective combustion chamber.
3. Incorrectly sized combustion chamber.
4. Lack of sufficient baffling.
5. Dirty heating surfaces.

407

6. Excessive draft.
7. Boiler or furnace overfired.
8. Unit unsuited to automatic firing.
9. Draft regulator improperly adjusted.

When the carbon dioxide (CO_2) content of the flue gas is too low (less than 8%), heat is lost up the chimney and the unit operates inefficiently. This condition is usually caused by one of the following:

1. Underfiring the combustion chamber.
2. Burner nozzle is too small.
3. Air leakage into the furnace or boiler.

When the carbon dioxide content is too high, the furnace operation is generally characterized by excess smoke and/or pulsations and other noises. A high carbon dioxide content is usually caused by insufficient draft or an overfired burner.

The carbon dioxide reading is also taken through a hole drilled in the flue pipe with a CO_2 indicator (Fig. 12-17). The CO_2 indicator is used as shown in Fig. 12-18. The results are indicated by a test liquid on a scale calibrated in %CO_2.

A correct draft is essential for efficient burner operation. Insufficient draft can make it almost impossible to adjust the oil

Fig. 12-17. Bacharach fyrite CO_2 indicator.

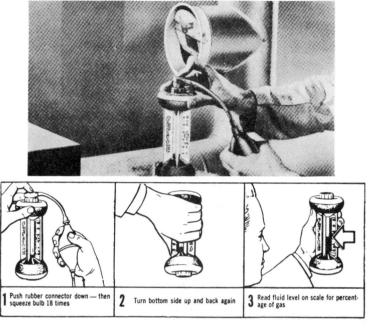

1 Push rubber connector down — then squeeze bulb 18 times	**2** Turn bottom side up and back again
3 Read fluid level on scale for percentage of gas	

Courtesy Bacharach Instrument Co.

Fig. 12-18. CO₂ test.

burner for its highest efficiency. Excessive draft can reduce the percentage of carbon dioxide in the flue gases and increase the stack temperature.

For the most efficient operating characteristics, the overfire draft generally should be not less than 0.02 in. wg. Smoke and odor often occur when the overfire draft falls below 0.02 in. wg.

It may be necessary to adjust the barometric draft regulator to obtain the correct overfire draft. If it is not possible to adjust the overfire draft for a −0.01 to 0.02 in. wg., install a mechanical draft inducer between the chimney and the barometric draft regulator.

The primary air band should be adjusted to a 0+ smoke or until a hard clean flame is visible. A clean flame is preferred to one with high carbon dioxide. Adjust the overfire draft for a −0.01 to a −0.02 in. wg. An excessive overfire draft condition will cause high stack temperature and inefficient operation. A too low or

409

positive draft over the fire will usually cause the flue gases and fume to seep into the space upon start up or shutdown.

The flue-pipe draft in most residential oil burners is between 0.04 and 0.06 in. water. This is sufficient to maintain a draft of 0.02 in. in the firebox.

The furnace or boiler draft is measured with a draft gauge as shown in Fig. 12-19. A hole is drilled in either the fire door (over-fire draft measurement) or flue-pipe (flue-pipe draft measurement), and the unit is run for approximately 5 minutes. The draft tube is then inserted into the test hole, and the gauge is read (Fig. 12-20).

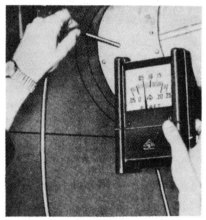

Fig. 12-19. Bacharach model MZF draft gauge.

Courtesy Bacharach Instrument Co.

BLOWERS AND MOTORS

Both direct-drive and belt-drive blowers are used with oil-fired furnaces. These blowers and motors are identical to those used in gas-fired furnaces. Read the section "Blowers and Motors" in Chapter 11 (Gas-Fired Furnaces) for additional information.

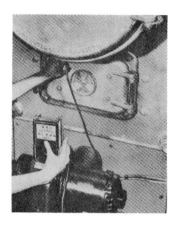

(A) Overfire draft.

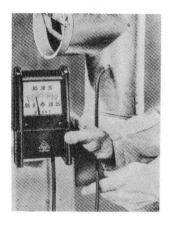

(B) Flue pipe draft.

Courtesy Bacharach Instrument Co.

Fig. 12-20. Overfire and flue pipe draft tests.

AIR DELIVERY AND BLOWER ADJUSTMENT

It is sometimes necessary to adjust the blower speed to produce a temperature rise through the furnace that falls within the limits stamped on the furnace nameplate. Blower adjustment procedures are described in Chapter 11 (Gas-Fired Furnaces).

AIR FILTER

A forced-warm-air, oil-fired furnace is supplied with either a disposable air filter or a permanent (washable) one. Never use a filter with a gravity warm-air furnace because it will obstruct the airflow.

A disposable air filter should be inspected on a regular basis and replaced when dirty. Always replace the air filter with one of the same size and type of filter media. This information is usually found on a label attached to the filter.

411

A permanent air filter must also be regularly inspected. When it is dirty, it must be removed and cleaned. The usual method is to vacuum it and then to wash it in a soap or detergent and water solution.

Additional information about air filters is found in Chapter 11 (Gas-Fired Furnaces) and Chapter 14 of Volume 3 (Air Cleaners and Filters).

AIR CONDITIONING

A furnace used in conjunction with a cooling unit should be installed in parallel or on the upstream side of the evaporator coil to avoid condensation on the heating element. In a parallel installation, dampers or comparable means should be provided to prevent chilled air from entering the furnace.

Duct sizing should include an allowance for air conditioning, even though it may not be initially installed with the furnace. Air conditioning involves a greater volume of air than heating.

All ductwork located in unconditioned areas (attics, crawl spaces, etc.) *downstream* from the furnace should be insulated to prevent unwanted heat gain.

INSTALLATION AND MAINTENANCE CHECKLIST

A properly installed and maintained furnace will operate efficiently and economically. The following installation checklist is offered as a guide to the installer.

1. Be sure there is at least 0.01 in. wg draft over the fire.
2. Check for sufficient combustion and ventilation air.
3. Eliminate downdraft or back draft.
4. Provide sufficient space for service accessibility.
5. Check all field wiring.
6. Supply-line fuse or circuit breaker must be of proper size and type for furnace.
7. Line voltage must meet specifications while furnace is operating.

8. Ductwork must be checked for proper balance, velocity, and quietness.
9. Check the fuel line for leaks.
10. Cycle the burner.
11. Check the limit switch.
12. Check the fan switch.
13. Make final adjustments to the fire.
14. Adjust the blower motor for desired speed.
15. Make sure air filter is properly secured.
16. Make sure all access panels have been secured.
17. Pitch air conditioning equipment condensate lines toward a drain.
18. Check thermostat heat anticipator setting.
19. Check thermostat for normal operation. Observe at least five ignition cycles before leaving the installation.
20. Clear and clean the area around the furnace.

Other installation and maintenance recommendations are given in various sections of this chapter.

TROUBLESHOOTING AN OIL-FIRED FURNACE

The list that follows contains the most common operating problems associated with oil-fired furnaces. Each problem is given in the form of a symptom, the possible cause, and a suggested remedy. This list is intended to provide the operator with a quick reference to the cause and correction of a specific problem.

Symptom and Possible Cause   *Possible Remedy*

Change in Size of Fire

(a) Dirty nozzle.   (a) Clean or replace.
(b) Low pressure.   (b) Adjust at pump.
(c) Plugged strainer.   (c) Clean.
(d) Cold oil.   (d) Adjust pressure.

413

Symptom and Possible Cause   *Possible Remedy*

No Oil Flow

(a) Oil level below intake line in the supply tank.

(a) Fill tank.

(b) Clogged strainer.

(b) Remove and clean strainer.

(c) Clogged nozzle.

(c) Clean or replace.

(d) Air leak in intake line.

(d) Tighten fittings and plugs; check valves.

(e) Restricted intake line (high vacuum reading).

(e) Replace kinked tubing; check valves, filters.

(f) Two-pipe system air bound.

(f) Check bypass plug.

(g) Single-pipe system air bound.

(g) Loosen gauge port and drain oil until foam is gone.

(h) Slipping or broken coupling.

(h) Tighten or replace coupling.

(i) Frozen pump shaft.

(i) Replace

Oil Spray but No Ignition

(a) Dirty electrodes.

(a) Clean or replace.

(b) Improper spacing.

(b) Reset.

(c) Cracked porcelain.

(c) Replace.

(d) Dead transformer.

(d) Replace.

(e) Loose connection.

(e) Tighten.

(f) Faulty relay.

(f) Replace

Burner Motor Does Not Start

(a) Defective thermostat.

(a) Replace.

(b) Fuse burned out.

(b) Replace.

(c) Limit control open.

(c) Check setting and correct.

(d) Contact dirty or open on primary relay.

(d) Clean or replace relay.

Symptom and Possible Cause	*Possible Remedy*
(e) Relay transformer burned out.	(e) Replace relay.
(f) Motor stuck or burned out or overload protector out.	(f) Replace if burned out.
(g) Primary relay off on safety.	(g) Push reset button.

Noisy Operation

(a) Bad coupling alignment.	(a) Loosen fuel unit or motor.
(b) Loose coupling.	(b) Tighten setscrews.
(c) Air in oil line.	(c) Bleed oil line; look for leaks.
(d) Pump noise.	(d) Continued running some-times works in gears. If not, replace.
(e) Hum vibration.	(e) Isolate pipes from struc-tural members.
(f) Combustion noise.	(f) Adjust noise.
(g) Furnace too small.	(g) Check heat.loss to be sure furnace properly sized.
(h) Burner noisy.	(h) Check mounting and position; adjust air.
(i) Blower noisy.	(i) Oil bearing. Tighten shaft collars; adjust belt ten-sion; align and tighten pulleys; position rubber isolators.

Burner Will Not Run Continuously

(a) Lockout timing too short.	(a) Replace primary control.
(b) Poor flame due to too much air; too little oil.	(b) Check nozzle, air adjust-ment, oil pressure, and size of nozzle.

415

Symptom and Possible Cause   *Possible Remedy*

(c) Water or air in oil.   (c) Look for leak in supply.

(d) Control wired wrong.   (d) Check and rewire.

Pulsation

(a) Air adjustment.   (a) Readjust air.

(b) Pressure over fire.   (b) Correct draft to 0.02 in. W.C. negative.

(c) Dirty or improperly set electrodes.   (c) Clean and reset; wire primary control for continuous ignition.

(d) Too much oil impingement.   (d) Check nozzle and pump pressure; check nozzle size and angle and position of drawer assembly.

Short Cycling of Fan

(a) Fan control setting.   (a) Set lower turn on (115°F).

(b) Input too low.   (b) Check burner input.

(c) Temperature rise too low due to excessive speed of blower.   (c) Slow blower down and check ventilation.

Short Cycling on Limit Control

(a) Limit setting low.   (a) Reset to maximum.

(b) Input too high.   (b) Check burner input.

(c) Temperature rise too high due to blower running too slow.   (c) Increase blower speed.

(d) Temperature rise too high due to restricted returns or outlets.   (d) Open dampers or add additional outlets or returns.

(e) Fan control setting too high.   (e) Reset lower (115°F).

(f) Control out of position.   (f) Place cad cell in proper position.

Symptom and Possible Cause   *Possible Remedy*

High Fuel Consumption

(a) Input too high.
(b) Flue loss too great.

(a) Check burner input.
(b) Measure CO^2 and flue-gas temperature; if loss is more than 25%, reset air, check input, and speed up blower. Check static pressure in return and outlet plenum and correct to recommended values.

Not Heating

(a) Low input.
(b) Insufficient air circulating.

(a) Check nozzles and input.
(b) Speed up blower. Check size and location of ducts and outlets. Set fan control and blower for continuous air circulation.

CHAPTER 13

Coal-Fired Furnaces

Coal-fired furnaces burn coal or coke. Some are designed to burn coal in combination with one or more other fuels, such as gas, oil, and/or wood (see "Multifuel Furnaces" in this chapter). Both gravity and forced-warm-air furnaces are available for use with solid fuels (Figs. 13-1 and 13-2).

Coal furnaces are either hand fired or fired with a coal stoker. A stoker is an automatic coal feeding device that carries the coal from the storage bin to the furnace as needed.

The fire in a coal furnace must be kept burning throughout the heating season. The fire *must* be attended at *regular* intervals. The supply of fuel, method of feeding, and regulation are always based on weather conditions.

Solids-fuel, forced-warm-air furnaces are rated in accordance with equations provided by the National Warm Air Heating and Air Conditioning Association. Check the latest edition of their publication *Commercial Standard for Solid-Fuel-Burning Forced Air Furnaces.*

419

Courtesy Oneida Heater Co., Inc.

Fig. 13-1. Oneida coal-fired, forced-warm-air furnace.

PLANNING SUGGESTIONS

The first step in planning a heating system is to calculate the maximum heat loss for the structure. This should be done in accordance with procedures described in the manuals of the National Warm Air Heating and Air Conditioning Association or by a comparable method. This is very important because the data will be used to determine the size (capacity) of the furnace selected for the installation.

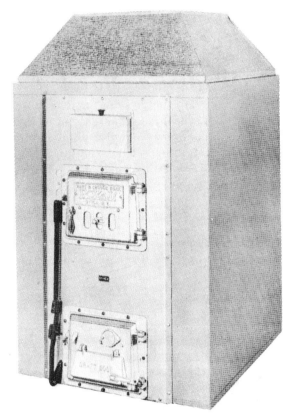

Fig. 13-2. Oneida coal-fired, gravity warm-air furnace.

Location and Clearance

A coal-fired furnace should be located a safe distance from any combustible materials. Consult the local codes and regulations for the required clearances.

Never obstruct the front of the furnace. Access must be provided to the fire and ashpit doors in order to operate the furnace.

A centralized location for the furnace usually results in the best operating characteristics because long supply ducts and the heat loss associated with them are eliminated.

421

Installation Recommendations

Make certain you have familiarized yourself with all local codes and regulations that govern the installation of coal-fired furnaces and coal stokers. Local codes and regulations take precedence over national standards.

Coal-fired furnaces are shipped disassembled with complete assembly instructions. The installer must assemble the furnace in accordance with these instructions. If the furnace is fired by a coal stoker or contains some automatic controls, the installer must connect the electrical service from the line voltage main. The low-voltage thermostat must also be connected.

Duct Connections

Detailed information concerning the installation of an air-duct system is contained in the following two publications of the National Fire Protection Association:

1. *Installation of Air Conditioning and Ventilating Systems of Other Than Residence Type* (NFPA No. 90A).
2. *Residence Type Warm Air Heating and Air Conditioning Systems* (NFPA No. 90B).

Additional information about duct connections can be found in Chapter 7 of Volume 2 (Ducts and Duct Systems). The comments in Chapter 11 (Gas-Fired Furnaces) about furnace duct connections and air distribution ducts also apply to ducts used with coal-fired, forced-warm-air furnaces.

Electrical Wiring

A wiring diagram specifying the electrical connections between the various controls should be sent by the stoker manufacturer.

As shown in Fig. 13-3, a room thermostat operating through the stoker time relay starts and stops the stoker in response to temperature conditions. The time relay operates the stoker to keep the fire alive during periods when heat may not be required by the room thermostat. A stoker time relay should include a device to shut down the stoker immediately after a shutdown call from the thermostat. This action eliminates fuel waste by preventing the overshooting of room temperature.

A high-limit control is used to protect the system against excessive temperatures. This control takes the form of an air switch in a warm-air heating system.

A snap switch should be installed in the wiring between the stoker time relay and the stoker motor to open the stoker motor circuit when the fire is being cleaned (Fig. 13-3).

A wiring diagram for a combination wood-oil furnace is shown in Fig. 13-4. Note that the wood controls and oil controls are wired as separate systems. This is a common practice on furnaces designed to burn more than one fuel.

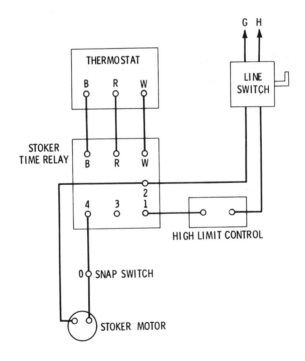

Fig. 13-3. Snap switch located between stoker time relay and stoker motor. The snap switch is used to open the stoker motor circuit when cleaning the fire.

423

Ventilation and Combustion Air

The total draft requirement of a coal-fired furnace is greater than furnaces that burn gas or oil because the chimney draft must overcome the resistance of the fuel bed.

Both a primary and secondary air supply are necessary for the combustion of solid fuels. The primary air passes through the fuel

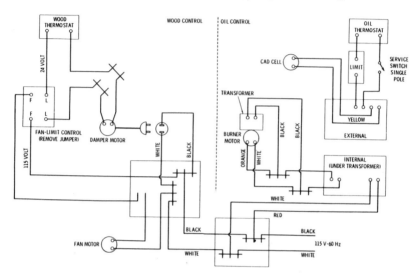

Fig. 13-4. Wiring diagram for an Oneida wood-oil furnace.

bed generally on an upward path from the ashpit. The secondary air is usually admitted through slots in the furnace fire door and passes *over* the fire to complete the combustion process (Fig. 13-5).

On residential furnaces, the secondary air slots in the fire door should be kept *closed*. There is usually enough air leakage to admit a suitable amount of secondary air without having to open the fire door. An excessive amount of secondary air (particularly after the fire has taken hold) will actually reduce the efficiency of the furnace.

The draft will determine the rate at which the fuel is burned.

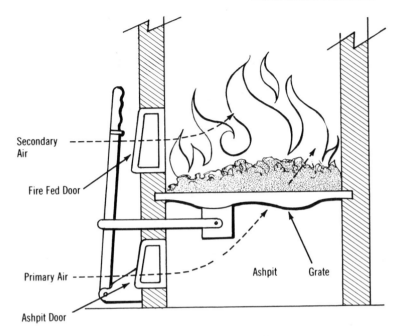

Fig. 13-5. Entry points for primary and secondary air.

The draft in furnaces using coal or coke will depend upon the following factors:

1. Type of fuel (anthracite or bituminous, coke, etc.).
2. Size of fuel.
3. Fuel-bed thickness.
4. Ash and clinker accumulation.
5. Flue resistance to the flow of gas.
6. Soot accumulation in flue.
7. Grate area.

The rate at which the fuel is burned will depend upon the heating, load demand of the specific location. The burning rate can be varied by regulating the draft. One of the most effective methods of doing this is by placing a damper in the flue outlet (Fig. 13-6). This method proportionally reduces both primary and secondary air.

The use of an ashpit damper is required for low combustion

rates. A cold, air check damper is necessary when there is excessive chimney draft.

Venting

Provisions must be made for venting the products of combustion to the outside in order to avoid contamination of the air in the living or working spaces of the structure. Masonry chimneys and low-heat Type A prefabricated chimneys are the most common methods of venting coal-fired furnaces.

Fig. 13-6. Damper placed in flue outlet.

Flue Pipe

Coal-fired furnaces require flue passages that are larger than those used with other fuels. This is due to the far greater volume of smoke, soot, and other products of combustion associated with burning solid fuels.

CHIMNEY

The chimney used with a coal-fired furnace should be of sufficient height and area to meet the requirements of the furnace. For best results, only the furnace should be connected to the chimney. Read the section "Chimneys" in Chapter 11 (Gas-Fired Furnaces).

Chimney Troubleshooting

The chimneys used with coal-fired furnaces are basically the same as those used with gas-fired and oil-fired units. Read the section "Chimney Troubleshooting" in Chapter 11 (Gas-Fired Furnaces) for a description of common chimney problems and suggested remedies.

BASIC FURNACE COMPONENTS

The basic components of a coal-fired, forced-warm-air furnace are:

1. Automatic controls.
2. Furnace grate.
3. Heat exchanger.
4. Coal stoker.
5. Blower and motor.

Each of these components is described in the sections that follow. Additional information is contained in Chapter 10 (Furnace Fundamentals) and the various chapters in which furnace controls and fuel-burning equipment are described.

Automatic Controls

The automatic controls of a coal-fired furnace which uses a stoker to feed the coal are shown in Fig. 13-7. This control system is very similar to the one used for an oil burner except for the automatic timer included with the stoker relay and transformer. During the heating season, the fire of a coal furnace must burn continually. The timer is a device designed to operate the coal stoker for a few minutes every hour or half hour in order to keep the fire alive during those periods when little heat is required.

A stack thermostat or similar device is recommended for use with stokers in areas subject to electric power failures. An electric power failure will shut down the stoker. If the shutdown period is long enough, the fire will die for lack of fuel. When the electricity comes on again, the stoker will feed coal to a cold firepot. A stack thermostat or light-sensitive electronic device will monitor the stack heat or fire and prevent the stoker operation when the fire is out.

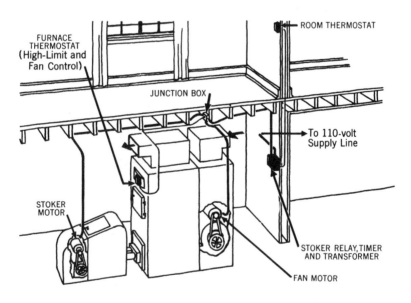

Courtesy U.S. Department of Agriculture

Fig. 13-7. Automatic controls for a coal-fired furnace fed by a stoker.

A blower fan control is generally included with most forced-warm-air furnaces. Many manufacturers will also provide a hand-operated draft control with a coal-fired furnace. Electrically operated dampers for draft control are generally available, but at extra cost.

The operation of a furnace is controlled by the room thermostat. In heating systems in which a coal-fired furnace is used, the thermostat opens or shuts dampers to increase or decrease the supply of air to the fire. When the supply of air is increased, the fire becomes hotter, and the amount of heat generated by the furnace is increased. Decreasing the supply of air to the fire has the opposite effect. Older coal-fired furnaces were not controlled by a room thermostat. Hand-operated dampers were used instead.

Furnace Grate

An essential part of any coal-fired furnace is the metal grate on which the fuel is burned (Fig. 13-8). The grate should be designed to allow sufficient primary air to pass upward through the fuel bed for the combustion process. The ashes will drop to the ashpit below through the same openings when the grate is shaken.

There should be a metal heat transfer surface (heat exchanger) of sufficient size above the fire to transfer the heat from the combustion process to the water or air in the heating system.

Coal Stoker

A coal stoker is a device used to automatically feed coal to a coal-fired furnace or boiler. See Chapter 3 of Volume 2 (Coal Firing Methods) for additional information about coal stokers.

Hand-Firing Methods

The method used to hand-fire a furnace will depend largely upon the type of solid fuel used in it. Coke and the various types of coals each have their own special hand-firing methods. Some of these firing methods are described in Chapter 3 of Volume 2 (Coal Firing Methods).

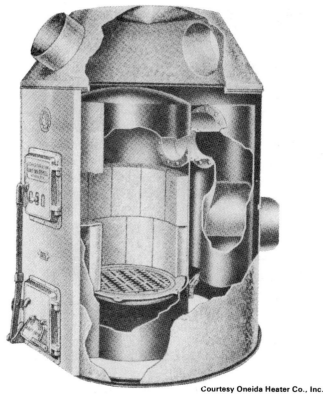

Courtesy Oneida Heater Co., Inc.

Fig. 13-8. Cutaway view of a coal-fired furnace showing grate and combination chamber details.

Blower and Motor Assembly

The blowers used on coal furnaces are the same centrifugal
types found on gas, oil, and electric forced-warm-air furnaces.
The blowers are usually installed on either side of the furnace
(Fig. 13-1) or at the back (Fig. 13-9). The latter arrangement is
especially recommended for wood-burning and multifuel furnaces. A detailed description of blowers and blower motors is
included in Chapter 11 (Gas-Fired Furnaces).

Blower Adjustment

Instructions for making blower adjustments can usually be
obtained from the furnace manufacturer. A brief description of

Courtesy XXth Century Heating & Ventilating Company

Fig. 13-9. Wood- and coal-burning furnace.

methods used to make blower adjustments is included in Chapter 11 (Gas-Fired Furnaces).

Air Filter

The number, size, and type of air filter will be recommended by the furnace manufacturer in the specifications. See also Chapter 14 of Volume 3 (Air Cleaners and Filters).

Air Conditioning

One method of adding summer air conditioning to a coal-fired heating installation is to install a separate and independent air conditioning system. This is expensive because it includes the equipment and separate ducts, but it avoids many complications.

If you are considering the idea of adding air conditioning at some future date, you should select a solid-fuel furnace capable of meeting your cooling needs. For example, Onedia All-Fuel and Two-in-One furnaces are equipped with blowers large

431

enough to handle up to 4 tons of air conditioning and electronic air cleaning.

MULTIFUEL FURNACES

Furnaces designed for use with more than one fuel are available from some manufacturers for use in residential warm-air heating systems. Both gravity and forced-warm-air units are manufactured.

The Onedia All-Fuel furnace will burn any combustible material (coal, oil, gas, or wood). It is designed to burn coal and wood, oil and wood, or gas and wood at the same time. The cast-iron grates allow the burning of anthracite coal, bituminous coal, or wood.

The furnace shown in Fig. 13-9 burns wood and bituminous or anthracite coal with a maximum output of 125,000 Btuh (95,000 Btuh with anthracite). This furnace can be used as the sole heating unit or with the existing furnace in either an in-line system (without the blower) or in a parallel system (Figs. 13-9 and 13-10).

WOOD-FIRED FURNACES

Furnaces are available that burn wood in combination with other fuels (see above) or that burn *only* wood (Figs. 13-11 and 13-12). One standard cord of good hardwood is approximately equal to the heating equivalent of 200 gal. of fuel oil or 1 ton of coal.

The burning of wood requires a furnace built specifically for this purpose. The Daniels gravity warm-air furnaces illustrated in Figs. 13-11 and 13-12 are constructed of steel with fire doors and rectangular firepots big enough to burn large chunks of wood.

The so-called chunk furnace in Fig. 13-9 has a plate grate with round holes and slotted holes. Each time the furnace is fired, the wood coals are pulled to the front of the grate and stirred up so that there is a draft up through the holes in the plate. By closing the draft door located in the large fire door and opening the draft door in the ashpit door, a draft is forced up through the wood coals to complete their burning. For general firing, the ashpit draft door is closed and the fire is regulated by the draft door in the fire (feed) door.

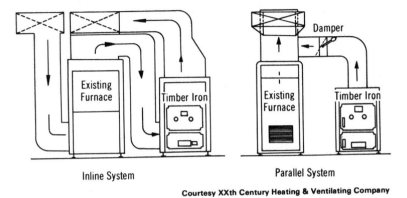

Inline System   Parallel System

Courtesy XXth Century Heating & Ventilating Company

Fig. 13-10. Inline and parallel furnace installations.

Courtesy Sam Daniels Co.

Fig. 13-11. Blower installation on a wood-fired, forced-warm-air furnace.

Forced-warm-air, wood-fired furnaces are also available for use in residential heating systems. The Daniels furnace shown in Fig. 13-13 is available in capacities ranging from 100,000 to 400,000 Btu (at bonnet). These units contain three large filters, and the air is circulated by means of a large, low-velocity fan. One advantage of this type of furnace is the provision for gravity heat. When a power failure occurs, the furnace can provide gravity heat by removing the cold-air panels.

Courtesy Sam Daniels Co.

Fig. 13-12. Wood-fired chunk furnace.

MAINTENANCE AND OPERATING INSTRUCTIONS

The following recommendations are offered as a guide to the maintenance and operation of a coal-fired furnace:

1. Carry a deep or high fire by keeping the pot full. Let coals

Fig. 13-13. Pipeless chunk furnace.

come up to the feed door and higher, and then slope back. Do not disturb the fire by frequent feeding in driblets or poking or shaking through the day. Feed, shake, or clear the grate at regular intervals.

2. When feeding coal to the furnace or shaking the fire, keep the choke damper open and the draft door closed.

3. In severe weather, give the furnace most careful attention late at night. Clear the grate until it is bright underneath, and fill the pot full.

4. In mild weather (running with checked fire), feeding the fire in the morning and at night is usually enough. In extreme weather, it may be necessary to feed more often.

5. Do not leave the feed door open in order to check the fire (in mild weather or at night). Correct adjustment of the

435

damper control will regulate the fire to supply just the right amount of heat needed.

6. Do not allow ashes to bank up under the grate in the ashpit. Grate bars are hardy, but it is possible to warp them by carelessness. Taking up ashes once a day is the best rule, even if few ashes have fallen into the pit. If convenient, spray the ashes with a water hose before taking them up.

7. Keep the heating surfaces and flues clean. Soot will reduce heating efficiency, and the furnace will require much more fuel to produce the required heat than would be the case if it were clean.

8. Do not overshake or poke the fire in mild weather. Shake it enough to make room for a little more fuel. Never poke a hard coal fire from above.

9. Clean the fire regularly especially in cold weather. This will result in fuel savings and increased heating efficiency. Do not slice or poke the fire from the top. Use the feed door for removing large clinkers. Do all cleaning through the clinker door or by shaking the dumping grate.

10. If the fire does not burn evenly over the entire grate, or if it seems sluggish in starting up after the grates have been thoroughly shaken, a thorough cleaning is required. This is accomplished by allowing the fire to burn down until a thin layer of fuel is left on the grate. Shake well and remove all the slate and clinkers left on the grate with a slice bar and hoe. To start the fire, first add a thin layer of fresh coal and feed the full charge of fuel only after this layer is burning briskly.

11. The grate will sometimes get covered with slate and clinkers massed together when a poor-quality coal is burned. The best way to correct the situation is to dump the grate and build an entirely new fire.

12. Hard clinkers lodged between the grate bars should be removed with a poker or slice bar.

TROUBLESHOOTING A COAL-FIRED FURNACE

The list that follows contains the most common operating problems associated with coal-fired furnaces. Each problem is given in the form of a symptom, the possible cause, and a suggested remedy. This list is intended to provide the operator with a quick reference to the cause and correction of a specific problem.

Symptom and Possible Cause   *Possible Remedy*

Inadequate Fire

(a) Insufficient draft.

(a) Clean ash pit and remove obstruction to primary air supply; adjust damper control.

(b) Dirty furnace.

(b) Clean furnace.

(c) Dirty flue.

(c) Clean flue.

(d) Poor-quality fuel.

(d) Replace with better-quality fuel.

(e) Grate clogged with slate and clinkers.

(e) Dump grate and rebuild fire; dislodge by poking gently with poker.

Coal Stoker Stops

(a) Power off.

(a) Check main power switch, fuses, and correct.

(b) Obstruction in feed screw.

(b) Remove obstruction.

(c) Dirty fire.

(c) Clean fire.

Stoker Motor Fails to Start

(a) No electrical power.

(a) Check main power switch, fuses, and correct.

(b) Overload.

(b) Push reset button on transmission; push reset button on stoker.

(c) Blown fuses.

(c) Replace fuses.

(d) Limit control contacts open.

(d) Let furnace cool off.

437

Symptom and Possible Cause   *Possible Remedy*

Stoker Operates Continuously

(a) Controls out of adjustment.

(a) Readjust controls or call local sales representative for service.

(b) Dirty fire.

(b) Clean fire.

(c) Fire out.

(c) Rebuild fire.

(d) Dirty furnace.

(d) Clean furnace.

Excessive Coal in Firebox

(a) Stoker feeding too much coal.

(a) Reduce coal feed rate.

(b) Insufficient air.

(b) Open manual draft.

(c) Stoker windbox full of siftings.

(c) Clean out windbox.

(d) Accumulation of clinkers in fire.

(d) Clean fire.

Excessive Fuel Use

(a) Dirty furnace.

(a) Clean furnace.

(b) Dirty flue.

(b) Clean flue.

(c) Improper draft.

(c) Adjust dampers for correct rate of combustion.

Not Enough Heat

(a) Thermostat set too low.

(a) Raise setting.

(b) Thermostat improperly located.

(b) Relocate thermostat.

(c) Thermostat out of calibration.

(c) Recalibrate or replace.

(d) Lamp or other heat source too near to the thermostat.

(d) Remove heat source.

(e) Dirty air filter.

(e) Clean or replace.

438

Symptom and Possible Cause	*Possible Remedy*
(f) Limit set too low.	(f) Reset or replace.
(g) Fan speed too slow.	(g) Check motor and fan belt, and tighten if too loose.

Blower Won't Run

(a) Power not on.	(a) Check power switch; check fuses and replace if necessary.
(b) Fan control adjustment too high.	(b) Readjustment or replace.
(c) Loose wiring.	(c) Check and tighten.
(d) Defective motor overload, protector, or motor.	(d) Replace motor.

Blower Won't Stop

(a) Manual fan on.	(a) Switch to automatic.
(b) Fan switch defective.	(b) Replace.
(c) Short in wiring.	(c) Check wiring and correct.

Rapid Fan Cycling

(a) Fan switch differential too low.	(a) Readjust or replace.
(b) Blower speed too high.	(b) Readjust to lower speed.

Noisy Blower and Motor

(a) Fan blades loose.	(a) Replace or tighten.
(b) Belt tension improper.	(b) Readjust to allow 1-in. slack.
(c) Pulleys out of alignment.	(c) Realign.
(d) Bearings dry.	(d) Lubricate.
(e) Defective belt.	(e) Replace.
(f) Belt rubbing.	(f) Reposition.

439

CHAPTER 14

Electric-Fired Furnaces

Electric heating is the only heat produced almost as fast as the thermostat calls for it. It is almost instantaneous because there are no heat exchangers to warm up. The heating elements start producing heat the moment the thermostat calls for it.

Because there is no flame with electric heat, there is no need to vent smoke or flue gases to the outside. Furthermore, there is no chimney loss with electric heat. It is 100 percent efficient, compared with an efficiency of 80 percent in furnaces using other fuels.

Electric-fired furnaces are available in upflow, downflow, or horizontal flow models, and in a wide range of sizes. For example, Carrier electric furnaces are available in 15 standard models from 5 to 35 kW in 5-kW increments. Other manufacturers offer a similar range of models (Figs. 14-1 and 14-2).

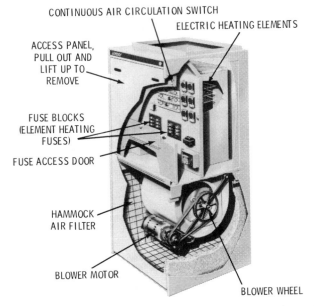

CONTINUOUS AIR CIRCULATION SWITCH

ELECTRIC HEATING ELEMENTS

ACCESS PANEL, PULL OUT AND LIFT UP TO REMOVE

FUSE BLOCKS (ELEMENT HEATING FUSES)

FUSE ACCESS DOOR

HAMMOCK AIR FILTER

BLOWER MOTOR

BLOWER WHEEL

Courtesy Lennox Air Conditioning and Heating

Fig. 14-1. Lennox model E10 upflow electric furnace.

An electric-fired furnace should be listed by the Underwriters' Laboratories for construction and operating safety. Furnaces approved by the agency are marked "UL Approved."

Electric-fired furnaces should be installed in accordance with local codes and regulations, the National Electrical Code, and recommendations made by the National Fire Protection Association.

ELECTRICAL POWER SUPPLY

Contact the local power company and make certain adequate electrical service is available for the furnace load *plus* all other appliances that will be on the line.

Check the National Electrical Code and the local code requirements. All wiring (including sizing) *must* comply with the requirements of these codes. When there is any conflict, the local

HEATING ELEMENT

FUSE BLOCK

GROUND TERMINAL

FAN SPEED
TERMINAL BOARD

LOW-VOLTAGE BOARD

FAN CAPACITOR

BLOWER

FILTER

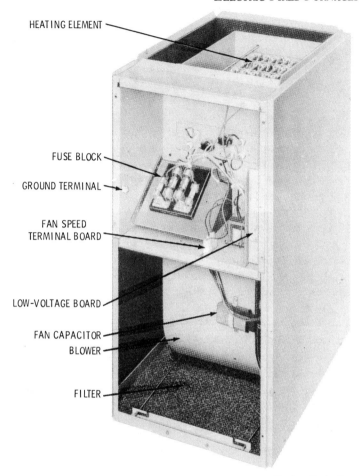

Courtesy Carrier Corp.

Fig. 14-2. Carrier forced-warm-air electric furnace.

codes and regulations take precedence. The manufacturer's requirements are also important. For example, Janitrol discourages the use of aluminum wire although it is acceptable to the National Electrical Code. Use of aluminum wire in this case could jeopardize the furnace warranty.

Be sure to consult the manufacturer's wiring diagrams for elec-

443

trical power requirements. Read the instructions carefully, and be sure you understand them thoroughly before you begin work.

PLANNING SUGGESTIONS

If an electric-fired furnace is being planned for a new structure, the maximum heat loss for each heated space must be calculated in accordance with procedures described in the manuals of the National Warm Air Heating and Air Conditioning Association or by a comparable method. This is very important because these data will be used to determine the size (capacity) of the furnace selected for the installation.

Do not consider an electric heating system unless the structure is properly insulated. This insulating will be more extensive than that used with other types of heating systems. For example, ceilings should have a *minimum* of 6 in. of blanket or loose-fill insulation, and cavities between studs in exterior walls should be completely filled with insulation. A description of the insulation requirements for a structure in which an electric heating system is used is included in Chapter 9 (Electric Heating Systems).

If the heating or heating/cooling installation is to be approved by either the FHA or VA, heat loss and heat gain calculations should be made in accordance with the procedures described in *National Environmental Systems Contractors Manual J*.

Location and Clearance

An electric-fired, forced-warm-air furnace should be located as near as possible to the center of the heat distribution system. Centralizing the furnace eliminates the need for one or more exceptionally long supply ducts. Long supply ducts are uneconomical because they are subject to a certain amount of heat loss. The number of elbows should be kept to a minimum for the same reason.

Electric furnaces are not vented because electric heat is not produced by the combustion process. No flue gases or other toxic products of the combustion process occur with electric heat. As a result, a chimney and flue pipe are not required, and it is not necessary to consider these factors when locating the furnace.

A clearance of 24 to 30 in. in front of the furnace access panel should be provided for servicing and repairs. There is no minimum clearance requirement for ductwork and combustible materials. Electric furnaces may be installed with zero clearance between the cabinet and combustible materials because the heat from the furnace is not produced by a flame.

INSTALLATION RECOMMENDATIONS

New electric furnaces for residential installation are shipped from the factory with all internal wiring completed. These furnaces are also generally shipped as a preassembled unit. In order to install the new furnace, the electric service from the line voltage main and the low-voltage thermostat must be connected. Directions for making these connections are found in the furnace manufacturer's installation instructions.

Familiarize yourself with all local codes and regulations that govern the installation of an electric-fired furnace. Local codes and regulations take precedence over national standards.

Check the insulation of the structure to determine if it is properly insulated for electric heat. The insulation should be installed in accordance with recommendations in *All Weather Comfort Standard of Electrically Heated and Air Conditioned Homes* (Electric Heating Association).

The furnace should be mounted on a level surface. If the unit is not level, it may develop serious vibrations. An insulating material can be placed under the furnace in most installations to reduce sound vibrations when the unit is operating. A noncombustible base is recommended for counterflow models.

Duct Connections

Detailed information concerning the installation of an air-duct system is contained in the following two publications of the National Fire Protection Association:

1. *Installation of Air Conditioning and Ventilating Systems of Other Than Residence Type* (NFPA No. 90A).

2. *Residence Type Warm Air Heating and Air Conditioning Systems* (NFPA No. 90B).

Additional information about duct connections can be found in Chapter 7 of Volume 2 (Ducts and Duct Systems). The comments made in Chapter 11 (Gas-Fired Furnaces) about furnace duct connections and air distribution ducts apply for the most part to ducts used with electric-fired, forced-warm-air furnaces.

BASIC COMPONENTS

An electric-fired, forced-warm-air furnace will generally consist of the following basic components (Fig. 14-3):

1. Automatic controls.
2. Heating elements.
3. Safety controls.
4. Blowers and motors.
5. Air filter(s).

Each of these components is described in the sections that follow. Additional information is contained in Chapter 10 (Furnace Fundamentals) and the various chapters in which furnace controls are described.

AUTOMATIC CONTROLS

The automatic controls used in an electric heating system are designed to ensure its safe and efficient operation. Detailed descriptions of these controls are found in Chapter 4 of Volume 2 (Thermostats and Humidistats). This section is primarily concerned with outlining the operating principles of the automatic controls used with an electric furnace. These controls include:

1. Room thermostat.
2. Thermostat heat anticipator.
3. Timing sequences.

In a central heating system, the wall-mounted room thermostat is the control that governs the normal operation of the furnace.

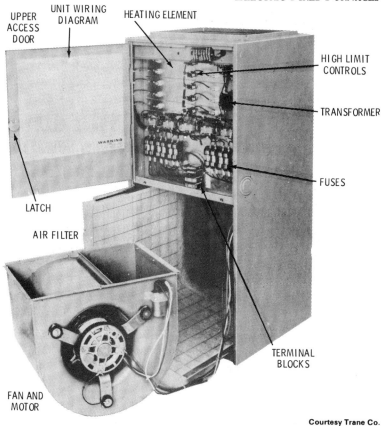

UPPER ACCESS DOOR

UNIT WIRING DIAGRAM

HEATING ELEMENT

HIGH LIMIT CONTROLS

TRANSFORMER

FUSES

LATCH

AIR FILTER

TERMINAL BLOCKS

FAN AND MOTOR

Courtesy Trane Co.

Fig. 14-3. Principal components of an electrical furnace.

The operating principle is simple. The temperature selector on the thermostat is set for the desired temperature. When the temperature in the room falls below this setting, the thermostat will call for heat and cause the first heating circuit in the furnace to be turned on. There is generally a delay of about 15 seconds before the furnace blower starts. This prevents the blower from circulating cool air in the winter. After about 30 seconds, the second heating circuit is turned on. The other circuits are turned on one by one in timed sequence.

When the temperature reaches the required level, the thermostat opens. After a short time, the first heating circuit is shut off.

447

The others are shut off one by one in timed sequence. The blower will continue to operate until the air temperature in the furnace drops below a specified temperature.

A typical room thermostat will have a fan switch, a system switch, and a temperature selector (Fig. 14-4). The temperature selector (a dial or lever device) on the thermostat is used to select the desired temperature. The actual operation of the heating system is governed by the positions of the fan and system switches. The switch positions and their functions are listed in Tables 14-1 and 14-2.

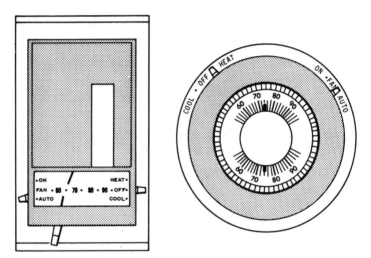

Fig. 14-4. Typical room thermostats used in central electric heating systems.

Most room thermostats contain a *heat anticipator*. This is a device designed to assist the thermostat in controlling closer to the desired temperature range (Fig. 14-5).

When timing sequences are used (see below), the current flowing through the first time-delay sequencer (relay) must also flow through the heat-anticipator. In order to obtain satisfactory operation, the heat-anticipator setting must be equal to the current draw of the sequencer.

The furnace manufacturer will generally recommend the set-

Table 14-1. Thermostat Operation in Single-Stage Heating/Two-Stage Cooling; Two-Stage Heating/Single-Stage Cooling; Two-Stage Heating/Two-Stage Cooling

Thermostat Switch Setting		Function
Fan	**System**	
Auto	Off	System completely shut down.
On	Off	Blower only, continuous operation. Provides air circulation when no cooling or heating is desired.
Auto	Cool	Blower and cooling system cycle on and off as thermostat demands.
On	Cool	Blower runs continuously; cooling system cycles on and off as thermostat demands.
Auto	Heat	Blower and furnace cycle on and off as thermostat demands.
On	Heat	Blower runs continuously; furnace cycles on and off as thermostat demands.
Auto	Auto	Unit cycles for either cooling or heating as per demand of thermostat.
On	Auto	Blower runs continuously; cooling and heating cycle as per demand of thermostat.

Courtesy Fedders Corporation

ting for the heat-anticipator adjustment for each size unit. For example, the setting recommended for a Trane Model EUADH 07 electric furnace is 0.45. This thermostat adjustment will vary depending upon the type of time-delay sequencer used, the furnace manufacturer, and the size of the furnace. This may be illustrated by the recommended heat-anticipator settings given by Coleman for its 10-kW, 15-kW, and 20-kW furnace models (Table 14-3). All Coleman 25-kW models require a heat-anticipator setting of 0.60.

After you have adjusted the heat anticipator to the suggested setting, operate the furnace several hours and observe the results. If there is *insufficient* heat, it may be caused by short furnace cycles. This can be corrected by moving the heat-anticipator pointer to a slightly higher setting. If there is too much heat, then

449

Table 14-2. Thermostat Operation in Single-Stage Heating/Cooling

Thermostat Switch Setting		Function
Fan	System	
Auto	Off	System completely shut down.
On	Off	Blower only, continuous operation. Provides air circulation when no cooling or heating is desired.
Auto	Cool	Blower and cooling system cycle on and off as thermostat demands.
On	Cool	Blower runs continuously; cooling system cycles on and off as thermostat demands.
Auto	Heat	Blower and furnace cycle on and off as thermostat demands.
Heat	On	Blower runs continuously; furnace cycles on and off as thermostat demands.

Courtesy Fedders Corporation

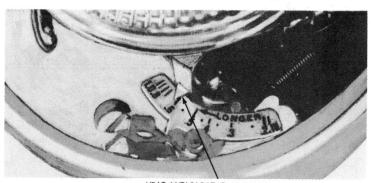

HEAT ANTICIPATOR ADJUSTMENT

Courtesy Trane Co.

Fig. 14-5. Thermostat heat anticipator.

long furnace cycles are overheating the structure. This can be corrected by moving the heat-anticipator pointer to a slightly lower setting. After making these thermostat adjustments, allow the furnace to operate several hours to determine if further adjustment will be required. Additional information about the

Table 14-3. Recommended Heat-Anticipator Settings for Coleman 10-, 15-, and 20-kW Electric Furnaces

Sequencer	10 kW	15 kW	20 kW
White-Rogers	0.15	0.15	0.10
Klixon	0.15	0.30	
Honeywell	0.40	0.40	

Courtesy The Coleman Company, Inc.

thermostat heat anticipator is contained in Chapter 11 (Gas-Fired Furnaces).

HEATING ELEMENTS

A heating element must operate at a temperature well above its surroundings to deliver heat at a useful high rate. If it is a radiant heating element, it may operate red hot or nearly white hot over a relatively long period of time.

Wires with a high resistance, such as iron, chromium, nickel, manganese, and their alloys, are commonly used for heating elements. The heat output of a wire can be varied by changing its material composition, or by changing its size or diameter. A smaller-diameter wire will have a higher resistance than a larger-diameter wire.

The heating elements in an electric furnace are resistance coils made from high-temperature, chrome nickel, heat-generating wire. Most manufacturers design their furnaces so that the entire heating element assembly can be removed for easy maintenance or repair (Fig. 14-6).

Open elements are used in noncentral heating units, such as radiant or convective heaters, where it is desirable for the element to operate at a relatively high temperature. These elements reach operating temperatures very rapidly when energized. The exact operating temperature depends on the material used in the element and the type of heat desired. A radiant heater, for example, would probably operate at higher temperatures than would a wall-mounted convective heater.

Another type of element used in noncentral heating is the

451

HEATING ELEMENT SUPPORT ROD

HEATING
ELEMENTS

FAN CONTROL   HIGH LIMIT CONTROL

Fig. 14-6. Heating elements and controls.

encapsulated, or completely enclosed, element. The simplest form of encapsulated element is ceiling cable, which has a layer of plastic insulation over it that can withstand the heat. Ceiling cable is designed to operate at very low temperatures.

A more complicated enclosed element is that used in baseboard convectors. An outer sheath of ceramic or metal protects the resistance wire from damage, corrosion, or deterioration. The heat-dissipating fins add surface area to increase the rate of heat transfer to the surrounding medium (air or water).

TIMING SEQUENCES

When all the heating elements in an electric furnace turn on simultaneously, there is a momentarily excessive demand on the power supply. The result is a temporary interruption in the electric service. This problem of power drains and surges can be eliminated by timing the heating elements so that they start one at a time in predetermined increments. Sequencing relays are used for this purpose.

SAFETY CONTROLS

Most electric-fired furnaces are equipped with a variety of different safety controls to protect the appliance against current overloading or excessive operating temperatures. These safety controls are:

1. Temperature limit controls.
2. Secondary high-limit control.
3. Furnace fuses.
4. Circuit breakers.
5. Control voltage transformer.
6. Thermal overload protector.

Temperature-Limit Controls

In the line control diagram of a Trane EUADH 07A model electric furnace shown in Fig. 14-7, each heating element is shown with a high-limit control device located between the heating element and the time-delay relay. These high-limit control devices are designed to limit the outlet air temperature on Trane electric furnaces to 200°F. If the temperature of the outlet air should exceed 200°F, the high-limit control will open and interrupt the supply of electric power to the heating element.

Secondary High-Limit Control

Secondary high-limit protection is provided by a fusible link in each electric heating element (Fig. 14-8). This device is designed to shut off the current when temperatures in the furnace become excessive. It functions as a backup system in case of limit switch failure.

Furnace Fuses

Furnace fuses are used to provide protection against possible overload conditions and to ensure correct, safe field wiring. Each

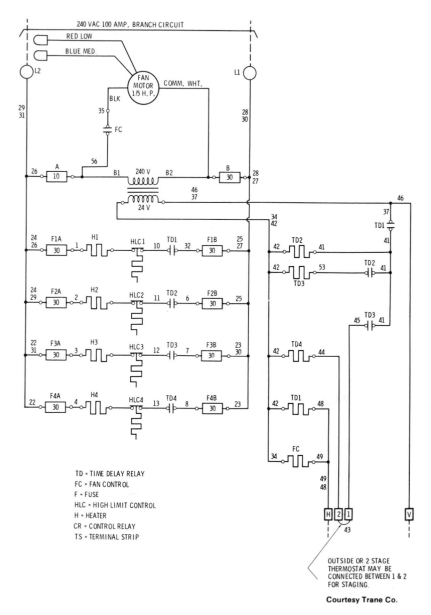

TD = TIME DELAY RELAY
FC = FAN CONTROL
F = FUSE
HLC = HIGH LIMIT CONTROL
H = HEATER
CR = CONTROL RELAY
TS = TERMINAL STRIP

OUTSIDE OR 2 STAGE
THERMOSTAT MAY BE
CONNECTED BETWEEN 1 & 2
FOR STAGING.

Courtesy Trane Co.

Fig. 14-7. Line-control wiring diagram of a Trane electric furnace.

heating element circuit is protected by two branch fuses as shown in Fig. 14-9. These are sized to limit the current draw of each heater and are designed to open on a short-circuit or an overloaded-circuit condition. For overload protection, the blower motor and relay are also safeguarded with a separate fuse.

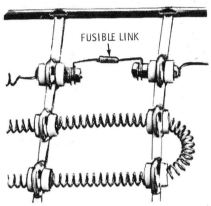

FUSIBLE LINK

Courtesy International Heating and Air Conditioning Corp.

Fig. 14-8. Fusible link connection in electric heating element.

Circuit Breakers

Some electric furnaces are equipped with circuit breakers and a terminal board. The wiring diagram for the Coleman 25-kW electric furnace illustrated in Fig. 14-10 indicates that both the terminal board and circuit breakers are located in the power supply feed line to the furnace.

When it is desirable to run branch circuits to the circuit breakers (bypassing the terminal board), the jumper wires between the circuit breakers and the wiring are connected as shown in Figs. 14-11 and 14-12.

Transformer

A control voltage transformer is used to limit the amount of output current. Limiting the amount of output current permits the use of open control wiring.

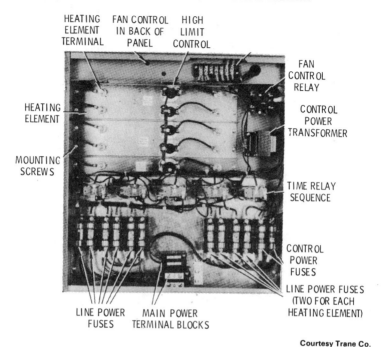

HEATING ELEMENT TERMINAL

FAN CONTROL IN BACK OF PANEL

HIGH LIMIT CONTROL

FAN CONTROL RELAY

HEATING ELEMENT

CONTROL POWER TRANSFORMER

MOUNTING SCREWS

TIME RELAY SEQUENCE

CONTROL POWER FUSES

LINE POWER FUSES (TWO FOR EACH HEATING ELEMENT)

LINE POWER FUSES

MAIN POWER TERMINAL BLOCKS

Courtesy Trane Co.

Fig. 14-9. Electric furnace heating-section control panel showing location of line power fuses.

Thermal Overload Protector

The fan motor is protected against locked rotor or overheated conditions by thermal overload devices. When these conditions occur, the fan motor circuit is automatically opened and the motor is shut off.

ELECTRICAL WIRING

All internal furnace wiring is done at the factory before it is shipped. At the site, the following two types of electrical connections are required to field wire the unit:

1. Line voltage field wiring.
2. Control voltage field wiring.

456

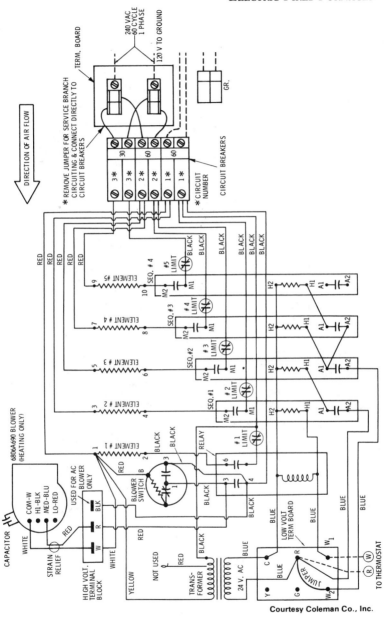

Courtesy Coleman Co., Inc.

Fig. 14-10. Wiring diagram for the Coleman 25-kW electric furnace.

457

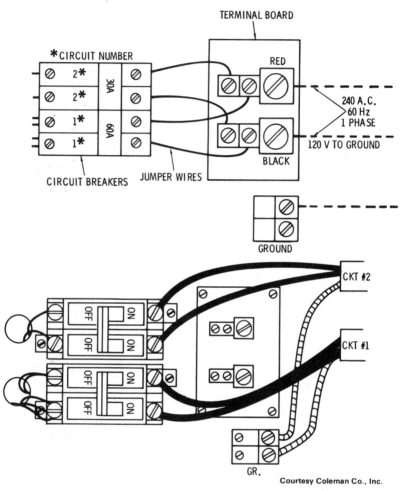

Fig. 14-11. Electrical wiring for a Coleman 15- or 20-kW electric furnace with circuit breakers.

A typical example of line voltage field wiring is shown in Fig. 14-13. Line voltage wiring involves the connection of the furnace to the building power supply. Line voltage wiring runs directly from the building power panel to a fused disconnect switch. From there, the wiring runs to terminals L_1 and L_2 on the power-supply terminal block.

458

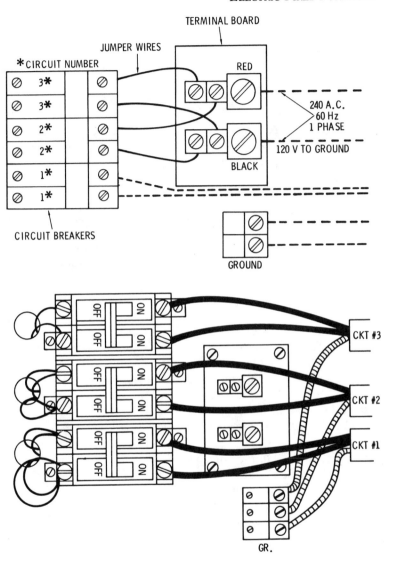

Fig. 14-12. Electric wiring for a Coleman 25-kW electric furnace with circuit breaker.

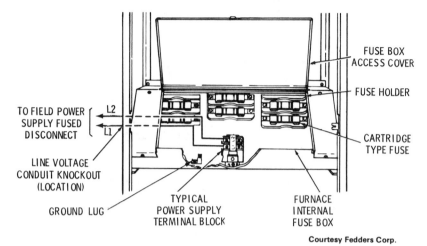

Courtesy Fedders Corp.

Fig. 14-13. Typical line voltage field wiring shown for a Fedders electric furnace.

The unit must be properly grounded either by attaching a grounded conduit for the supply conductors (knockouts in the side panels of Fedders electric furnaces are provided for this purpose) or by connecting a separate wire from the furnace ground lug to a suitable ground.

The external control voltage circuitry consists of the wiring between the thermostat and the low-voltage terminal block located in the control voltage section of the furnace. Instructions for control voltage wiring are generally shipped with the thermostat.

Some typical examples of control voltage field wiring used with Fedders electric furnaces are shown in Figs. 14-14 and 14-15. Control voltage field wiring connections used with Coleman electric furnaces are shown in Fig. 14-16.

The National Electric Code requires that furnaces larger than 10 kW be supplied with branch circuit fusing. Power connections on units of this size are usually made to lugs on the fuse blocks.

BLOWERS AND MOTORS

The blowers and motors used with electric-fired, forced-warm-air furnaces are identical to those used in gas-fired furnaces. Read the section "Blowers and Motors" in Chapter 11 (Gas-Fired Furnaces) for additional information.

AIR DELIVERY AND BLOWER ADJUSTMENT

It is sometimes necessary to adjust the blower speed to produce a temperature rise through the furnace that falls within the limits stamped on the furnace nameplate. Blower adjustment procedures are described in full detail in Chapter 11 (Gas-Fired Furnaces).

AIR FILTER

The air filters used in electric-fired, forced-warm-air furnaces are either permanent types that can be removed and cleaned on a periodic basis or replaceable, throwaway filters.

More detailed information concerning furnace air filters is contained in Chapter 14 of Volume 3 (Air Cleaners and Filters). Also read the section on air filters in Chapter 11 (Gas-Fired Furnaces). The filters used in the various types of forced-warm-air furnaces are essentially the same regardless of the fuel used to fire the unit.

AIR CONDITIONING

A furnace should be installed in parallel or on the upstream side of the cooling unit to avoid condensation in the heating section. Parallel installation will require dampers or some other means to prevent cool air from entering the furnace.

The air conditioning component of a typical electric heating and cooling system generally consists of an outdoor condensing unit, indoor coils, and a cabinet to house the cooling coils. Most furnace manufacturers provide detailed instructions for adding air conditioning to the heating unit. The important thing to remember is to size the ducts for the larger volume of air used in air conditioning.

Additional information about air conditioning can be found in Chapters 7 and 8 of Volume 3.

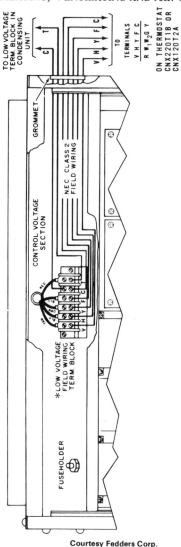

Fig. 14-14. Control voltage wiring for two-stage heating, single-stage cooling applications.

Courtesy Fedders Corp.

MAINTENANCE AND OPERATING INSTRUCTIONS

Maintenance and operating instructions will normally be provided for the furnace by the manufacturer. If no instructions are

462

available, try contacting a field representative or writing the company for a duplicate copy of the owner's manual.

Always shut off the electrical power supply to the furnace before attempting to service it. This is very important because there is the possibility of a fatal electric shock. Be sure to open all furnace fused disconnect switches before servicing.

A clean air filter is important to the efficient operation of the

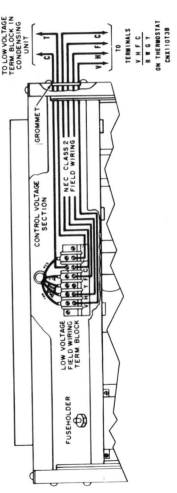

Fig. 14-15. Control voltage wiring for single-stage heating, single-stage cooling applications.

Courtesy Fedders Corp.

463

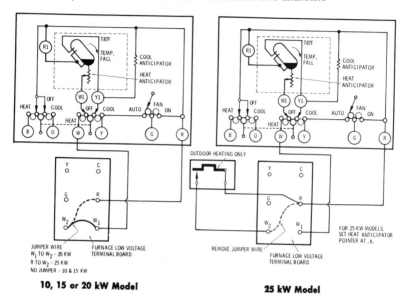

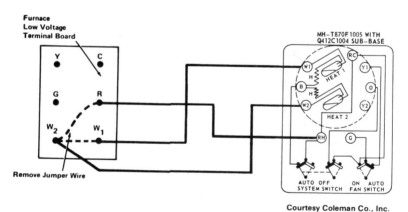

Courtesy Coleman Co., Inc.

Fig. 14-16. Three connections for indoor and outdoor heating (only) thermostats for a Coleman model 6806 electric furnace.

furnace. Air filters should be removed and inspected on a periodic basis (preferably every six months). Clean or replace the filters as required.

Inspect the heating element and heating control wiring to make certain connections are tight and clean. Check for any breaks or cracks in the wire insulation. These can cause shorting. All terminal block wiring connections should also be tight and clean.

Check the furnace wiring diagram to make certain the fuses are of the correct type and amperage.

Some fan motors are permanently lubricated and will not require further attention. Others have ports for oiling and require periodic lubrication. See "Blowers and Motors" in Chapter 11 (Gas-Fired Furnaces).

A combined heating and cooling system is designed to operate year-round without being shut down. The only changeover required is that of the room thermostat and periodic maintenance.

When an electric furnace is installed *without* an air conditioning unit, the furnace should be shut down at the end of the heating season. The procedure for doing this will be found in the furnace owner's manual. It is a simple operation that generally consists of opening the main fused disconnect switch (or switches) in the power-supply lines serving the furnace.

TROUBLESHOOTING AN ELECTRIC FURNACE

Any appliance may sometimes fail to operate efficiently because of a malfunction somewhere in the equipment. The problems most commonly associated with electric-fired furnaces are listed below.

Symptom and Possible Cause   *Possible Remedy*

Unit Fails to Operate

(a) Defective furnace transformer.	(a) Replace.
(b) Blown or defective transformer fuse.	(b) Replace fuse.
(c) Defective thermostat.	(c) Replace thermostat.
(d) Improperly set room.	(d) Change thermostat to proper setting.

Symptom and Possible Cause   *Possible Remedy*

(e) Open fused disconnect switch.

(e) Correct problem and close furnace circuit.

(f) Blown or faulty furnace fuse.

(f) Replace fuse.

Fan Operates with Low or No Heat

(a) Blown or faulty heater element fuse.

(a) Replace fuse.

(b) Defective time-delay sequencer.

(b) Replace sequencer.

Heats without Fan Operation

(a) Faulty fan-control relay.
(b) Defective fan motor.
(c) Faulty fan motor wiring or loose connections.
(d) Defective fan motor run capacitor.

(a) Replace.
(b) Repair or replace.
(c) Repair or replace wiring.
(d) Repair or replace.

Individual Heater Fails to Operate

(a) Blown or faulty heater circuit fuse.

(a) Replace fuse.

(b) Defective high-limit control.

(b) Replace.

(c) Defective time-delay sequencer.

(c) Replace.

(d) Faulty heater element.

(d) Replace.

Fan Operates on Heating, Not on Cooling

(a) Defective cooling-cycle control relay.

(a) Replace.

Symptom and Possible Cause	*Possible Remedy*
(b) Improperly connected or faulty room thermostat.	(b) Make proper connections to thermostat.
(c) Defective or improper fan motor connections.	(c) Repair or replace.

CHAPTER 15

Boilers and Boiler Fittings

A boiler is a device used to supply steam or hot water for heating, processing, or power purposes. This chapter is primarily concerned with a description of the low-pressure steam and hot-water boilers used in the heating systems of residences and small buildings.

The basic construction of both low-pressure steam and hot-water, space heating boilers fired by fossil fuels consists of: (1) an insulated steel jacket enclosing a lower chamber in which the combustion process takes place; and (2) an upper chamber containing cast-iron sections or steel tubes in which water is heated or converted to steam for circulation through the pipes of the heating system (Figs. 15-1 to 15-4).

The design and construction of the lower chamber depends upon the type of fuel used to fire the boiler. It serves as a combustion chamber for coal- and oil-fired boilers and as a com-

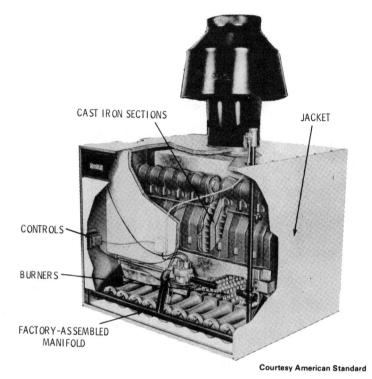

CAST IRON SECTIONS

JACKET

CONTROLS

BURNERS

FACTORY-ASSEMBLED
MANIFOLD

Courtesy American Standard

Fig. 15-1. Gas-fired boiler for a forced-hot-water heating system.

partment for housing the gas burner assembly on gas-fired boil-ers. These gas burner assemblies are generally designed for easy removal so that they can be periodically cleaned or serviced (Fig. 15-5).

Oil burners are externally mounted with the burner nozzle extending into the combustion chamber. This is also true of gas conversion burners. Gas burner assemblies, on the other hand, are located inside the lower chamber of the boiler.

The cast-iron sections or steel tubes in the upper chamber of the boiler contain water that circulates through the pipes in the heating system either in the form of steam or hot water. The heat from the combustion process in the lower chamber of the boiler is transferred through the metal surface of the cast-iron sections or steel tubes to the water contained in them, causing a rise in

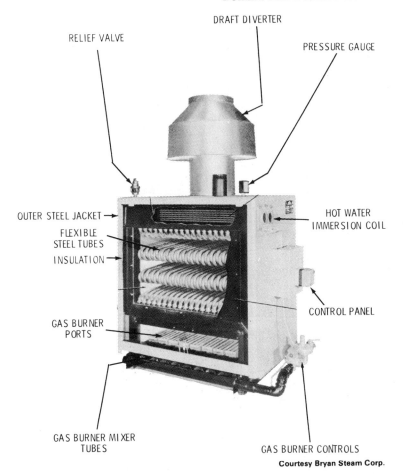

RELIEF VALVE

DRAFT DIVERTER

PRESSURE GAUGE

OUTER STEEL JACKET

FLEXIBLE STEEL TUBES

INSULATION

HOT WATER IMMERSION COIL

GAS BURNER PORTS

CONTROL PANEL

GAS BURNER MIXER TUBES

GAS BURNER CONTROLS

Courtesy Bryan Steam Corp.

Fig. 15-2. Basic components of a Bryan steel-tube, gas-fired boiler used in hot-water heating systems.

temperature. The amount of water contained in these passages is one of the ways in which steam and hot-water, space heating boilers are distinguished from one another. In hot-water, space heating boilers, these passages are completely filled with water; whereas in low-pressure steam boilers only the lower two-thirds are filled. In the latter, the water is heated very rapidly, causing steam to form in the upper one-third. The steam, under pressure,

471

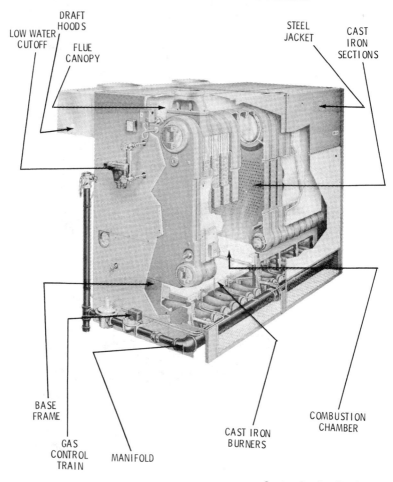

DRAFT
HOODS

LOW WATER
CUTOFF

FLUE
CANOPY

STEEL
JACKET

CAST
IRON
SECTIONS

BASE
FRAME

GAS
CONTROL
TRAIN

MANIFOLD

CAST IRON
BURNERS

COMBUSTION
CHAMBER

Fig. 15-3. Gas-fired-steam-heating boiler.

rises through the supply pipes connected to the top section of the boiler.

A boiler jacket contains a number of different openings for pipe connections and the mounting of accessories. The number and type of openings on a specific boiler jacket depends upon the type of boiler (e.g., steam or hot water). Among the different openings to be found on a boiler jacket are the flue connection,

472

water feed (supply) connection, inspection and cleanout tapping, blowdown tapping, relief valve tapping, control tapping, drain tapping, expansion tank tapping, and return tapping. There are also gas and oil burner connections. Fig. 15-6 illustrates the arrangement of control tappings for American Standard gas-fired steam and hot-water boilers.

Most (but not all) of the controls on low-pressure steam and hot-water space heating boilers fired by the *same* fuel are similar in design and function; there are exceptions. For example, a few

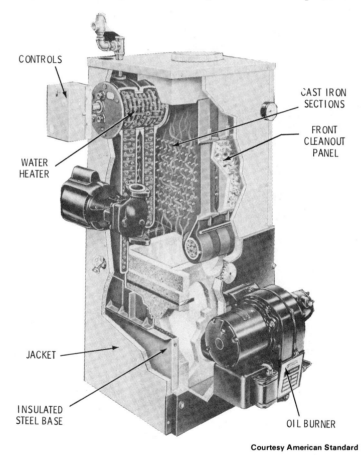

CONTROLS

CAST IRON SECTIONS

FRONT CLEANOUT PANEL

WATER HEATER

JACKET

INSULATED STEEL BASE

OIL BURNER

Fig. 15-4. Oil-fired boiler for a forced-hot-water heating system.

473

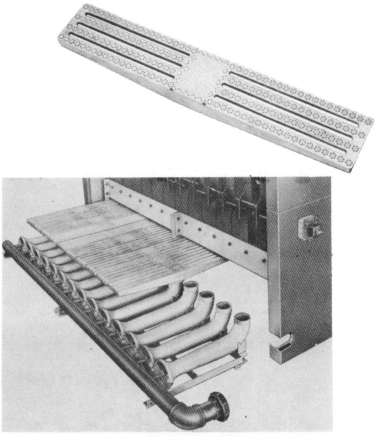

Fig. 15-5. Gas burner boiler heads and support frames. Support frames are mounted on casters to facilitate removal of burner for cleaning.

474

location on boiler	size (inches)	steam boilers	water boilers
A	6	supply	supply
B	4	return	return
C	3	safety valve	relief valve
D	1¼	blow-off	not required
E	1	water feeder	not required
F	¾	pressure limit	temperature limit
G	¾	drain	drain
H	½	pressure gauge	temperature and altitude gauge
J	½	gauge glass and low water cut-off	not required
K ■	⅜	try-cocks	not required

■ tappings available on special order only.

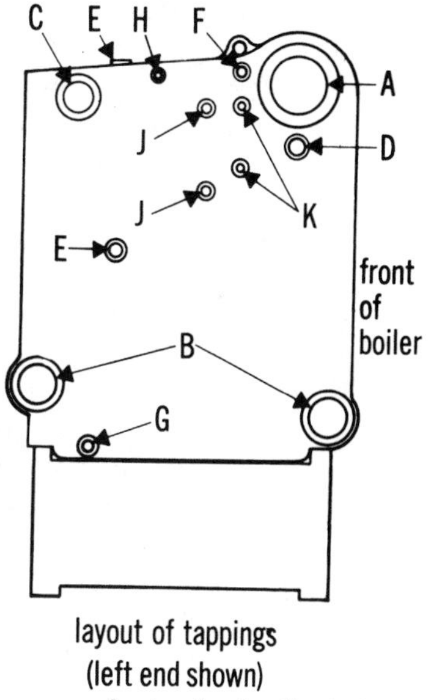

Fig. 15-6. Arrangement of control tappings.

layout of tappings
(left end shown)

Courtesy American Standard

boiler controls and fittings are designed to be specifically used on steam boilers; others are found only on hot-water, space heating boilers. The various boiler controls and fittings are described in the appropriate sections of this chapter.

475

BOILER RATING METHODS

The construction of low-pressure steel and cast-iron heating boilers is governed by the requirements of the ASME Boiler and Pressure Vessel Code. This is a nationally recognized code used by boiler manufacturers, and any boiler used in a heating installation should clearly display the ASME stamp. State and local codes are usually patterned after the ASME Code.

The location of the identification symbols used by the ASME is specified by the code and determined by the type of boiler. For example, on a water-tube boiler, it appears on a head of the steam-outlet drum near and above the manhole opening. On vertical fire-tube boilers, the stamp bearing the identification symbol should appear on the shell above the fire door and handhole opening. Other types of boilers (e.g., scotch marine and superheaters) have their own specified location for the identification symbol stamp.

The ASME Boiler and Pressure Vessel Code applies only to boiler construction, specifically to maximum allowable working pressures, not to its heating capacity. A number of different methods are used to rate the heating or operating capacity of a boiler. The boiler manufacturers have developed their own ratings, but these are generally used along with rating methods available from the following organizations:

1. Institute of Boiler and Radiator Manufacturers (IBR).
2. Steel Boiler Institute (SBI).
3. American Gas Association (AGA).
4. Mechanical Contractors Association of America (MCAA).

The Institute of Boiler and Radiator Manufacturers has adopted a code for rating most types of sectional cast-iron boilers. Steel boilers are rated by the Steel Boiler Institute rating code. The Mechanical Contractors Association of America has devised a method for rating boilers not covered by either the SBI or IBR Codes. Finally, gas-fired boilers are rated in accordance with methods developed by the American Gas Association.

Other organizations that rate boilers include the Packaged Fire-tube Branch of the American Boiler and Affiliated Industries (ratings for modified scotch marine boilers) and numerous international groups.

In terms of its heating capacity, the rating of a boiler can be expressed in square feet of equivalent direct radiation (EDR) or thousands of Btuh. Sometimes a boiler horsepower rating is also given, but this has proven to be misleading.

For steam boilers, 1 sq. ft. of equivalent direct radiation (EDR) is equal to the emission of 240 Btuh. For a water boiler, 1 sq. ft. EDR is considered equal to the emission of 150 Btuh.

A boiler horsepower (bhp) is the evaporation of 34.5 lb. of water into dry steam from and at 212°F. For rating purposes, 1 bhp is considered as the heat equivalent of 140 sq. ft. of steam radiation per hour. In some cases bhp ratings are obtained by dividing steam SBI ratings by 140.

A boiler is rated according to its operating or heating capacity, but this rating will vary in accordance with the *type* of load used as the basis for the rating. The three types of connected loads used to determine the rating of a boiler are:

1. Net load.
2. Design load.
3. Gross load.

Net load refers to the actual connected load of the heat-emitting units in the steam or hot-water heating system. *Design load* includes the net-load rating plus an allowance for piping heat loss. Finally, *gross load* will equal the net load and the piping heat loss, plus an additional allowance for the pickup load.

BOILER HEATING SURFACE

The boiler *heating surface* (expressed in square feet) is that portion of the surface of the heat transfer apparatus in contact with the fluid being heated on one side and the gas or refractory being cooled on the other side. The *direct* or *radiant* heating surface is the surface against which the fire strikes. The surface that comes in contact with the hot gases is called the *indirect* or *convection* surface.

The *heating capacity* of any boiler is influenced by the amount and arrangement of the heating surface and the temperatures on either side. The arrangement of the heating surface refers to the ratio of the diameter of each passage to its length, as well as its contour (straight or curved), cross-sectional shape, number of passes, and other design variables.

BOILER EFFICIENCY

The *boiler efficiency* is the ratio of the heat output to the calorific value of the fuel. Boiler efficiency is determined by a number of different factors, including the type of fuel used, the method of firing, and the control settings. For example, oil- and gas-fired boilers have boiler efficiencies ranging from 70 to 80 percent. A hand-fired boiler in which anthracite coal is used will have a boiler efficiency of 60 to 75 percent.

TYPES OF BOILERS

The boilers used in low-pressure, steam and hot-water space heating systems can be classified in a number of different ways. Some of the criteria used in classifying them will include:

1. Construction material.
2. Construction design.
3. Boiler position.
4. Number of passes of the hot gases.
5. Length of travel of the hot gases.
6. Type of heating surface.
7. Type of fuel used.

Most boilers are either cast iron or steel construction. *Very few* are constructed from nonferrous materials. Cast-iron boilers generally display a greater resistance to the corrosive effects of water than steel ones do, but the degree of corrosion in steel boilers can be significantly reduced by chemically treating the water.

Cast-iron boilers can be classified on the basis of their construction design as either round boilers with horizontal pancake sections or square (or rectangular) boilers with either horizontal or vertical sections. A coal-fired, round, vertical boiler commonly used in older heating installations is shown in Fig. 15-7.

A square-shaped boiler with horizontal section design is shown in Fig. 15-8. The heating surface of each cast-iron section is

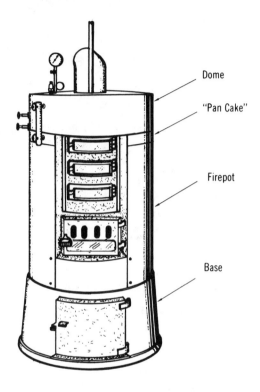

Dome

"Pan Cake"

Firepot

Base

Fig. 15-7. Coal-fired round vertical boiler.

exposed at right angles to the rising flue gases (Fig. 15-9). The water travels in a zigzag path from section to section in a manner similar to the flow of water in a steel tube boiler (Fig. 15-10).

Vertical cast-iron sections are available in a number of different designs, depending upon the manufacturer (Fig. 15-11). A typical arrangement of these sections is shown in Fig. 15-12.

Steel boilers may be classified with respect to the relative position of water and hot gases in the tubular heating surface. In

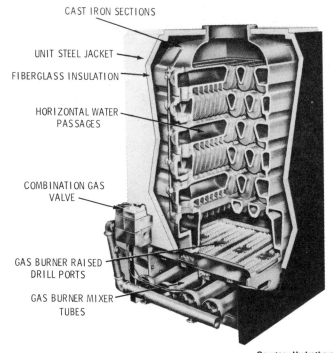

CAST IRON SECTIONS

UNIT STEEL JACKET

FIBERGLASS INSULATION

HORIZONTAL WATER PASSAGES

COMBINATION GAS VALVE

GAS BURNER RAISED DRILL PORTS

GAS BURNER MIXER TUBES

Courtesy Hydrotherm, Inc.

Fig. 15-8. Horizontal section design of a Hydrotherm gas-fired, hydronic, cast-iron boiler.

fire-tube boilers, for example, the hot gases pass within the boiler tubes and the water necessary to produce the steam that circulates around them. In water-tube boilers, the reverse is true.

Flexible steel tubes are used in Bryan steam and hot-water, space heating boilers for the rapid circulation of the water around the heat rising from the fire (Figs. 15-13 and 15-14).

Boilers can also be classified according to the number of passes made by the hot gases (e.g., one pass, two passes, three passes) (Fig. 15-15).

The *length* of travel of the hot gases is another method used for classifying boilers. The efficiency of a boiler heating surface depends, in part, upon the ratio of the cross-sectional area of the passage to its length.

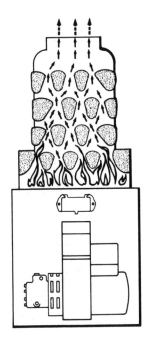

Courtesy Hydrotherm, Inc.

Fig. 15-9. Direction of heat travel.

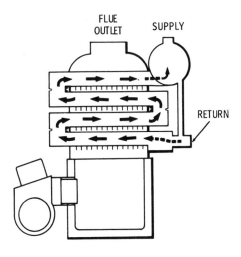

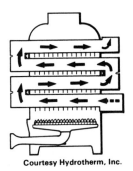

Courtesy Hydrotherm, Inc.

Fig. 15-10. Direction of water flow.

481

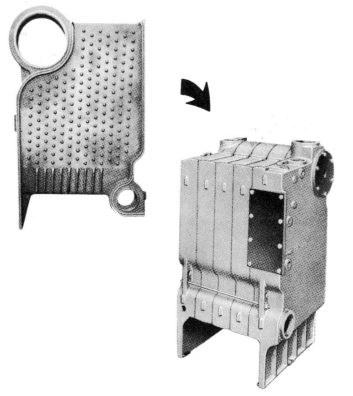

Fig. 15-11. Vertical cast-iron section.

Among the various fuels used to fire boilers are oil, gas (natural and propane), coal, and coke. Conversion kits for converting a boiler from one gas to another are available from some manufacturers. Changing from coal (or coke) to oil or gas can be accomplished by using conversion chambers and making certain of other modifications (see Chapter 16, Boiler and Furnace Conversions).

Electricity can also be used to fire boilers. One advantage in using electric-fired boilers is that the draft provisions required by boilers using combustible fuels is not necessary.

The classification criteria described above are selective and limited to the more common types in use. Considering the multi-

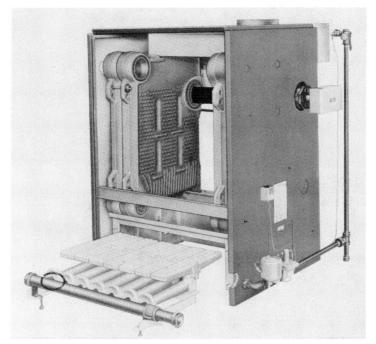

Fig. 15-12. Cast-iron sections in a hydronic gas-fired boiler.

Fig. 15-13. Flexible steel water tube.

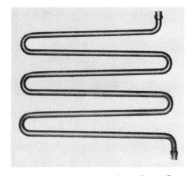

483

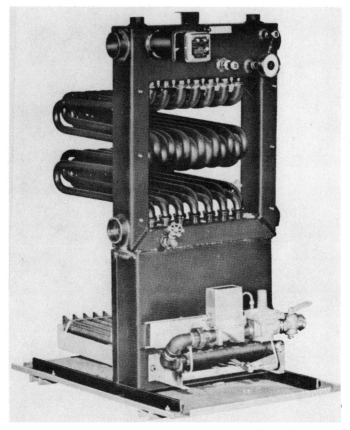

Fig. 15-14. F-series, gas-fired boiler with insulation and outer steel jacket removed.

plicity of boiler types and designs available, it is extremely difficult to establish a classification system suitable for all of them.

Gas-Fired Boilers

Gas-fired, steam and hot-water, space heating boilers generally consist of a number of closely placed cast-iron sections or steel tubes with a series of gas burners (i.e., a gas burner assembly) placed beneath them. The flue gases pass upward between the sections or tubes to the flue collector.

Draft losses are kept low in these boilers because the pressure at which the gas is supplied is generally sufficient to draw in the amount of air necessary for combustion. The fact that there is no fuel-bed resistance, as is the case with coal-fired boilers, also contributes to low draft loss.

The draft in gas-fired boilers is generally nullified by the diverter; consequently, the resistance offered by the boiler passages is not an important variable. When there is a problem of excessive draft, it can be controlled by installing a sheet-metal baffle in the flue connection at the boiler (see "Control of Excessive Draft on Gas-Fired Boilers" in this chapter).

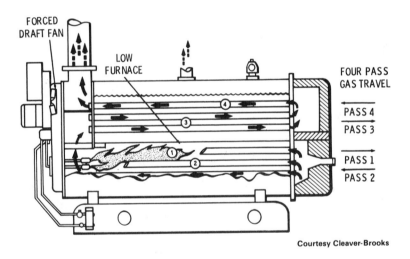

Courtesy Cleaver-Brooks

Fig. 15-15. A steam boiler with four-pass travel.

Most of the controls and accessories used to operate gas-fired boilers are described in considerable detail in Chapter 2 (Gas Burners), Chapter 5 (Gas and Oil Controls), and Chapter 6 of Volume 2 (Other Automatic Controls). The exact placement of these controls and accessories on the steel boiler jacket may differ slightly from one manufacturer to another, but not significantly. The major difference will be in the types of controls and accessories used to govern the temperature, pressure, and flow-rate of the heat-conveying medium; and this is determined by

485

whether it is a steam or hot-water, space heating boiler (see "Steam Boiler Fittings and Accessories" and "Hot-Water Boiler Fittings and Accessories" in this chapter). An exploded view of a Hydrotherm gas-fired boiler illustrating the location of various controls and accessories is shown in Fig. 15-16.

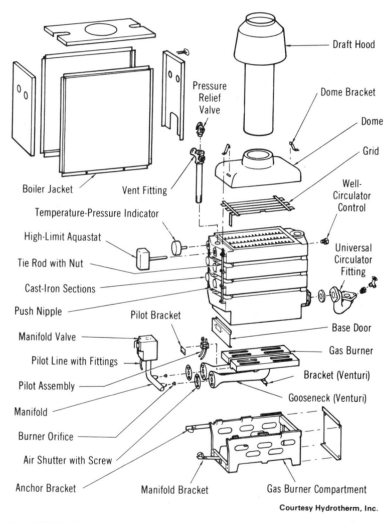

Courtesy Hydrotherm, Inc.

Fig. 15-16. Exploded view of a gas-fired, hot-water, space-heating boiler.

Oil-Fired Boilers

An oil-fired boiler contains a heat transfer surface consisting of either cast-iron sections or steel tubes and a special combustion chamber shaped to meet the requirements of an oil burner.

The oil burner is a device designed to mix fuel oil with air under controlled conditions and to deliver the mixture to the combustion chamber for burning. The burner is mounted outside the chamber. The oil burners used in residential heating boilers are usually high-pressure atomizing burners, although other types are used on occasion.

It is possible to convert a coal-fired boiler to oil by redesigning the combustion chamber. However, boilers specifically designed to use oil as a fuel have proven to be more efficient and economical than coal-fired boilers converted to oil.

Read Chapter 1 (Oil Burners), Chapter 5 (Gas and Oil Controls), and Chapter 6 of Volume 2 (Other Automatic Controls) for a description of the controls and accessories used to operate oil-fired boilers. Also read the appropriate sections in this chapter ("Steam Boiler Fittings and Accessories" and "Hot-Water Boiler Fittings and Accessories"). An exploded view of a Hydrotherm oil-fired boiler illustrates the location of the various controls and accessories (Fig. 15-17).

Coal-Fired Boilers

A typical coal-fired boiler contains a grate for the fuel bed located beneath metal, heat transfer surfaces (e.g., cast-iron sections). The hot gases resulting from combustion rise through the passages transferring their heat to the water contained in them.

The controls and accessories used to operate a coal-fired boiler are described in Chapter 3 of Volume 2 (Coal Firing Methods). The controls and accessories used to govern the temperature, pressure, and flowrate of the heat-conveying medium are described in this chapter (see "Steam Boiler Fittings and Accessories" and "Hot-Water Boiler Fittings and Accessories").

A coal-fired boiler requires more draft than gas- or oil-fired boilers because of the resistance of the fuel bed. Special care must be given to the design of the chamber surrounding the grate to ensure sufficient volume for a proper mixture of fuel and air.

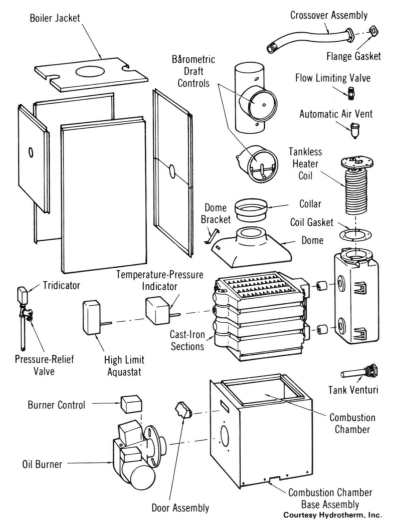

Fig. 15-17. Exploded view of an oil-fired, hot-water, space-heating boiler.

Provisions must also be made for introducing sufficient air through the grate (and fuel bed) for the combustion process.

Before the introduction of oil burners, coal was generally used for fuel and is still largely used especially for heating plants already built and in cases where the owner cannot afford the expense of converting to oil. Moreover, some cast-iron boilers designed to burn coal are not suited to burn oil, which adds to the expense of a conversion job. Finally, the increasing worldwide shortage of oil will find it more and more necessary to find applications for low polluting uses of coal as a heat source. An important use for coal will be steam and hot-water heating systems equipped with devices to reduce the polluting effect of burning coal.

Coal-burning boilers are usually arranged for hand firing but may be fitted for automatic stoker operation. In many automatic stokers, fuel is carried from the hopper through a feed tube by means of a rotating worm. Intermittent action of the worm agitates the fuel bed, prevents arching of the coal in the retort, and enures that the incoming air reaches every part of the fire at all times. An auxiliary air connection between the feed tube and the windbox prevents gas accumulating in the tube and eliminates any tendency to "smoke back" through the hopper.

An underfeed stoker is one in which the fuel is fed upward from underneath. The action of a screw or worm carries the fuel back through a retort from which it passes upward as the fuel is consumed, the ash being finally deposited on dead plates on either side of the retort from which it can be removed.

Only part of the fuel being burned is actually burned in the fuel bed. Under the influence of high temperature created in the fuel bed and lack of sufficient air, unburned gases are released above the retort and tuyeres; and unless these gases are mixed with air and burned inside of the combustion chamber, they will leave the boiler in the unburned condition, carrying a large percentage of the heat originally contained in the coal.

Nearer the outside of the fuel bed, the fuel burns less violently and much more air is passed through the fuel bed than is necessary. It is this excess air that must be mixed with the unburned gases issuing from the central point of the fuel bed if high-combustion efficiency is to be realized.

Coke can also be used to fire boilers, but it requires different handling. In order to be completely consumed, coke needs a greater volume of air per pound of fuel than coal and therefore requires a stronger draft, which is increased by the fact that it can burn economically only in a thick bed.

489

Because less coke is burned per hour per square foot of grate than coal, a larger grate and a deep fire-pot are required to accommodate the thick bed of coke.

The quick-flaming combustion that characterizes coal is not produced by coke because the latter fuel contains very little hydrogen; however, a coke fire is more even and regular.

Electric-Fired Boilers

Compact electric-fired boilers enjoy widespread use in forced-hot-water heating systems. A typical example of one of these boilers is shown in Fig. 15-18. Heat is generated by electric heating elements immersed in water contained in a waterproof cast-iron shell. Although small in size, these boilers are capable of generating as high as 90,000 Btuh; enough heat for the average eight-room house.

The basic components of an electric-fired boiler are shown in Fig. 15-19. An automatic air vent (1) located above the cast-iron boiler shell (8) is used for bleeding off trapped air from the water system. Safety devices include a water pressure-relief valve (2) for reducing system pressure, an adjustable limit control (3) that allows selection of maximum boiler water temperature for system design, and preset pressure controls (4) to guarantee safe operation within the prescribed pressure range.

The limit control is an immersion device that shuts off the boiler if the temperature exceeds a predetermined setting. The preset pressure controls consist of a high-pressure switch and a preset safe-fill switch. The high-pressure switch is designed to deenergize the boiler if the pressure reaches 28 psi. This switch is reinforced by the relief valve, which is preset to relieve system pressure at 30 psi. The preset safe-fill switch prevents the electric heating elements from being energized unless pressure in the system is 4 psi. This prevents any chance of the heating elements burning out should an attempt be made to operate the boiler while dry.

The circulator (5), cast-iron boiler shell (8), and expansion (9) are all enclosed within the outer steel jacket (10) of the boiler. The electric heating elements (6) are mounted inside the ends of the boiler casting. A drain valve (11) mounted below the boiler casting is designed for the connection of standard hose fittings.

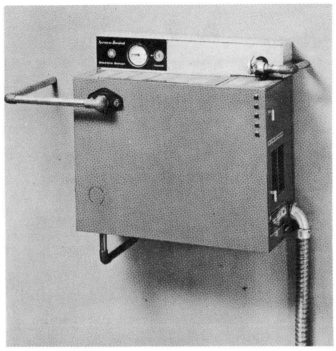

Fig. 15-18. An electric-fired boiler for hot-water heating.

All controls are prewired and mounted in place by the manufacturer. The only electrical connections that need to be made during the installation of the boiler are those to the main power supply and room thermostat.

The sequencing relay switch (7) provides for an incremental loading on the electric service line. This reduces line voltage fluctuation and prevents power surge during energizing.

Minimum installation clearances for the boiler described above are shown in Fig. 15-20. Two air openings of 108 sq. in. (6″ × 18″) each are required for closet installation.

A typical piping arrangement for a series-loop, hot-water heating system using an electric-fired boiler of this capacity is shown in Fig. 15-21. An outline of the procedure used for filling this system is as follows:

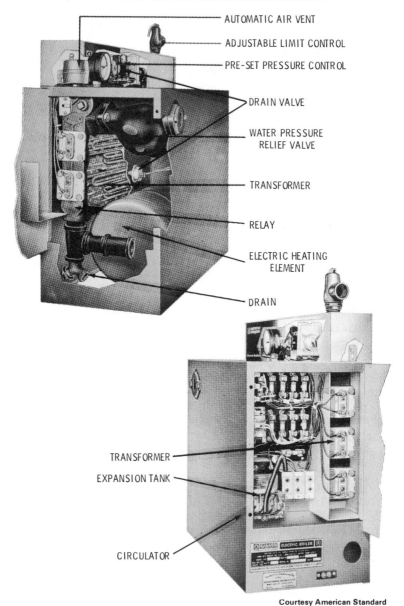

AUTOMATIC AIR VENT

ADJUSTABLE LIMIT CONTROL

PRE-SET PRESSURE CONTROL

DRAIN VALVE

WATER PRESSURE RELIEF VALVE

TRANSFORMER

RELAY

ELECTRIC HEATING ELEMENT

DRAIN

TRANSFORMER

EXPANSION TANK

CIRCULATOR

Courtesy American Standard

Fig. 15-19. Basic components of an electric-fired boiler.

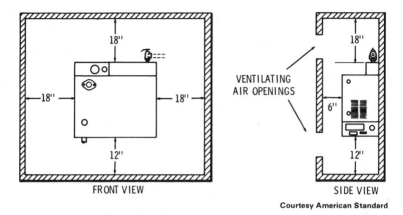

FRONT VIEW   SIDE VIEW

Courtesy American Standard

Fig. 15-20. Minimum installation clearances.

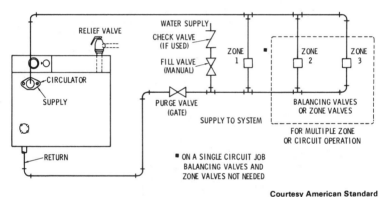

Courtesy American Standard

Fig. 15-21. Typical piping arrangement for a multizoned hot-water heating system.

1. Close all zone valves except zone to be purged.
2. Open boiler drain valve.
3. Open fill valve.
4. Close purge valve.
5. Vent air from boiler by manually opening relief valve.
6. Open second valve to be purged, and close first zone.
7. After second zone is purged, open third zone valve and close second zone.

493

8. Repeat the precedures described in Steps 6 and 7 for as many zones as are in the system.
9. When all zones have been purged, close boiler drain cock.
10. When water appears at relief valve, release lever and allow to close.
11. Continue filling until pressure gauge reads approximately 12 psi. Close fill valve.
12. Open purge valves.

Where multiple circuits are used without zone valves, balancing valves should be installed for balancing. These can be shut off to purge each circuit individually. For a single-loop system, no additional vents, valves, drains or other accessories are required.

Smaller electric-fired boilers having outputs ranging from approximately 6000 to 20,000 Btu are available for individual apartment heating or for residential zoned heating through the use of several units. The basic components of one of these smaller boilers is shown in Fig. 15-22. The piping arrangement for a series-loop heating system using one of these boilers is illustrated in Fig. 15-23. The procedure for filling the system is similar to the one outlined above except for the elimination of Steps 6 to 9.

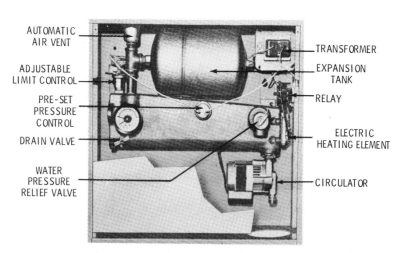

AUTOMATIC AIR VENT
ADJUSTABLE LIMIT CONTROL
PRE-SET PRESSURE CONTROL
DRAIN VALVE
WATER PRESSURE RELIEF VALVE
TRANSFORMER
EXPANSION TANK
RELAY
ELECTRIC HEATING ELEMENT
CIRCULATOR

Courtesy American Standard

Fig. 15-22. Basic components of an electric-fired boiler.

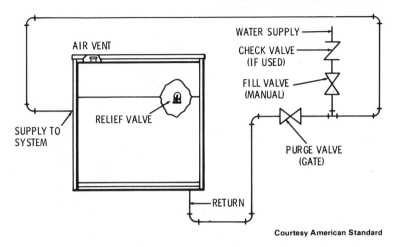

Fig. 15-23. Typical piping arrangment for a hot-water heating system.

STEAM BOILER FITTINGS
AND ACCESSORIES

The boilers used in steam heating systems are fitted with a variety of devices designed to ensure the safe and proper operation of the boiler. These boiler fittings and accessories can be divided by function into two basic categories: (1) indicating or measuring devices, and (2) controlling devices. Sometimes both functions are combined in a single unit.

Indicating devices include water gauges, pressure gauges, and similar types of equipment that provide information about the operating conditions in the boiler. They are designed to indicate temperatures, pressures, or a water level that fall outside the design limits of the boiler. Controlling devices include boiler equipment designed to cause changes in these conditions. For example, pressure-relief valves are used to relieve excess pressure in the boiler.

Typical fittings and accessories commonly found on steam boilers are illustrated in Figs. 15-24 and 15-25. A steam boiler is easily identified by an externally mounted low-water cutoff and pump control.

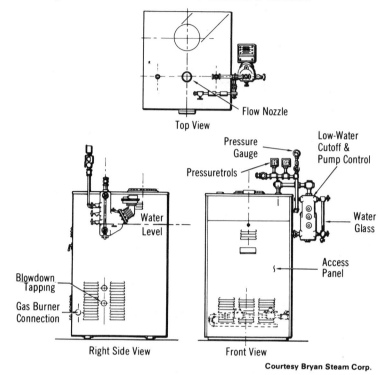

Courtesy Bryan Steam Corp.

Fig. 15-24. Gas-fired steam boiler fittings.

Steam Boiler, Pressure-Relief Valves

A steam boiler is equipped with a pressure-relief valve (or valves) that opens and releases excess steam at or below the maximum allowable working pressure of the boiler. These are pop safety-relief valves designed to comply with the requirements of the ASME Boiler and Pressure Vessel Code.

Pop safety valves exhaust the steam through holes drilled around the top of the spring housing. They function by "popping" wide open at the set pressure, remaining in that position until pressure has dropped the predetermined amount (commonly known as blowback, or blowdown), and then snapping shut instantly. Lift levers and drain holes in the discharge side of the valve are required by ASME codes.

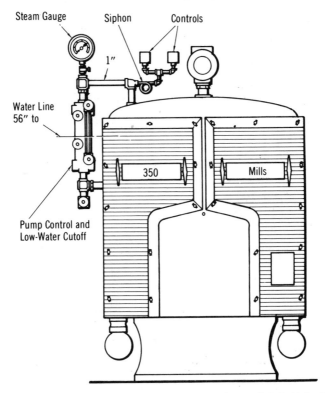

Steam Gauge   Siphon   Controls

1″

Water Line
56″ to

Pump Control and
Low-Water Cutoff

350   Mills

Fig. 15-25. Water column, piping, fittings, and steam controls for a gas-fired steam boiler.

On the low-pressure boilers used in residential steam heating systems, the pressure- (or safety-) relief valve is generally preset to open and release steam when a maximum pressure of 15 psi is reached in the boiler. The valve closes when the steam pressure once again falls below 15 psi.

Water Gauges

Water gauges are used on both steam and hot-water space heating boilers to indicate the water level. Some examples of different water gauges used on boilers are shown in Fig. 15-26.

497

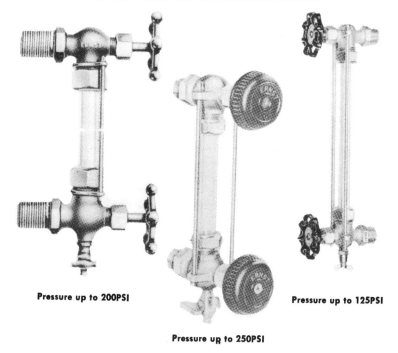

Pressure up to 200PSI

Pressure up to 250PSI

Pressure up to 125PSI

Courtesy Ernst Gage Co.

Fig. 15-26. Various types of water gauges.

Fig. 15-27 illustrates the typical location of a water gauge on a boiler.

A water gauge is used to *visually* check the level of water in the boiler. If the water level in the boiler is high enough, the glass tube will be approximately half full. If no water is showing in the tube, the boiler must be turned off and refilled to the proper level. Do *not* add the water until the boiler has had time to cool off.

The construction of a typical water gauge is shown in Fig. 15-28. The glass tube of the water gauge must be long enough to cover the safe range of water level in the boiler. The ends of the water gauge are connected to the interior of the boiler by fittings located above the safe high-water level and below the safe low-water level.

Water Columns

A water column (Fig. 15-29) is a boiler fitting that combines tyro cocks, a water gauge, and an alarm whistle in a single unit. A float in the column activates an alarm whistle when the water drops below a safe level.

Steam Gauge

The difference between the pressure found on the inside of the boiler and the pressure on the outside of it is indicated by a steam gauge (Fig. 15-30). This is a gauge pressure reading and should

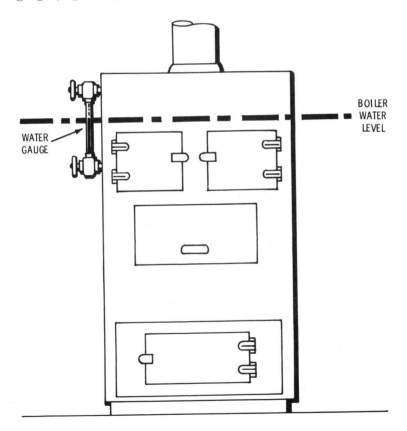

WATER
GAUGE

BOILER
WATER
LEVEL

Fig. 15-27. Typical location of a water gauge.

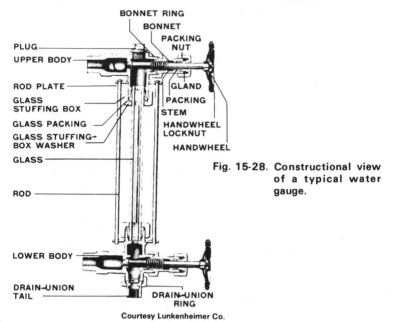

Fig. 15-28. Constructional view of a typical water gauge.

Courtesy Lunkenheimer Co.

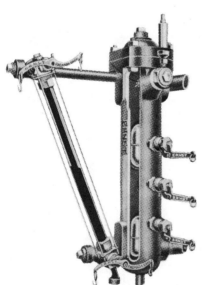

Fig. 15-29. Water column.

Courtesy Ernst Gage Co.

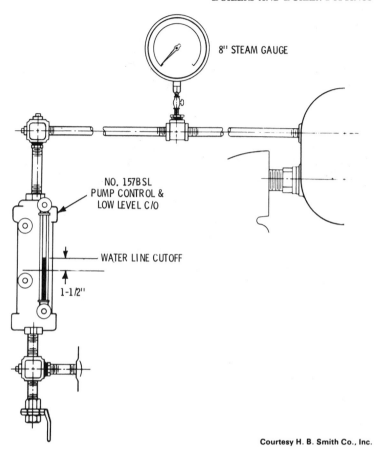

8" STEAM GAUGE

NO. 157B SL
PUMP CONTROL &
LOW LEVEL C/O

WATER LINE CUTOFF

1-1/2"

Courtesy H. B. Smith Co., Inc.

Fig. 15-30. Steam gauge.

not be confused with absolute pressure. A steam gauge is used to measure the steam pressure at the top of the boiler. Generally a reading of 12 psi (pounds per square inch) indicates a dangerous buildup of pressure in low-pressure steam heating boilers. The boiler should be shut down before the pressure exceeds this level.

If the steam gauge is operating properly, the needle (or pointer) will move with each change of pressure inside the boiler. Shut off the steam, and the gauge needle should drop to zero; turn on the steam, and the needle should rise to the correct read-

501

ing. It is very important that the steam gauge be regularly checked to ensure that it is operating properly.

The steam gauges used on boilers operate on the bent-tube principle; that is to say, the tendency of a bent or curved tube to assume a straight position when pressure is applied. As shown in Fig. 15-31, one end of the curved tube is attached in a fixed position to a pigtail (connector tube), which, in turn, is attached to the boiler. The gauge needle is mounted on a rack-and-pinion gear attached to the *free* end of the curve tube. The pressure in the curved tube causes its free end to move slightly in its effort to assume a straight position. This slight movement is multiplied by the rack-and-pinion gear, causing the needle to rotate and indicate the steam pressure.

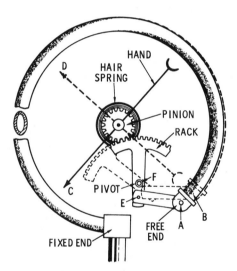

Fig. 15-31. Mechanism of a bent-tube steam gauge.

Steam Gauge Pigtails

A steam gauge pigtail (Fig. 15-32) is a length of tubing with one or more loops in it used to connect the steam gauge to the boiler. It functions as a protective device by preventing live boiler steam from coming into contact with working parts of the gauge. The steam is denied passage by condensation which forms in the loop

Fig. 15-32. Steam-gauge pigtail.

(or loops) of the pigtail. It is recommended that the steam gauge pigtail be filled with water *before* it is attached to the boiler. It should be attached to the boiler in a position where heat and vibration will be at a minimum.

Pressure Controllers

When the room thermostat calls for heat in a steam heating system, the electrical current flows to the operating controls of the automatic fuel-burning equipment (oil burner, gas burner, or coal stoker) through a high-limit switch. In a steam boiler control system, the high-limit switch is activated by pressure and is referred to as a *pressuretrol*.

Fig. 15-33 shows a pressuretrol operating controller designed to provide on-off and proportional control of steam boilers fired by proportional-type burners. It can be field adjusted to operate either in unison (burner starts at high fire) or in sequence (burner starts at some firing rate other than high fire). A common bellows assembly located in the bellows housing actuates the stop-start switch (on-off operation) and the wiper of the 185-ohm potentiometer (proportional operation) .

The pressure controller shown in Fig. 15-34 operates strictly as a high-limit pressure safety control on steam heating boilers. It breaks the electrical circuit on pressure rise. A variation of this controller (the Honeywell PA404B Pressuretrol Controller) is used for suspension-type unit heaters.

Direct control for a proportional motor operating an automatic burner is obtained by using the pressure controller illustrated in

503

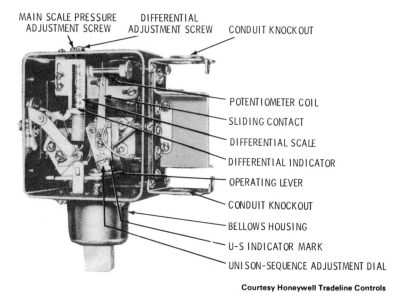

MAIN SCALE PRESSURE ADJUSTMENT SCREW

DIFFERENTIAL ADJUSTMENT SCREW

CONDUIT KNOCKOUT

POTENTIOMETER COIL

SLIDING CONTACT

DIFFERENTIAL SCALE

DIFFERENTIAL INDICATOR

OPERATING LEVER

CONDUIT KNOCKOUT

BELLOWS HOUSING

U-S INDICATOR MARK

UNISON-SEQUENCE ADJUSTMENT DIAL

Courtesy Honeywell Tradeline Controls

Fig. 15-33. Pressuretrol operating controller.

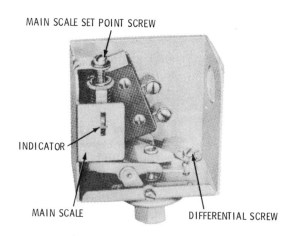

MAIN SCALE SET POINT SCREW

INDICATOR

MAIN SCALE

DIFFERENTIAL SCREW

Courtesy Honeywell Tradeline Controls

Fig. 15-34. Pressuretrol controller.

504

Fig. 15-35. This unit contains two potentiometers operating in unison, which makes it possible to simultaneously control two motors. It is also provided with an adjustable throttling range.

Pressure controllers are also available for vapor or vacuum systems. A pressure controller with a bellows-operated mercury switch is manufactured by Honeywell for use on vapor heating systems with pressures up to 4 psi. The pressure controller illustrated in Fig. 15-36 can be used as a boiler high-limit control with

PRESSURE ADJUSTMENT SCREW   THROTTLING RANGE ADJUSTMENT SCREW

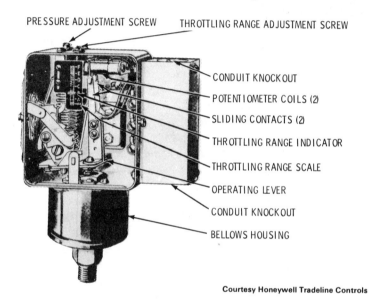

CONDUIT KNOCKOUT

POTENTIOMETER COILS (2)

SLIDING CONTACTS (2)

THROTTLING RANGE INDICATOR

THROTTLING RANGE SCALE

OPERATING LEVER

CONDUIT KNOCKOUT

BELLOWS HOUSING

Courtesy Honeywell Tradeline Controls

Fig. 15-35. Proportional Pressuretrol controller.

Fig. 15-36. Vacuumstat controller.

Courtesy Honeywell Tradeline Controls

cut-in and cut-out settings in the vacuum range. In such installations, the heating system must include a vacuum pump and a siphon loop.

Low-Water Cutoffs

Steam boilers must be provided with a water-level control device that will shut off the automatic fuel-burning equipment when the water level in the boiler drops to a level too low for safe operation. This water-level control device is referred to as a *low-water cutoff*. The two low-water cutoffs used on steam boilers are the float type and the robe type.

A float low-water cutoff device consists of a cutoff switch operating in conjunction with a float located in the boiler water or in a float chamber installed next to the boiler. The float is connected through a linkage to a switch that operates a feed water valve. As the water level falls, the float drops with it until it reaches a point at which the feed water switch is actuated. If the water level continues to fall, a second switch connected to the float by the linkage is actuated, and the automatic fuel-burning equipment is shut off.

A probe low-water cutoff device depends on the flow of a low electrical current to control the operation of the automatic fuel burning equipment. The electrical current flows from the probe through the water to keep the relay energized. When the water level falls, there will be a point at which the probe loses direct contact with the water. As a result, the contact is broken, and the flow of the electrical current is stopped. This, in turn, causes the relay to be energized, which results in shutting off the fuel-burning equipment. A probe low-water device *cannot* be used in the direct operation of feed water valves.

On some boilers, the low-water cutoff is combined with a water feeder to add water to the boiler when the water level falls below the safe operating limit. An example of one of these combined units is shown in Fig. 15-37. They are available with either single or dual switch assemblies. The single double-throw switch assembly provides a combination feeder and burner cutoff switch with an extra terminal for line voltage with single-pole, double-throw service (Fig. 15-38). The dual switch assembly is

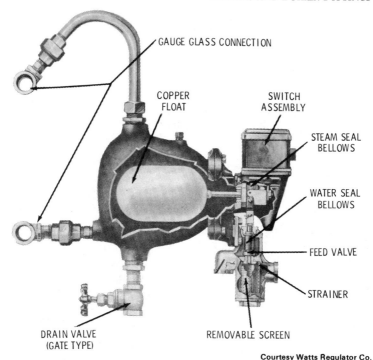

GAUGE GLASS CONNECTION

COPPER
FLOAT

SWITCH
ASSEMBLY

STEAM SEAL
BELLOWS

WATER SEAL
BELLOWS

FEED VALVE

STRAINER

DRAIN VALVE
(GATE TYPE)

REMOVABLE SCREEN

Courtesy Watts Regulator Co.

Fig. 15-37. Basic components of a boiler water feeder and cutoff.

used for line voltage burner service and for independent low- (or
high-) voltage alarm, feed valve, or pump starter service. When
an emergency condition occurs, the switch interrupts the current
to the burner and shuts it off. When the emergency has passed,
the water feeder takes over and feeds makeup water when
needed for normal operation. On units equipped with dual
switch assemblies, the alarm, feed valve, or pump starter switch
actuates just before the burner switch cuts the firing. A typical
installation of a Watts 60 LWD boiler water feeder and cutoff on
a steam boiler is illustrated in Fig. 15-39.

A low-water cutoff *without* a boiler water feeder is usually
adequate for providing automatic low-water safety protection
for most small boilers (Fig. 15-40). These units can be installed on
any boiler having gauge glass connections.

Direct installation low-water cutoffs are available for boilers

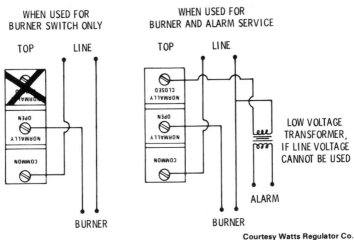

Courtesy Watts Regulator Co.

Fig. 15-38. A single-switch assembly and wiring diagram.

with limited space in the boiler jacket (Fig. 15-41). Each unit contains a switch assembly and float and is screwed into the boiler through a 2½-in. tapping.

Installing a Low-Water Cutoff

The Watts No. 89 low-water cutoff illustrated in Fig. 15-40 is suitable for low-pressure steam heating boilers with a maximum operating steam pressure of 15 lb. Although installation is relatively simple, the instructions should be carefully followed. This is a boiler safety device, and it must operate properly or the boiler can be damaged.

Before installing the Watts low-water cutoff, the gauge glass and cocks must be removed from the boiler. When you have done this, proceed as follows:

1. Install the ½-in. tee (with long and short nipples) on the float chamber by inserting the short nipple in the float chamber tapping in the end opposite the switch. Screw the tee tight with the long nipple pointing in the correct direction, depending on the location of the boiler (Fig. 15-42).

2. Insert the long nipple into the lower glass gauge tapping on the boiler.

3. Swing the entire float chamber until the nipple is made up tight. Line up the cutoff so that the top of the switch box is level (Fig. 15-43).

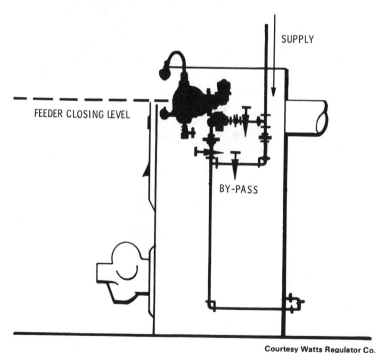

SUPPLY

FEEDER CLOSING LEVEL

BY-PASS

Courtesy Watts Regulator Co.

Fig. 15-39. Typical installation of a boiler water feeder and cutoff.

509

Courtesy Watts Regulator Co.

Fig. 15-40. A low-water cutoff.

4. Screw the nipple in the tee carrying the tubing connector into the upper gauge glass tapping and pull up tight. Install a compression coupling in the top float chamber tapping (Fig. 15-44).

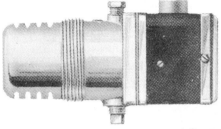

Courtesy Watts Regulator Co.

Fig. 15-41. Direct-installation, low-water cutoff.

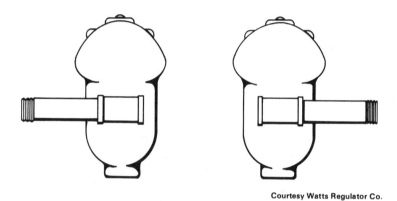

Courtesy Watts Regulator Co.

Fig. 15-42. Installing the short nipple in the float chamber topping.

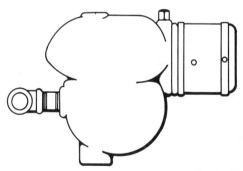

Courtesy Watts Regulator Co.

Fig. 15-43. Installing long nipple into lower-gauge glass topping in boiler.

511

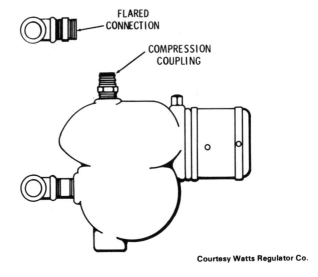

Fig. 15-44. Installing compression coupling.

5. Hold the tube bend in position and mark the tube on a level with the top of the top of the hex on the compression coupling. Cut off the tube at this mark (Fig. 15-45).

6. Slide the ring and nut over the end of the connector tube.

7. Slide the end of the tube into the compression coupling in the float chamber, and tighten both couplings.

8. Install a drain valve in the bottom float chamber tapping (Fig. 15-46).

9. Reinstall the gauge cocks in the end of the tees and replace the gauge glass (Fig. 15-46).

The drain valve should be opened once every month (or oftener) during boiler operation to flush out sediment from the float chamber.

Steam System, Vacuum-Relief Valves

A vacuum-relief valve (Fig. 15-47) is an effective means of providing protection against the buildup of excessive vacuum conditions in steam heating and steam processing systems. It is

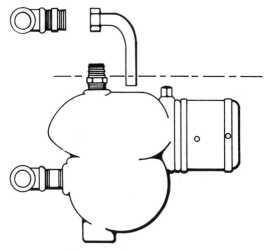

Fig. 15-45. Cutting tube on level with top of hex.

also used in combination with temperature-pressure-relief valves on water heaters.

Vacuum conditions can often occur after the supply line has been shut off. Steam condenses and a vacuum can be created that not only affects system operation but also can cause damage to equipment. The vacuum relief valve automatically admits air to the system in order to break up the vacuum. These valves operate under conditions up to a maximum temperature of 250°F. The valve disk should be constructed from a material capable of withstanding high temperatures, particularly when the valve is used in a steam processing system.

Steam Boiler Injectors

A steam boiler injector is a device used on some boilers to create accelerated steam circulation in a steam heating system. The essential parts of a boiler injector are shown in Fig. 15-48.

This device operates on the induction principle. In operation, steam from the boiler, entering the steam nozzle, passes through it, through the space between the steam nozzle and combining tube, and then out through the overflow. This produces a

513

vacuum that draws in the water through the water inlet. The incoming cold water condenses the steam in traversing the combining tube, and the water jet thus formed is driven at first out through the overflow. As the velocity of the water jet increases, sufficient momentum is obtained to overcome the boiler pressure, with the result that the water enters the delivery tube and passes the main check valve into the boiler.

The induction principle of the steam injector is the same as the operating principle of hydraulic boosters used in forced-hot-water heating systems with the exception that fast-flowing hot water is used instead of steam to obtain the inductive action to accomplish accelerated circulation.

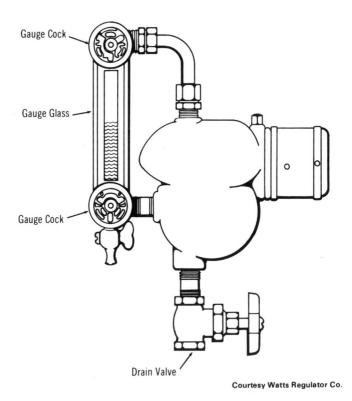

Courtesy Watts Regulator Co.

Fig. 15-46. Installing drain valve.

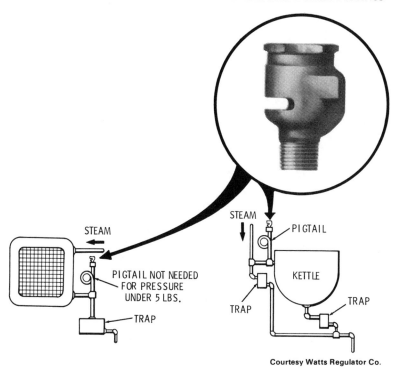

STEAM

STEAM ←

STEAM ↓

PIGTAIL

PIGTAIL NOT NEEDED FOR PRESSURE UNDER 5 LBS.

KETTLE

TRAP

TRAP

TRAP

TRAP

Courtesy Watts Regulator Co.

Fig. 15-47. Vacuum relief valve used in combination with a unit heater in a steam-heating system and with a jacketed kettle in a steam-processing system.

Try Cocks

Try cocks are small valves installed on steam boilers at the safe high-water level and at the safe low-water level (Fig. 15-49). When the boiler water column is inoperable, try cocks function as a backup system to determine the water level. The water level is determined by opening *slightly* first one try cock and then the other. The water level is indicated by the color of the plume escaping from the try-cock orifice. A steam plume is characteristically colorless and will indicate that the water level is too low (i.e., below the level of the try-cock orifice). Water, on the other hand, is characterized by a white plume. A false water-level reading will be obtained if the try cock is opened too wide when the

515

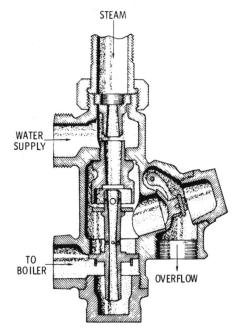

STEAM

WATER
SUPPLY

TO
BOILER

OVERFLOW

Fig. 15-48. Cross-sectional view of a typical injector.

Courtesy Ernst Gage Co.

Fig. 15-49. Try cocks.

516

level of the water is only slightly below the level of the try-cock orifice. The violent agitation of the water caused by a wide-open try cock will result in some of the water escaping, giving the false impression that the level of the water in boiler is at or above the try-cock fitting.

On large steam boilers, a third try cock is often installed between the other two. Sometimes a try cock on these larger boilers is at a level too high to reach. When this is the case, chain-operated, lever-type try cocks are installed.

Blowoff Valve

Sediments, contaminants, and other impurities found in water will settle out over a period of time and accumulate at a low point in the bottom of the boiler. These can be removed by means of a blowoff valve placed in a line connected to the lowest part of the boiler. The blowoff valve is opened periodically, and the accumulated sediments are drained off.

Fusible Plugs

A fusible plug is used on some boilers as a protection against dangerous low-water conditions. Examples of fusible plugs are shown in Fig. 15-50. They are generally made of bronze and

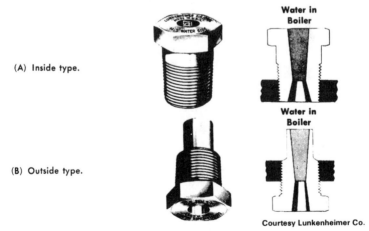

(A) Inside type.

(B) Outside type.

Courtesy Lunkenheimer Co.

Fig. 15-50. Fusible plugs.

filled with pure tin. When the temperature (and pressure) in the boiler builds up to about 450°F (the approximate melting point of pure tin), the tin core melts and relieves the pressure within the boiler.

Fusible plugs must comply with ASME standards and are available in several sizes for use on boilers having a steam pressure of less than 250 psi.

Steam Loop

A steam loop (Fig. 15-51) is a piping arrangement used in a steam heating system to return condensation to the boiler. The principal components of a steam loop piping arrangement are:

1. Riser.
2. Condenser.
3. Drop leg.
4. Check valve.

The *riser* carries a mixture of steam and water to a horizontal *condenser* where the steam portion of the mixture is condensed. The condensation of the steam causes a reduction of pressure,

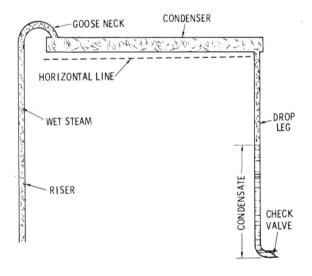

Fig. 15-51. Essential parts of a steam loop.

and it is this reduced pressure that provides the motive force for the steam and water mixture in the riser. The goose neck in the riser prevents any backflow of the steam. The condensation accumulates in the *drop leg* to a height such that its weight will balance the weight of the mixture in the riser. The *check valve* on the drop leg allows condensation to pass into the boiler but prevents steam in the boiler from backing into the pipe.

HOT-WATER BOILER FITTINGS AND ACCESSORIES

The boilers used in hot-water heating systems are also fitted with a variety of devices designed to ensure the safe and proper operation. Some of these fittings and accessories are similar to those used on steam boilers (e.g., pressure-relief valves and low-water cutoffs). They also share the same measuring and controlling functions described for steam boiler fittings and accessories. Figs. 15-52 and 25-53 illustrate their arrangement on a hot-water, space heating boiler.

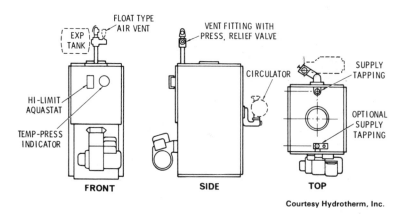

Fig. 15-52. Pipings, fittings, and controls for an oil-fired, hot-water, space-heating boiler.

ASME-Rated Water-Pressure-Relief Valves

Pressure-relief valves are used on hot-water, space heating boilers to relieve pressure created by two different conditions:

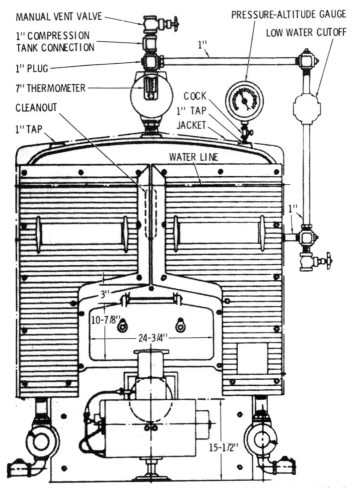

MANUAL VENT VALVE

1" COMPRESSION
TANK CONNECTION

1" PLUG

7" THERMOMETER

CLEANOUT

1" TAP

PRESSURE-ALTITUDE GAUGE

LOW WATER CUTOFF

1"

COCK

1" TAP

JACKET

WATER LINE

1"

3"

10-7/8"

24-3/4"

15-1/2"

Courtesy H. B. Smith Co., Inc.

Fig. 15-53. Piping, fittings, and controls for a gas-fired, hot-water, space-heating boiler.

(1) water thermal conditions and (2) steam pressure conditions. These are exclusively safety devices and should not normally be used as regulating valves or other units intended to regulate or control the flow pressure. Relief valves are intended to prevent personal injury and damage to property. These valves start to

open at the set pressure and require a certain percentage of over-pressure to open fully. As the pressure drops, they start to close and shut at approximately the set pressure.

As thermal conditions develop inside a hot-water, space heating boiler, pressures may be built up to the setting of the relief valve. When the pressure reaches the safety limit setting, the valve functions as a *water* relief valve and discharges the small amount of water which is expanded in the system (Fig. 15-54).

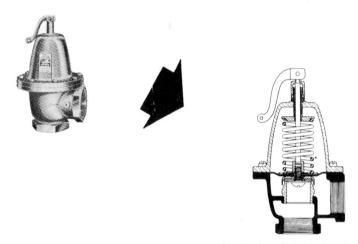

Fig. 15-54. Typical ASME pressure-relief valves.

If *both* water and steam are present in a hot-water, space heating boiler, the firing controls are probably malfunctioning or have completely broken down. This may result in runaway firing of the boiler, which could cause the boiler water to reach steam-forming temperatures, creating a steam pressure condition. Under these circumstances, the relief valve functions as a steam safety-relief valve. The steam is discharged at a rate or faster than the boiler can generate it, thus restoring system pressure to a safe level. Although these valves are steam rated and have an emergency Btu steam-discharge capacity if runaway firing conditions occur, they should *not* be continuous steam service.

In order to be completely effective, the safety-relief valves used on hot-water, space heating boilers must be designed to discharge the excessive water pressure created by thermal expansion and also the excessive steam pressure resulting from runaway emergency temperature conditions.

Altitude Gauges

An altitude gauge mounted on the boiler is used to indicate the water level in *open*, hot-water heating systems. The level (altitude) of the water is indicated on the gauge by the relative positions of two hands or arrows. One hand (usually black) is stationary and is permanently set when the boiler is filled. The movable hand (usually red or white) is initially set in the same position as the stationary hand after the boiler has been filled with water. Its position will change as the water level in the boiler changes. This movable hand indicates the true level of the water. Efficient operation is being provided as long as the movable hand is directly above the stationary one.

Fig. 15-55 shows a pressure gauge, thermometer, and altimeter (altitude gauge) combined in one unit. These Marsh Ther-Alti-Meters are used on residential hot-water, space heating boilers. They provide water pressure readings from 0 to 30 psi (red field to 50 psi), corresponding altitude from 0 to 70 ft., and water temperature from 60° to 320°F.

Courtesy Marsh Instrument Co.

Fig. 15-55. Pressure gauge, thermometer, and altimeter combined in one unit.

In *closed* hot-water heating systems, automatic valves are used to control boiler water level (see "Pressure-Relief Valves" and "Pressure-Reducing Valves" in this chapter).

Aquastats

An aquastat is a device commonly used on hot-water space boilers and some steam boilers to control temperature limit or to operate the circulator (hot-water, heating system pump). It is similar in function to the pressure control on a steam boiler or the fan and limit control on a forced-warm-air furnace.

Basically, an aquastat is an automatic switching device consisting of a metal- or liquid-filled, heat-sensitive element designed to detect temperature drop or rise of the boiler water. Aquastats can be strapped to the hot-water supply riser or mounted so that the heat-sensitive element is immersed in a boiler well.

The type of aquastat used in a heating installation generally depends upon whether it is designed to control temperature limit or to switch on the circulator. If the former is the case, the aquastat will close on temperature drop and open on temperature rise. The aquastat will close on temperature rise if it is used to operate the circulator.

An example of a strap-on type aquastat is the ITT General Controls L-53 Strap-On Hot Water Control illustrated in Fig. 15-56. This type of aquastat responds to water temperature changes as conducted through the supply-riser pipe wall to the temperature responsive base of the device. The enclosed switching is supplied in direct action (Fig. 15-57), reverse action (Fig. 15-58), and double action (Fig. 15-59) models. Consequently, these strap-on aquastats can be divided into three basic types, depending upon their operating principles: (1) direct-action types, (2) reverse-action types, and (3) double-action types.

A direct-action (N.C.) aquastat used as a high-limit control must always be located on the supply riser where it will be subjected to the maximum boiler water temperature. Its location on the riser will therefore have to be as close to the boiler as possible but ahead of any line shutoff valves.

Direct-action aquastats will open the circuit on temperature rise. Setting the adjustable scale pointer to a position on the scale

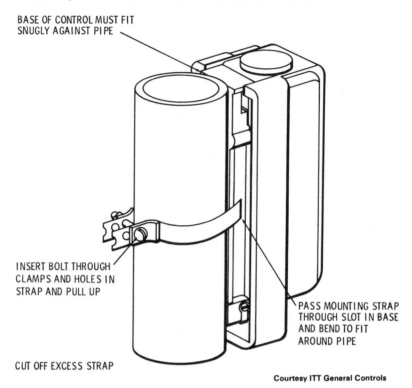

BASE OF CONTROL MUST FIT
SNUGLY AGAINST PIPE

INSERT BOLT THROUGH
CLAMPS AND HOLES IN
STRAP AND PULL UP

PASS MOUNTING STRAP
THROUGH SLOT IN BASE
AND BEND TO FIT
AROUND PIPE

CUT OFF EXCESS STRAP

Courtesy ITT General Controls

Fig. 15-56. Snap-on, hot-water control.

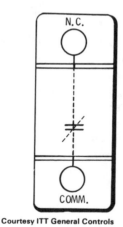

N.C.

COMM.

Courtesy ITT General Controls

Fig. 15-57. Direct-action switch
operation.

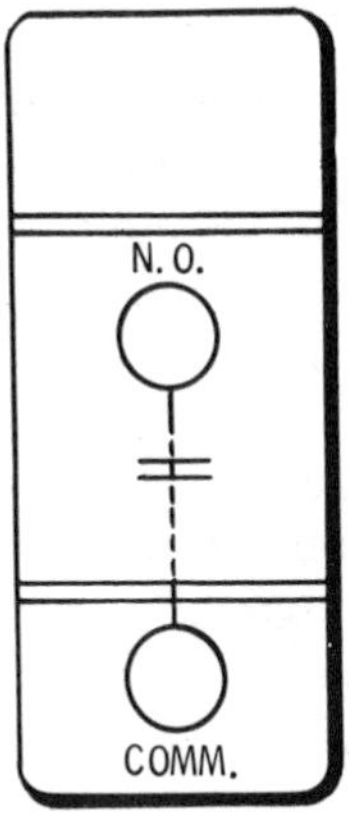

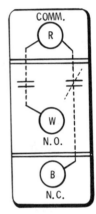

Fig. 15-58. Reverse-action switch operation.

Fig. 15-59. Double-action switch operation.

Courtesy ITT General Controls

will result in breaking the circuit (Fig. 15-60). The "cut-out" setting, when used as a high-limit control, should be as low as possible and still ensure proper heating in cold weather. An initial "cut-out" setting of 170°F is recommended for a gravity hot-

Fig. 15-60. Indicator can be set for a temperature higher than 200°F by removing the screws (A) and holding stop bracket to scale plate and removing the bracket (B).

Courtesy ITT General Controls

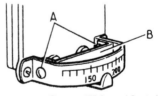

water, heating system. An initial setting of 200°F is suggested for a forced hot-water (hydronic) heating system. The mechanical differential of the control illustrated in Fig. 15-60 is nonadjustable and approximately 15°F.

A reverse-action (N.O.) aquastat should be mounted ahead of any valves or traps *on the return line* when it is used on a unit heater installation. It should be mounted on the largest riser from the boiler if it is used to prevent circulator operation when boiler water temperature is low. A reverse-action aquastat closes the circuit on temperature rise.

A double-action (SPDT) aquastat is used to start circulator operation with a single switch actuation and to function as an operating control to maintain boiler water temperature. This type of aquastat should be mounted on the largest riser ahead of any valve but at a point where it will be subjected to maximum boiler water temperatures. The double-action aquastat illustrated in Fig. 15-59 opens R to B contacts and closes R to W contacts on temperature rise at scale setting.

Another example of a strap-on aquastat is the Honeywell LA409A aquastat shown in Fig. 15-61. This is a nonimmersion, surface-mounted aquastat which functions as a high-limit or safety control device in hot-water and steam heating systems, and which requires no tapping of the boiler or draining of the system. This aquastat is also available in a reverse-action model to operate unit heater fans.

A strap-on aquastat can be mounted in any position. When mounting these aquastats, make absolutely certain that the pipe surface is clean and free of rust and corrosion. All rough and high

Fig. 15-61. An aquaset controller.

Courtesy Honeywell Tradeline Controls

spots should be filed smooth. Nothing should be allowed to interfere with the operation of the temperature responsive base of the control.

The Honeywell L6081A aquastat shown in Fig. 15-62 is used to control boiler water temperature in gas- and oil-fired hydronic heating systems. It has an immersion-type, liquid-filled, heat-sensing element that actuates two snap switches. One switch operates as a high-limit control, the other as a low-limit and/or circulator control.

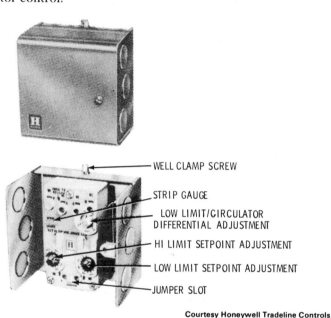

WELL CLAMP SCREW

STRIP GAUGE

LOW LIMIT/CIRCULATOR DIFFERENTIAL ADJUSTMENT

HI LIMIT SETPOINT ADJUSTMENT

LOW LIMIT SETPOINT ADJUSTMENT

JUMPER SLOT

Fig. 15-62. Multiple aquastat controller.

Low-Water Cutoffs

A low-water cutoff device, such as the one shown in Fig. 15-63, can also be used in a hot-water heating system to provide protection against a low-water-level condition in the boiler resulting from runaway firing caused by malfunctioning controls or a break in the return piping. The low-water cutoff device should be installed in the piping so that the raised line cast on the float

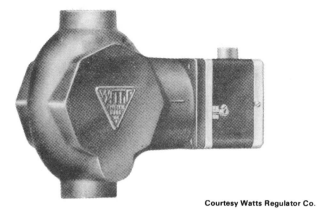

Fig. 15-63. Low-water cutoff used in a forced-hot-water heating system.

chamber body is on a level with the top of the boiler (Fig. 15-64). The drain valve located directly below the low-water cutoff should be opened periodically to flush mud and sediment from the float chamber. It is recommended that this be done at least once a month.

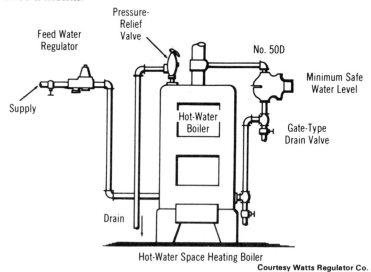

Fig. 15-64. Installation of a low-water cutoff in a forced-hot-water heating system.

Water-Pressure-Reducing Valve

Most forced-hot-water heating systems are equipped with a pressure-reducing valve designed to automatically feed water into the boiler when the pressure in the system drops below the valve setting. When the pressure returns to the minimum pressure setting, the valve automatically closes. Thus, the function of a water pressure-reducing valve is to keep the system automatically filled with water at the desired operating pressure (Fig. 15-65). These valves are also referred to as *pressure-reducing boiler feed valves* or *feed water pressure regulators*. Regardless of what they are called, their function remains the same: to deliver water to the boiler at the required pressure. To perform this function, they are *always* installed in the cold-water supply line. If they are part of a combination pressure-reducing and pressure-relief valve (dual control), the pressure-relief part of the combination valve is always located between the boiler and the pressure-reducing part.

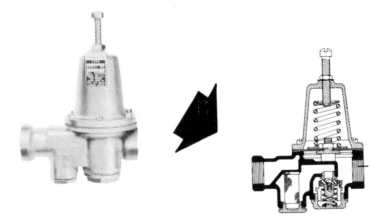

Courtesy A. W. Cash Valve Mfg. Co.

Fig. 15-65. Water pressure-reducing valve.

A manually operated feel valve was used on older hot-water heating systems to provide the same function as the automatic water-pressure-reducing valve. On these older systems, the manually operated feed valve should be *partly* opened and water added to the boiler when the movable (white or red) arrow on the altitude gauge drops below the setting of the stationary (black) arrow (see "Altitude Gauges" in this chapter).

Some pressure-reducing valves are equipped with a built-in check valve in the supply inlet to prevent a backflow of contaminated boiler water into the domestic water supply. This backflow drainage of boiler water generally occurs when the supply pressure falls below the feeder valve setting. Not only can this drainage cause possible damage to the boiler, it can also contaminate the domestic water supply. Some pressure-reducing valves are also equipped with integral strainers to trap foreign matter that may clog the valve. The strainer screen should be removed for cleaning at the beginning of each heating season. This can be done by removing the bottom plug from the strainer and withdrawing the screen. Clean and carefully replace screen and plug.

The principal components of a water-pressure-reducing valve are shown in Fig. 15-66. This particular valve is constructed with an integral strainer. It is also possible to use separate units in combination such as Fig. 16-67 illustrates. The two units are joined by a short threaded connection. A typical installation in which a water-pressure-reducing valve is used is shown in Fig. 15-68.

As mentioned before, a water-pressure-reducing valve should be installed in the water supply line to the boiler. It should also be installed at a level *above* the boiler and in a horizontal position. Before installing the valve, flush out the supply pipe to clear it of chips, scale, dirt, and other foreign matter that could interfere with its operation. Install a shutoff valve *ahead* of the regulator, and then connect the supply line to the inlet (usually marked "in" on the valve casting).

To fill the system, open the shutoff valve located ahead of the pressure-reducing valve. Water will flow into the system until it is full and under pressure. The shutoff valve must always be kept open when the system is in operation.

Water-pressure-reducing valves are usually set by the manufacturer to feed water to the boiler at approximately 15 lb. pressure. This pressure setting is sufficient for residences and houses up to a three-story building in size. For higher buildings in which the pressure may not be sufficient to lift the water to highest

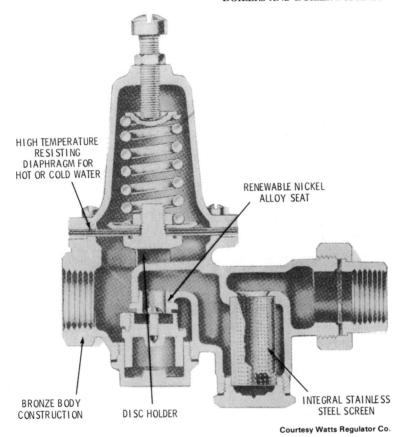

HIGH TEMPERATURE
RESISTING
DIAPHRAGM FOR
HOT OR COLD WATER

RENEWABLE NICKEL
ALLOY SEAT

BRONZE BODY
CONSTRUCTION

DISC HOLDER

INTEGRAL STAINLESS
STEEL SCREEN

Courtesy Watts Regulator Co.

Fig. 15-66. Basic components of a water pressure-reducing valve with integral stainless steel strainer.

radiation, it may be necessary to reset the water-pressure-
reducing valve for higher pressure. To do this, calculate the
number of feet from the regulator to the top of the highest radiation. Multiply this by 0.43 and add 3 lb. This will give the pressure needed to raise the water to the highest radiator and keep it
under pressure. Loosen the locknut on the valve, and turn the
adjustment screw clockwise slowly until the gauge indicates the
pressure calculated. When the desired pressure is obtained,
tighten the locknut.

531

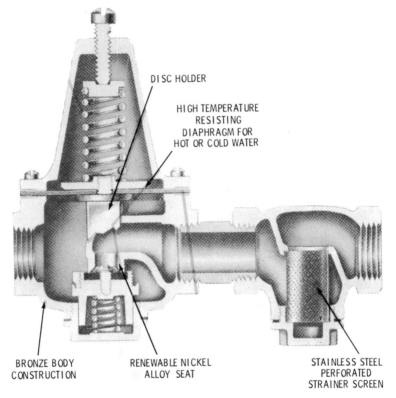

DISC HOLDER

HIGH TEMPERATURE
RESISTING
DIAPHRAGM FOR
HOT OR COLD WATER

STAINLESS STEEL
PERFORATED
STRAINER SCREEN

BRONZE BODY
CONSTRUCTION

RENEWABLE NICKEL
ALLOY SEAT

Fig. 15-67. Basic components of a water-pressure-reducing valve with separate strainer unit.

Combination Valves (Dual Controls)

Combination, or *dual control, valves* (Figs. 15-69 and 15-70) are designed for use in hot-water, space heating systems. These valves consist of a pressure-reducing and -regulating valve and a positive relief valve contained in one body. Sometimes an integral bypass valve is also included in the same unit.

The purpose of a combination valve is to provide pressure regulation and safety control, and to reduce boiler pressure and ensure automatic filling when conditions warrant. The valve is installed on the supply line of a boiler with the relief-valve section closest to the boiler.

The regulator side of the combination valve is designed to reduce the incoming water supply to the boiler to the required 14-lb. operating pressure for a one-, two-, or three-story house. The pressure in the system builds up due to thermal expansion when the boiler is fired. Under normal operating conditions, the expansion tank will absorb the additional pressure. However, if the expansion tank is waterlogged or if the system has no expansion tank, the relief side of the valve will open at 23 psi and drop the pressure back to within safe limits. If the pressure should drop below 14 psi, the regulator will open again and automatically refill the system. A built-in check valve prevents the backflow of contaminated boiler water into the potable boiler water supply.

Exploded views of combination valves are shown in Figs. 15-71 and 15-72. The Cash-Acme Type CBL Valve (Fig. 15-73) differs from the other two by having a built-in bypass valve. In all other respects, it is similar to the others except that it should be installed as close as possible to the top of the boiler (using close nipples). The principal advantages of having the bypass valve included are:

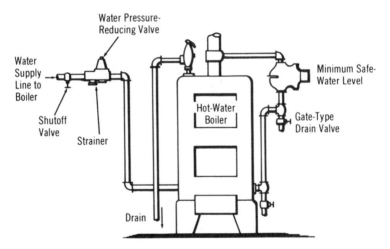

Courtesy Watts Regulator Co.

Fig. 15-68. Typical installation diagram showing the location of a water-pressure-reducing valve.

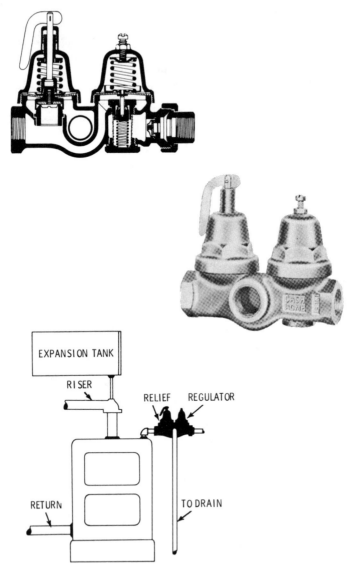

Fig. 15-69. A combination relief/regulator valve (Type CQ).

534

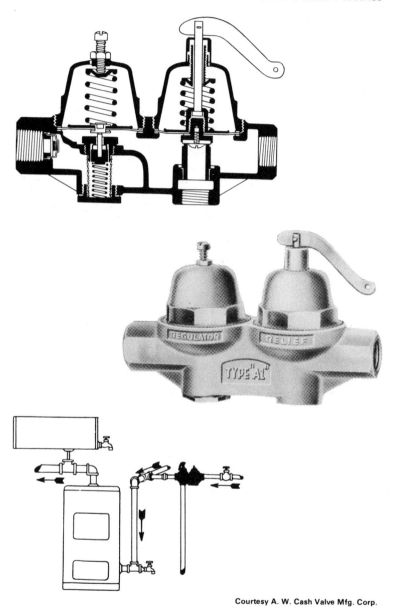

Courtesy A. W. Cash Valve Mfg. Corp.

Fig. 15-70. A combination relief/regulator valve (Type A-1).

535

RELIEF SECTION

1. Lever Pin
2. Lift Lever
3. Pressure Screw Cap
4. Pressure Screw
5. Lock Nut
6. Spring Chamber
7. Spring Button
8. Pressure Spring, stainless steel
9. Pull Rod Nut
10. Pull Rod
11. Lock Washer
12. Pressure Plate
13. Gasket, brass
14. Diaphragm
15. Composition Seat Shell
16. Composition Seat
17. Composition Seat Screw
18. Body Seat

24. Pressure Spring, black
25. Gasket, brass
26. Diaphragm Assembly
27. Cylinder
28. Body
29. Piston Assembly
30. Piston Spring
31. Strainer
32. Gasket, red fiber
33. Bottom Plug

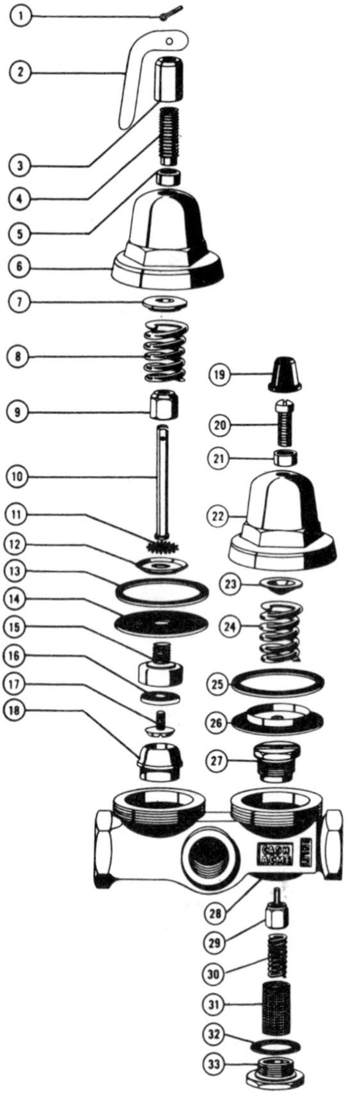

REGULATOR SECTION

19. Closing Cap
20. Pressure Screw
21. Lock Nut
22. Spring Chamber
23. Spring Button

Courtesy A. W. Cash Valve Mfg. Corp.

Fig. 15-71. An exploded view of Type CQ combination valve.

RELIEF SECTION

1. Lift Lever Pin
2. Lift Lever
3. Pressure Screw Cap
4. Pressure Screw
5. Lock Nut
6. Spring Chamber
7. Spring Button
8. Pressure Spring
9. Pull Rod Nut
10. Pull Rod
11. Pressure Plate
12. Diaphragm Gasket, brass
13. Diaphragms, bronze
14. Diaphragm Gasket, black
15. Seat Shell Gasket, gray
16. Seat Shell
17. Composition Seat Disc
18. Seat Disc Screw
19. Body Seat
20. Body

REGULATOR SECTION

21. Pressure Screw
22. Lock Nut
23. Spring Chamber
24. Spring Button

25. Pressure Spring
26. Diaphragm Gasket, brass
27. Diaphragm Assembly
28. Diaphragm Gasket, black
29. Cylinder
30. Piston Assembly
31. Piston Spring
23. Strainer Screen
33. Bottom Plug Gasket, red
34. Bottom Plug

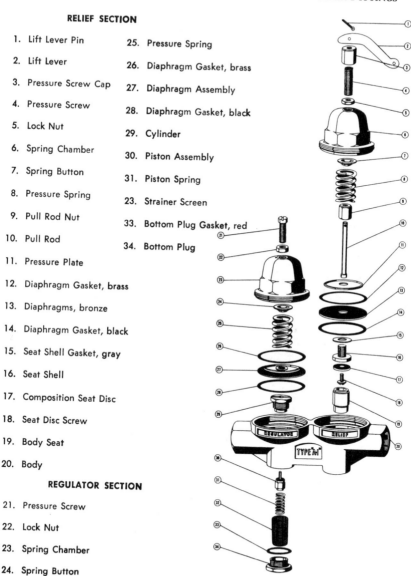

Courtesy A. W. Cash Valve Mfg. Corp.

Fig. 15-72. An exploded view of Type A-1 combination valve.

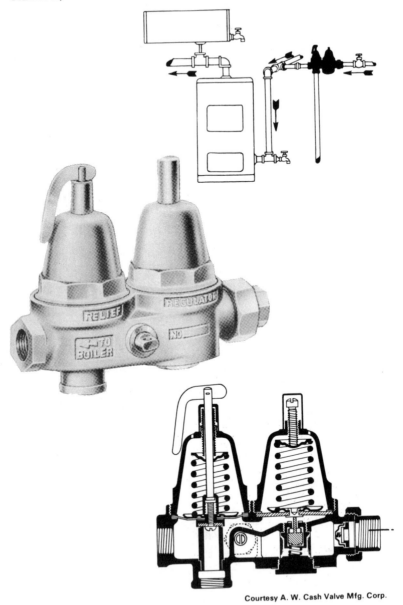

Courtesy A. W. Cash Valve Mfg. Corp.

Fig. 15-73. A combination relief/regulator valve (Type CBL).

1. Allows rapid filling of the system.
2. Permits an easy high-pressure test for leaks and system purging.
3. Passes first filling dirty water around the valve seat, ensuring a clean, good seating surface.
4. Permits the use of a wide opening, small seated regulator that prevents wire drawing and rapid wearing of the seat.

The most common types of problems encountered when using combination valves are the following:

1. Relief valve drips or cannot be shut off tightly.
2. Boiler pressure rises above the reducing valve setting.
3. Water escapes from "weep hole" on top of the regulator.
4. Water escapes from feeder side of the combination valve.
5. Foreign matter collects on the seat of the relief or regulator section of the valve.

Some of the problems listed above (particularly the first two) may *not* be the fault of the valve itself. For example, a dripping relief valve can be caused by an expansion tank being completely filled up with water, an undersized expansion tank, or a leak in the coil of a tankless or indirect water heater installed in the boiler. These possibilities should be checked *before* attempting to service or repair the valve. The procedures are outlined in this chapter.

If, by process of elimination, the problem can be traced to the valve, try tapping the side of the valve with a wrench. Sometimes a piece of foreign matter becomes lodged and causes the regulator piston to stick. A sharp tap with a wrench may dislodge it and allow the valve to function properly.

Foreign matter such as dirt, pipe scale, or chips often cause a valve to malfunction by lodging on a seating surface, or nicking or chipping the surface. Valve manufacturers often provide replacements or instructions for field servicing and repairing. The latter should be attempted only by a skilled and experienced worker with the proper tools and gauges.

Water seeping from the regulator "weep hole" or from the feeder side of a combustion valve is usually an indication that there is a rupture or leak in the diaphragm. Follow the procedures described in the preceding paragraph for repairing the valve. An occasional flushing of the relief side of a combination valve will reduce the possibility of the type of lime or scale buildup that can cause the valve to fuse shut.

Air Separators

An *air separator* is a device used to trap, separate, and eliminate air from the water in a hot-water, space heating boiler. Air separators are *not* installed on boilers used only for heating the domestic water supply.

The air separator shown in Fig. 15-74 is screwed into the boiler supply nozzle. The separator traps the air in the top of the boiler section where the water is hottest and where it travels a long

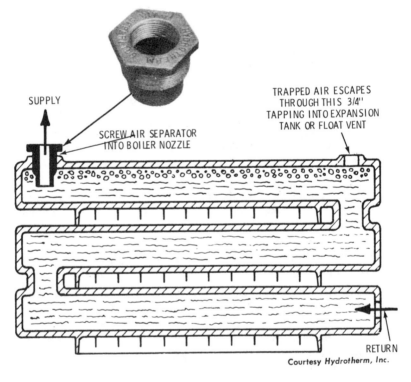

Fig. 15-74. Air separator.

horizontal path at low velocity, permitting the air bubbles to separate. The trapped air escapes through a ¾-in. tapping into an air-cushion expansion tank or through an automatic float vent on systems using diaphragm expansion tanks.

An air separator can only expel air that reaches the boiler. However, sometimes air pockets remain trapped in the piping or radiation, thereby impeding the water flow and reducing the heating performance of the boiler. Such air pockets can be eliminated by installing several different purge fittings in the heating system. A manual air vent is installed on the highest point of the system (automatic air vents are not generally required). (See below.)

Some boilers are constructed with built-in air separators, which make the addition of an external unit unnecessary. Fig. 15-75

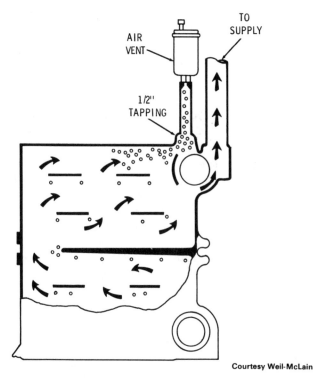

Fig. 15-75. Built-in air separator.

illustrates the working principles of one of these integral units. In this design, the air is diverted to an automatic air vent. A variation of this design provides for the diversion of air to an expansion tank. The manufacturer refers to this device as an *air eliminator*. Its function is similar to the air eliminator used on pipelines, but it differs in design and construction.

Purging Air from the System

Built-in or externally installed air separators are designed to separate air from the water in an *operating* boiler and vent it either into an expansion tank or to the atmosphere through an automatic air vent.

Air is also removed *from a heating system* by purging it from the piping connections to the boiler. This type of purging is standard when the system is being filled with water. Purging instructions are usually included in the owner's manual or the manufacturer's installation guide. Fig. 15-76 illustrates purging instructions for a Hydrotherm gas-fired boiler.

AIR SUPPLY AND VENTING

An adequate air supply must be provided for combustion to boilers that are fired by gas, oil, or coal, and the products of combustion must be vented to the outside atmosphere.

Combustion air (i.e., the air supply) is normally supplied through venting ducts or openings in the walls. The requirements for the air supply will depend upon the location and enclosure of the boiler.

Boilers installed in unconfined areas usually obtain adequate air for combustion by means of normal infiltration. However, normal air infiltration will prove to be inadequate for this purpose if the construction of the structure is unusually tight. Under these circumstances, provision must be made for the entry of additional air from outside the building. Unobstructed openings with a total free area of not less than 1 sq. in. per 5000 Btuh of the total input of the boiler will be necessary to provide an adequate air supply for combustion purposes.

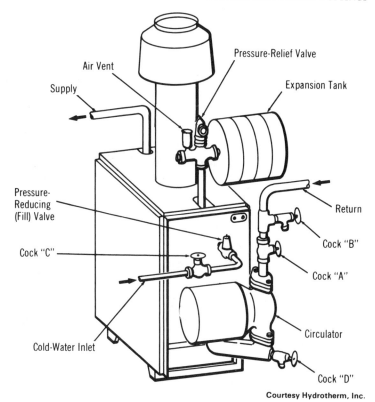

Courtesy Hydrotherm, Inc.

Fig. 15-76. Purging instructions for Hydrotherm HC series, gas-fired boiler.

Some boilers are installed in boiler rooms or enclosures supplied with combustion air from *inside* the structure. The air supply must enter and leave the boiler room through two openings in an *interior* wall or door. One opening is located near the ceiling, the other opening near the floor. Each opening should have a free area of not less than 1 sq. in. per 1000 Btuh of the total input of the boiler.

If a boiler is located in a boiler room or similar enclosure that receives its air supply from *outside* the building, the two air openings are located on an *exterior wall*. Each opening must have a free area of not less than 1 sq. in. per 4000 Btuh of the total input of the boiler.

543

The products of combustion must be vented to the outside atmosphere. In order to accomplish this, the boiler must be connected to a suitable venting system, which should include a flue pipe and a chimney or stack of adequate size and capacity.

The chimney height is usually governed by the height of the structure. As a general rule-of-thumb, the chimney should extend beyond the high-pressure area caused by the passage of 10 ft. of the flashing. If the chimney is too short, it will not extend beyond the high-pressure area caused by the passage of winds over the structure. As a result, a downdraft caused by the pressure difference in the high-pressure area will suck air down into the chimney, which will block the escape of flue gases and interfere with the combustion process. Figs. 15-77 and 15-78 illustrate these principles. The *correct* chimney height is shown by the dotted lines.

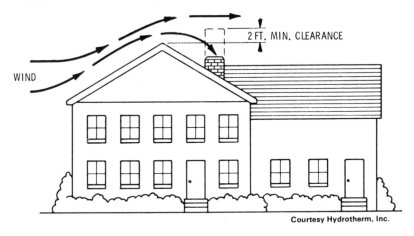

Courtesy Hydrotherm, Inc.

Fig. 15-77. Correct chimney design for a residence.

Fig. 15-79 shows a type of chimney construction referred to as a *Type A vent.* This chimney can be either masonry or factory-built construction, but it must be designed and constructed in accordance with the standards set forth in national codes. Type A vents are suitable for venting of all gas appliances. These are the *only* vents suitable for venting oil-burning equipment. Gas-fired boilers that produce temperatures at the draft diverter not in excess of 550°F may also use a *Type B vent* (Fig. 15-80), which

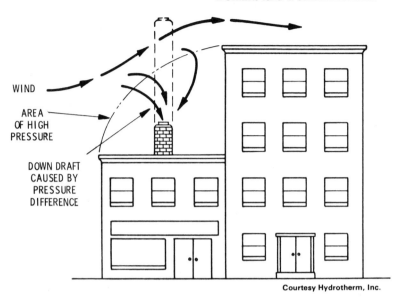

Courtesy Hydrotherm, Inc.

Fig. 15-78. Correct chimney design for a building.

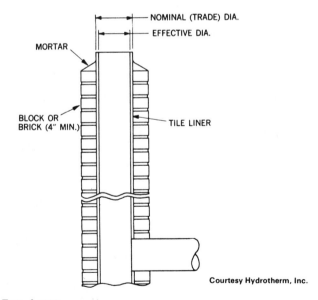

Courtesy Hydrotherm, Inc.

Fig. 15-79. Type A vent.

employs a double-wall metal vent pipe. A *Type C vent* (Fig. 15-81) is used to vent gas appliances in attic installations.

The boiler should be located so that the length of flue pipe connecting the boiler (or draft diverter) to the chimney is as short as possible. Additional information about chimneys and flues is contained in Chapter 11 (Gas-Fired Furnaces).

INDUCED DRAFT FANS

Sometimes a chimney is too small to provide enough updraft to convey the products of combustion from the boiler to the outside atmosphere. When it is evident that natural venting will be inadequate, a mechanical draft inducer can be used to increase the capacity of the chimney. These devices are used with both gas- and oil-fired boilers.

The two principal types of induced draft fans are: (1) direct-fan draft inducer and (2) induced-flow draft inducer. When a direct-fan draft inducer is used, the gases are drawn through an inlet located on the side, top, or bottom of the unit and discharged as shown in Fig. 15-82. The moving parts of an induced-flow draft inducer (Fig. 15-83) are located outside the path of the hot flue gases. Cool outside air is introduced through the blower section of the unit. This creates suction at the throat of the inducer body, which, in turn, increases the flow rate of the flue gases.

The draft created by the induced draft fan should closely match the demand. If the size of the draft inducer is correctly estimated, any dilution and excessive cooling of the flue gases (a condition that may result in condensation) will be greatly reduced. Many boiler manufacturers recommend the size draft inducer to be used with specific boiler models and provide data for making the appropriate calculations. Manufacturers of draft inducers are similarly helpful.

CONTROL OF EXCESSIVE DRAFT
ON GAS-FIRED BOILERS

Sometimes a boiler installation will have excessive draft conditions as a result of oversized chimneys or other factors. Draft

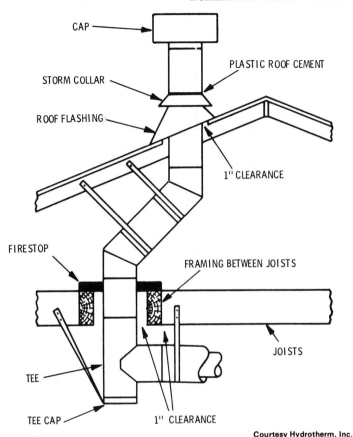

CAP

PLASTIC ROOF CEMENT

STORM COLLAR

ROOF FLASHING

1" CLEARANCE

FIRESTOP

FRAMING BETWEEN JOISTS

JOISTS

TEE

TEE CAP

1" CLEARANCE

Courtesy Hydrotherm, Inc.

Fig. 15-80. Type B vent.

conditions in excess of the draft design limits of the boiler will reduce the combustion efficiency of the boiler, resulting in higher fuel costs and possible pilot problems.

A suggested method of controlling the draft on such installations is to install a sheet-metal baffle, or restrictor, in the flue connection at the boiler (Fig. 15-84). This baffle may be inserted in a tee, as shown, and secured with stove bolts after the proper draft is obtained by measurements with a draft gauge. If the barometric control is installed in the side of the breeching by means of a draft control collar, the same method of baffling may

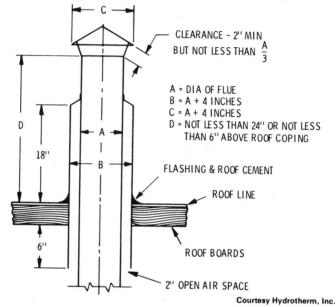

CLEARANCE - 2" MIN
BUT NOT LESS THAN $\frac{A}{3}$

A = DIA OF FLUE
B = A + 4 INCHES
C = A + 4 INCHES
D = NOT LESS THAN 24" OR NOT LESS
THAN 6" ABOVE ROOF COPING

FLASHING & ROOF CEMENT

ROOF LINE

ROOF BOARDS

2" OPEN AIR SPACE

Courtesy Hydrotherm, Inc.

Fig. 15-81. Type C vent.

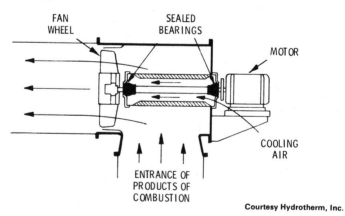

FAN
WHEEL

SEALED
BEARINGS

MOTOR

COOLING
AIR

ENTRANCE OF
PRODUCTS OF
COMBUSTION

Courtesy Hydrotherm, Inc.

Fig. 15-82. Direct fan draft inducer.

be used with the baffle restricting the breeching between the boiler and the barometric control. In this method of baffling, the breeching between the barometric control and breeching should not be restricted.

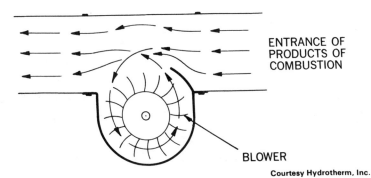

Courtesy Hydrotherm, Inc.

Fig. 15-83. Indirect fan draft inducer.

Sheet-Metal Baffle
Attached with Stove Bolts

To
Chimney

MG-2
Barometric
Double Acting

Measure Draft
at This Location

Suggested Baffle Dimensions			
Barometric Size Diameter	Cut Sheet		
	A	B	R
10"	10"	14"	5"
12"	12"	16¾"	6"
14"	14"	19½"	7"
16"	16"	22½"	8"
18"	18"	25"	9"
20"	20"	28"	10"
24"	24"	33½"	12"

The above dimensions for baffle may vary due to different types of flue connections and tee sizes.

Courtesy Bryan Steam Corp.

Fig. 15-84. Control of excessive draft of gas-fired boilers.

The baffle installation is correct if the gate of the barometric control is approximately half open while the gas burner is operating and the draft measurement is at or below the operating design of the boiler. The draft limit for a boiler can be obtained

549

from the manufacturer's specifications. For example, Bryan atmospheric gas boiler burner units are designed to operate at very low draft, never to exceed 0.01 in. W.C.

Always check with the local gas company regarding the use of baffles. Caution should be exercised to ensure against overrestricting the flue. Always be certain that there is a very slight draft after the baffle is secured in position.

TANKLESS WATER HEATERS

Some boilers are manufactured with the option of using a *tankless water heater*. This device consists of an immersion coil inserted in a steam or hot-water, space heating boiler to provide domestic and commercial service hot water. It is called a "tankless" water heater because no storage tank is used to store the heated water during periods of low demand.

The immersion coil is made of small-diameter copper tubes that are either straight with U-bends at the end (Fig. 15-85) or

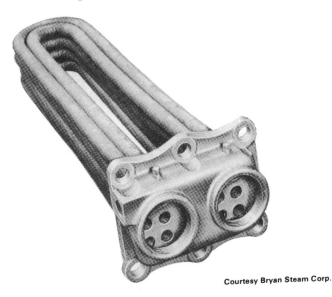

Courtesy Bryan Steam Corp.

Fig. 15-85. Heat exchanger coil.

550

formed in the shape of a spiral (Fig. 15-86). Typical locations on steam and hot-water, space heating boilers are illustrated in Fig. 15-87. As shown, the tankless heater is installed in the nipple port on cast-iron, hot-water boilers. On steam boilers, it is installed in

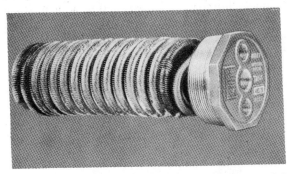

Courtesy H. B. Smith Co., Inc.

Fig. 15-86. Heat exchanger coil used in gas-fired boilers.

STEAM HEATER

WATER HEATER

Courtesy Weil-McLain

Fig. 15-87. Location of tankless water heater on oil-fired, hot-water boiler.

the left-hand side of a special back section at a point well below the water line. A uniform water temperature is maintained by a thermostatic three-way mixing valve installed in the supply line leading from the immersion coil.

Because these immersion coils function as a heat exchanger type device and operate on an indirect heating principle, they are also variously referred to as an *indirect water heater, indirect heat exchanger,* or *heat exchanger coil.*

In other applications, immersion coils provide hot-water radiation from a steam boiler, hot water for snow melting, heated water for pools, and industrial process water.

Additional information about water heaters, including suggestions for estimating the hot-water allowance, can be found in Chapter 4 of Volume 3 (Water Heaters).

LEAKING COILS

On occasion, the coil of a tankless or indirect water heater installed in a boiler will leak, allowing high-pressure water into the boiler water and resulting in a rise in the boiler pressure. This, in turn, can cause the relief valve to open and fail to close again tightly, or it can cause the boiler pressure to rise above the setting of the pressure-reducing valve. These symptoms are often wrongly attributed to a valve malfunction when the actual cause may be a leaking coil.

You can determine whether the problem is a leaking coil by taking the following steps:

1. Shut off the feed valve and/or break the connection to the coil.
2. Check the pressure reading in the boiler.
3. Wait about 8 hours, and check the boiler pressure reading again.

If the pressure reading has remained approximately the same over the 8-hour period, it is a strong indication that the coil is leaking. If the boiler pressure continues to rise, the problem may lie elsewhere in the system.

BLOWING DOWN A BOILER

Foam sometimes forms on the surface of the boiler water and is usually indicated by drops of water appearing with the steam. This condition is caused by the presence of oil, dissolved salts, or similar organic matter in the water. One method of eliminating this problem is by draining off part of the water in the boiler and adding an equal amount of fresh, clean water. Another method consists of blowing the foam from the water with a specially connected pipe or hose. The boiler should have a blowdown tapping for this purpose. The first method is the easiest. The second one (i.e., blowing down the boiler) requires considerable experience.

BOILER OPERATION AND MAINTENANCE

Always follow the manufacturer's recommendations for boiler operation and maintenance. If you do not have these instructions (e.g., an owner's manual), you should contact the local dealer for advice or write to the manufacturer.

The sections that follow contain recommendations for operating and maintaining boilers. Because many of these recommendations specifically pertain to either steam or hot-water, space heating boilers, they were divided and listed accordingly. The final three sections ("Boiler Water," "Cleaning Boilers," and "Insulating Boilers") contain general recommendations for all boilers.

Steam Boilers

Operating and maintaining a steam boiler differs in certain respects from the procedures followed for hot-water, space heating boilers. Many of these differences are evident in the recommendations that follow.

1. Keep a steam boiler filled to the water level recommended by the manufacturer when not in use.
2. Check the water level in the boiler before starting it. The heating surface can be damaged if the water level is too low.
3. Keep the water level at the center of the water gauge glass

553

during operation. If the water level is too low, use the manual feed valves to add more water. These valves are found in most systems.

4. Always add water to the boiler gradually. If at all possible, avoid adding water to a hot boiler. *Never* add water to an operating cast-iron boiler. Cold water fed rapidly into such boilers may come in contact with the hot surface of the cast-iron heat exchanger, causing it to crack. Shut off the boiler and wait until it has cooled down before adding water. After the boiler has had sufficient time to cool, slowly add water through the cold-water feed line.

5. Check the low-water cutoff at regular intervals. Sediment or rust accumulating under the float of the low-water cutoff can cause it to malfunction. As a result, a drop in the boiler water level would not register properly.

6. Check all boiler accessories to make certain they are functioning properly. Movable parts should be inspected and oiled regularly.

7. Never allow a low-pressure steam boiler to exceed the upper safe pressure limits recommended by the manufacturer (usually 3- to 4-lb. pressure).

Hot-Water, Space Heating Boilers

Recommendations for operating and maintaining hot-water space heating boilers are as follows:

1. Keep the boiler *and the pipes* in the heating system filled with water when not in use. Keeping the pipes filled with water reduces the possibility of rust and corrosion.

2. Check the water level in the boiler before starting it. The heating surface can be damaged if the water level is too low.

3. Always add water to a boiler gradually. Never add water to a hot boiler. Shut the boiler down and allow it to cool first.

4. Check all boiler accessories to make certain they are functioning properly. Movable parts should be inspected and oiled regularly. Such maintenance should also include the pump in a forced, hot-water heating system.

5. When operating a hot-water, space heating boiler, make certain all flow valves are open.

is provided for the steam to form at the top of the boiler. Too little water in a hot-water, space heating boiler can be caused by the limit control moving down to a lower setting.

3. *Too much time is required to get up steam in a steam boiler.* This common problem can be traced to the following possible causes:

   a. Too little or badly arranged heating surface.
   b. Heating surfaces covered with soot.
   c. Heating passages too short.
   d. Poor fuel or fuel firing.
   e. Poor draft.
   f. Boiler too small.
   g. Boiler defective.

4. *Boiler is slow to respond to the operation of the dampers.* Slow response to damper operation can be caused by any of the following:

   a. Air leakage into the chimney or stack.
   b. Poor fuel or fuel firing.
   c. Boiler too small.
   d. Clinkers on grate or ashpit full of ashes (coal-fired boilers).

5. *Water line is unsteady.* The problem of an unsteady water line may simply be due to connecting the water column to an extremely active section of the boiler. Therefore, it is extremely unlikely that the actual water level in the boiler can be read from the water column. Other possible causes of this problem are:

   a. Dirt or grease in the water.
   b. Varying pressure differences on the system.
   c. Excessive boiler output.

6. *Water disappears from the gauge glass.* The complete loss of water from the gauge glass could be due to priming (i.e., water globules being carried over into the steam). Other causes include:

a. Foaming.
b. Pressure drop too great in return line.
c. Improper water gauge connection.
d. Valve closed in the return line.

7. *Water is carried over into the steam main.* This problem is usually caused by one of the following:

a. Priming or foaming.
b. Water line is too high.
c. Outlet connections from boiler too small.
d. Steam liberating surface too small.
e. Boiler output excessive.

8. *Flues require cleaning too frequently.* A frequent buildup of soot and dirt in flues can be caused by any of the following:

a. Combustion rate too slow.
b. Poor draft.
c. Smoky combustion.
d. Excess air in firebox.

9. *Low carbon dioxide.* This condition can generally be traced to one of the following causes:

a. Air leakage between cast-iron sections.
b. Improper conversion job.
c. Problem with burners.

10. *Smoke from boiler fire door.* The following conditions may be the cause of this problem:

a. Dirty or clogged flues.
b. Incorrect setting of dampers.
c. Poor or defective draft in the chimney.
d. Incorrect reduction in the breeching size.

BOILER REPAIRS

Repairs to the boiler itself should be done by an experienced and skilled worker. They should never be attempted while the boiler is under pressure. Shut down the boiler and allow it to cool.

Most structural repairs should be regarded as only temporary until the section or part can be replaced with a new one. Minor cracks or leaks can be repaired as follows:

1. Minor cracks in the sections of a cast-iron boiler can be repaired by welding or brazing.
2. Boiler leaks can be *temporarily* sealed with a synthetic compound similar to the type used to seal automobile radiators. Permanent repairs should be made by an experienced worker.

BOILER INSULATION

All uninsulated boilers suffer a certain amount of heat loss. In some respects, this loss of heat can be beneficial because it does warm the basement and the floor above. However, if the boiler is also used to provide domestic or commercial service hot water, this heat loss can be uneconomical. This is particularly true for the summer months.

Steel-jacketed boilers are generally insulated by the manufacturer. For example, the boiler illustrated in Fig. 15-3 is insulated with a 1-in.-thick layer of fiberglass insulation attached to the inside surface of the steel jacket. However, other types of boilers, particularly the old cast-iron round boilers, do require an external layer of insulation to reduce the amount of heat loss. *Only* a noncombustible material such as asbestos cement, asbestos paper, or magnesia blocks should be used for this purpose.

Magnesia blocks or sheets can be cut to fit the contours of a boiler and wired to the surface. The cracks and insulating material are then covered with asbestos cement. Corrugated asbestos paper can be used in place of magnesia block, but it should be applied in three layers to obtain the desired thickness (about 1 ¼ in.). Each layer should be covered with asbestos cement.

BOILER INSTALLATION

All new boilers are shipped with a complete set of installation instructions. Usually these instructions will also contain an inventory list of the contents in the crates and cartons. *Always* check the contents off against this list before you do anything else. Missing or damaged parts should be immediately reported to the manufacturer or his local representative.

A boiler should always be located so that the connecting flue pipe between the boiler (or draft diverter) and the chimney is as short as possible. Another important consideration is the recommended minimum clearances between the boiler and combustible materials. These factors plus the design considerations of the system will dictate where the boiler is located.

The manufacturer's installation guide will also contain instructions on how to light and operate the boiler. The instructions will differ in accordance with the automatic fuel-burning equipment used (e.g., oil burner and gas burner).

After completing the installation of the boiler, inspect all the controls to make certain they are operating properly. Start and stop the burner or stoker several times by moving the room thermostat setting.

An important point to remember is that *any and all* requirements of national standards and codes, as well as local authorities having jurisdiction, *always* take precedence over the installation instructions provided by the manufacturer. For example, the latest edition of the American National Standard for Installation of Gas Appliances and Gas Piping Z21.30 should be consulted when installing gas-fired boilers.

WATER CHILLERS

One method of adding central air conditioning to a structure heated by either hot water or steam is to install a water chiller. A *water chiller* is a device consisting of a compressor, a condensor, a thermal expansion valve, and a refrigerant evaporator coil enclosed in a steel shell. The water in the system is pumped through the shell, where it is cooled by the evaporator coil and is circulated through the radiators or convectors for cooling and dehumidification.

The boiler and water chiller are available packaged in a single unit, or they may be installed as separate units. In a hot-water heating system, the same piping can be used to carry both the hot water and chilled water to the radiators or convectors. The room radiators or convectors are individually controlled to permit temperature adjustments. Steam piping should not be used to carry the chilled water.

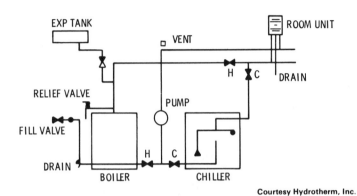

Fig. 15-89. Piping connections for boiler and water chiller. Valve H opened on heating with valve C closed. Reverse procedure on cooling.

Few residential water chillers are in use. The most popular method of applying central air conditioning to a steam or hot-water, space heating system is to install a split system.

When a water chiller is installed as a separate unit, the chilled water should be piped in parallel with the boiler. Appropriate valves should be installed to prevent the chilled water from entering the boiler (Fig. 15-89).

CHAPTER 16

Boiler and Furnace Conversions

Sometimes it becomes necessary to convert an existing boiler for furnace to another fuel. This is usually the case when the fuel for which it was originally designed has become too expensive relative to others or does not meet the desired level of efficiency. During the early 1950's, when coal was becoming more and more expensive as a domestic heating fuel and before oil and gas burners were offered as integral parts of heating units by manufacturers, converting boilers and furnaces was much more widespread than it is today.

A number of manufacturers have made available gas-fired and oil-fired burners specifically designed for boiler and furnace conversions. It would be impossible within the space limitations of this chapter to cover all the different models of conversion burners offered by these manufacturers. Consequently, this chapter will emphasize all the *ramifications* involved in converting a boiler or furnace from one fuel to another instead of examining the specifications and operating characteristics of one or more conversion burners.

PREPARATION FOR CONVERSION

Before constructing a combustion chamber and installing a burner, the heating system should be carefully checked for defects and cleanliness. A boiler or furnace in need of repairs will not give satisfactory results after the burner is installed. Be certain that *all* flue passages are cleaned so that the maximum amount of heat generated is absorbed by the boiler. Soot and ash are good insulators, and both are always undesirable.

All doors should fit tightly, and all other openings or cracks should be tightly cemented shut. The stack from the furnace to the chimney should have tight joints. Dampers should not close off more that 80 percent of the cross-sectional area of the stack. Stacks should be inspected for leaks and obstructions of any kind.

Remove the old fuel-handling parts (e.g., the coal-handling parts in a coal-fired furnace or boiler) and thoroughly clean the interior surfaces, removing all soot, scale, tars, and dirt.

Any cracks in the heat exchanger should be repaired or replaced. On boilers, any water leaks should be sealed, broken gauges replaced, and loose boiler doors repaired or replaced. Air leaks along the floor should be grouted.

No positive catches should be used on firing doors. File off any catches so that the firing door will open easily to relieve pressure. A spring-loaded door holder is recommended.

If a conversion burner with a direct spark ignition system is installed, no provision need be made for top venting or horizontal or downdraft furnace. Check the local codes for further information.

Fig. 16-1 illustrates some of the typical problems encountered when converting a coal-fired *boiler* to oil. It is highly doubtful that a boiler would have all these problems at the same time, but for your convenience they are included in the illustration to serve as a reference.

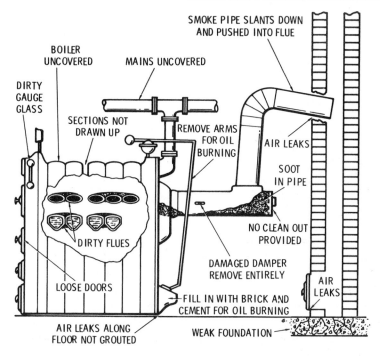

Fig. 16-1. Boiler problems to be remedied before converting and installing an oil burner.

BASIC COMBUSTION CHAMBER REQUIREMENTS

An essential requirement for combustion chambers is that the flame be in the presence of refractory materials so that it will not come into contact with any of the relatively cold heating surfaces of the boiler.

When the flame is burned in suspension in a combustion chamber of refractory material, the refractory walls will reflect the heat back into the flame and thereby increase its temperature. This increase in flame temperature due to reflected heat greatly increases the rapidity of combustion and thereby makes possible the burning of every particle of fuel with the zone combustion.

The combustion chamber walls should be high enough so that the flame will not come in contact with the relatively cool walls of the boiler. At the burner end, the refractory wall need not be higher than the grate line of the boiler.

Finally, a well-designed combustion chamber should provide rapid heating at the beginning of each firing cycle.

COMBUSTION CHAMBERS FOR CONVERSION GAS BURNERS

Conversion gas burners are designed for firing into a refractory lined combustion chamber constructed in the ashpit of a boiler or furnace. It is recommended that builtup combustion chambers used for this purpose be constructed of 2300°F insulating firebrick and cemented with insulating firebrick mortar.

Recommended *minimum* wall thicknesses and floor areas of these combustion chambers are based on the maximum rated Btu capacity (Table 16-1).

The height of combustion chamber walls will usually be determined by the grate line. Build the side and front walls about 2 in. above the grate line. The grate lugs and base of the water legs of boilers should be covered by about 3 or 4 in. to avoid heating sections that may be filled with sediment. Carry the back wall one or two brick courses higher than the side or front walls and allow it to overhang in order to deflect hot gases from impingement on the heat exchanger surfaces. Use a *hard* firebrick for the overhang section because the high-velocity gases moving against it tend to erode a softer brick.

If the combustion chamber is to be placed directly on the floor of the ashpit, a layer of asbestos millboard (½ -in. minimum thickness) should underlay the combustion chamber. Ordinary brick may be used if the floor of the combustion chamber is raised off the floor of the ashpit. The remaining open space should be filled with vermiculite or some other suitable loose insulation.

Table 16-1. Recommended Minimum Wall Thicknesses of Combustion Chambers Using Conversion Gas Burners

Input Btuh	Floor Area Sq. In	Preferred Width and Length	Recommended Minimum Wall Thickness	Recommended Minimum Floor Construction
100,000	180	12 × 15	2½ insulating	2½ insulating
150,000	200	12 × 16	firebrick plus	firebrick plus ½
200,000	220	13 × 17	backup of 1½	asbestos of mag-
250,000	235	13 × 18	or more loose	nesia block
300,000	260	13 × 20	insulation	
350,000	270	14 × 21		
400,000	330	15 × 22		2½ insulating
500,000	400	15 × 27	4½ insulating	firebrick plus 2¼
600,000	460	15 × 31	firebrick	common brick or
700,000	520	15 × 35		hard firebrick, plus ½ asbestos or magnesia block
800,000	600	18 × 33		
900,000	650	18 × 36		4½ insulating
1,000,000	700	22 × 32	4½ insulating	firebrick plus ½
1,100,000	750	22 × 34	firebrick plus	asbestos or
1,200,000	800	22 × 36	backup of 1½	magnesia block
1,600,000	1152	26 × 44	or more loose	or 2¼ common
2,000.000	1440	29 × 50	insulation	brick or hard
2,500,000	1760	32 × 55		firebrick
3,000,000	2100	35 × 60		
3,500,000	2470	38 × 65		
4,000,000	3120	40 × 68		

Courtesy Magic Servant Products Co.

COMBUSTION CHAMBERS FOR CONVERSION OIL BURNERS

A combustion chamber for a conversion oil burner should be constructed of lightweight 2000°F insulating firebrick. With lightweight refractory materials, there is no long smoky delay waiting for the firebox to reach the temperature necessary for complete combustion.

Different types of combustion chamber construction are shown in Figs. 16-2 and 16-3. The shape of the combustion chamber will be determined by the shape of the boiler or furnace.

567

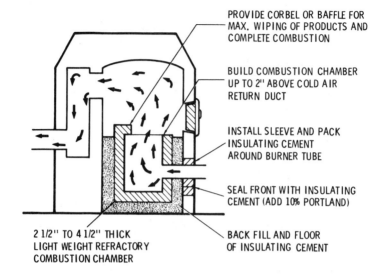

PROVIDE CORBEL OR BAFFLE FOR MAX. WIPING OF PRODUCTS AND COMPLETE COMBUSTION

BUILD COMBUSTION CHAMBER UP TO 2" ABOVE COLD AIR RETURN DUCT

INSTALL SLEEVE AND PACK INSULATING CEMENT AROUND BURNER TUBE

SEAL FRONT WITH INSULATING CEMENT (ADD 10% PORTLAND)

BACK FILL AND FLOOR OF INSULATING CEMENT

2 1/2" TO 4 1/2" THICK LIGHT WEIGHT REFRACTORY COMBUSTION CHAMBER

Courtesy Stewart-Warner

Fig. 16-2. Recommendations for constructing a combustion chamber in a furnace.

WHEN FLUE OUTLET AND TARGET WALL ARE ON SAME END, USE CORBEL AND LOCATE COMBUSTION CHAMBER FOR MAX. CONTACT OF PRODUCTS WITH CROWN SHEET

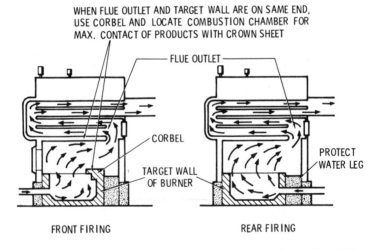

FLUE OUTLET

CORBEL

TARGET WALL OF BURNER

PROTECT WATER LEG

FRONT FIRING

REAR FIRING

Courtesy Stewart-Warner

Fig. 16-3. Combustion chambers for boilers.

The combustion chamber area (space) is equal to the inside width of the boiler times the length times the distance from the combustion chamber floor to the crown sheet. It is recommended that 1 cu. ft. be provided for each boiler horsepower, or 3 cu. ft. per gph oil-fired.

The floor area of the combustion chamber should be based on 100 sq. in. per gallon of oil (see Table 16-2 and Figs. 16-4 and 16-5). The minimum side wall height should conform to the recommended measurements listed in Table 16-2.

An overhang in the form of stepped corbels will deflect hot gases from impingement on the heat exchanger surfaces, increase combustion chamber temperatures, and thereby promote good combustion.

High-flame-retention burners provide an intense flame that eliminates the need for refractory combustion chambers above 3.0 gph, and has greatly reduced the need for such chambers in the 1.0- to 3.0-gph range.

Table 16-2. Combustion Chamber Floor Area Dimensions
(100 sq. in. per gph)

gph	Round I.D.	Rectangular W × L	Nozzle Height "N"	Min. Side Height "H"
0.50	8"	7" × 8"	5"	12"
0.75	9½"	8" × 9"	5"	12"
0.85	10½"	9" × 9½"	5"	12"
1.00	11½"	9" × 11"	5½"	13"
1.10	12"	10" × 11"	5½"	13"
1.25	12"	10½" × 12"	6"	14"
1.35	13"	11" × 12"	6"	14"
1.50	14"	12" × 12½"	6½"	15"
1.75	15"	13" × 13½"	6½"	15"
2.00	16"	14" × 14"	6½"	16"
2.25	17"	14" × 16"	7"	16"
2.50	18"	14" × 18"	7"	17"

Courtesy Stewart-Warner

CONSTRUCTION MATERIALS

The construction materials used in building combustion chambers for conversion burners should consist of the best grade of

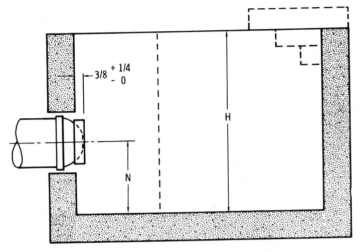

Courtesy Stewart-Warner

Fig. 16-4. Side view of combustion chamber.

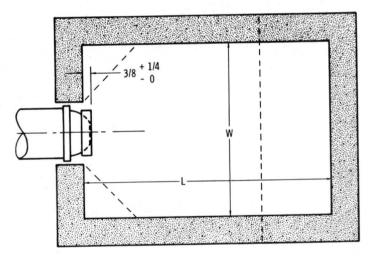

Courtesy Stewart-Warner

Fig. 16-5. Top view of combustion chamber.

insulating firebrick and a good, high-temperature firebrick cement. The shape of the bricks is also important, with arch or circled brick of 4½-in. wall thickness preferred. The cement-filled joints between the bricks should be not more than ⅛ in.

Always use the cement furnished or recommended by the brick manufacturer for cementing the insulating firebrick. The cement should be thinned to the consistency of a very thick cream before it is used. The usual method for applying the cement is to dip each brick into the cement and set it in place as you lay each course.

The insulating firebricks are usually backed by a second insulating layer, which consists of one of the following materials:

1. Asbestos.
2. Common brick.
3. Magnesia block insulation.
4. Hard firebrick.
5. Expanded mica products (e.g., vermiculite or zonolite).
6. Dry sand.
7. Dry sand mixed with an expanded mica product.

Approved insulating firebricks recommended for use in constructing combustion chambers are:

1. Babcock & Wilcox No. K-23 and No. K-26.
2. A.P. Green No. G-23 and No. G-26.
3. Armstrong Cork No. A-23 and No. A-26.
4. Johns-Manville No. JM-23 and No. JM-26.

The spaces between the outer edges of the firebrick should be filled with high-temperature cement and small pieces of firebrick to obtain firm construction and prevent infiltration of vapors through the wall.

BUILDING A COMBUSTION CHAMBER

Fig. 16-6 illustrates the construction of a typical custom-built combustion chamber. The steps involved in constructing this chamber are as follows:

1. Place split brick A on the ashpit floor to form the floor of the combustion chamber.
2. Mix the refractory binder in a shallow pan to the consistency of a heavy cream.
3. Dip the bottom and sides of brick B in the binder and place it in a circular position on the combustion chamber floor. You should start at the back and work toward the front of the chamber.

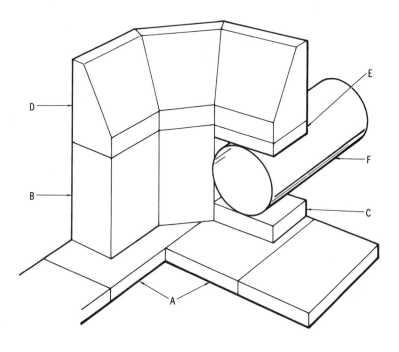

Fig. 16-6. Construction of a custom-built chamber.

4. Follow the same procedure with the rest of the bricks in the tier (i.e., those containing brick B).
5. When the tier is finished, tighten it in position with bands suitable for this purpose.

6. Fill behind the first tier with asbestos or rock wool.
7. Install top tier *D* in the same manner described in Steps 1 to 5; however, use only one band to tighten it.
8. Cap the chamber with a mixture of 3 parts asbestos and 1 part fire clay, and trowel smooth.
9. Place metal sleeve *F* in the opening, and cement it tightly in place.
10. Insert the draft tube tin metal sleeve, and pack with asbestos rope or rock wool.

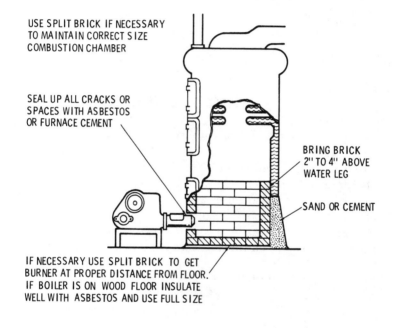

Fig. 16-7. Combustion chamber for an older cast-iron boiler.

Fig. 16-7 shows an old, round, cast-iron boiler that has been converted to oil. Note the hight of the combustion chamber wall. The procedure for setting up the oil burner to the combustion chamber is illustrated in Fig. 16-8.

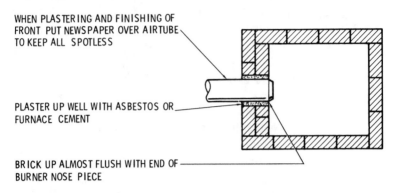

WHEN PLASTERING AND FINISHING OF
FRONT PUT NEWSPAPER OVER AIRTUBE
TO KEEP ALL SPOTLESS

PLASTER UP WELL WITH ASBESTOS OR
FURNACE CEMENT

BRICK UP ALMOST FLUSH WITH END OF
BURNER NOSE PIECE

Fig. 16-8. Procedure for setting up burner to combustion chamber.

VENTILATION REQUIREMENTS

If the existing boiler or furnace is located in an open area (basement or utility area) and the ventilation is relatively unrestricted, there should be sufficient supply of air for combustion and draft-hood dilution. On the other hand, if the heating unit is located in an enclosed furnace room or normal air infiltration is effectively reduced by storm windows or doors, then certain provisions must be made to correct this situation. Fig. 16-9 illustrates the type of modification that should be made in a furnace room to provide an adequate supply of air. As shown, two permanent grilles are installed in the walls of the furnace room, each of a size equal to 1 sq. in. of free area per 1000 Btuh of burner output. One grille should be located near the ceiling, and the other near the floor.

If the boiler or furnace is located in an area of particularly tight construction, the heating unit should be *directly* connected to an outdoor source of air. A permanently open grille sized for *at least* 1 sq. in. of free area per 5000 Btuh of burner output should also be provided. Connection to an outside source of air is also recommended if the building contains a large exhaust fan.

A conversion burner equipped with a spark ignition system is not suited for use in furnaces or boilers located in areas subjected to sustained reverse draft or where large ventilating fans operate in combination with insufficient makeup air. There is always the danger of drawing flue gases into the structure. (See "Installing a Conversion Gas Burner " in this chapter.)

FLUE PIPE AND CHIMNEY

The *flue pipe* is a pipe connecting the smoke outlet of the furnace or boiler with the flue of a chimney (Figs. 16-10 and 16-11). It should *not* extend beyond the inner liner of the chimney and should *never* be connected with the flue of an open fireplace. Furthermore, flue connections from two or more sources should never enter the chimney at the same level from opposing sides.

Flue pipe should be constructed from corrosive resistant metal and be designed with as few sharp turns as practical. The straightest and shortest possible passage should be provided for the existing flue gases.

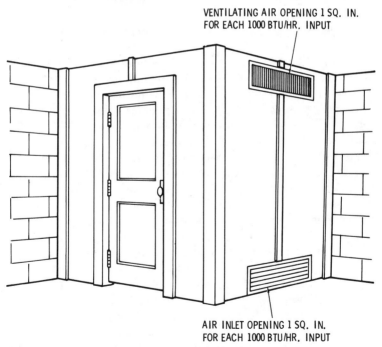

VENTILATING AIR OPENING 1 SQ. IN.
FOR EACH 1000 BTU/HR. INPUT

AIR INLET OPENING 1 SQ. IN.
FOR EACH 1000 BTU/HR. INPUT

Fig. 16-9. Air openings necessary to supply air for combustion when furnace is installed in an enclosed room.

575

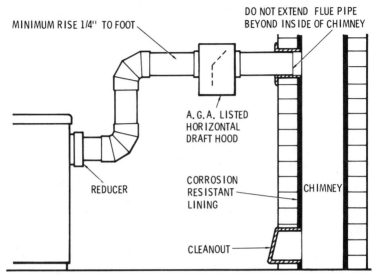

MINIMUM RISE 1/4" TO FOOT

DO NOT EXTEND FLUE PIPE
BEYOND INSIDE OF CHIMNEY

A.G.A. LISTED
HORIZONTAL
DRAFT HOOD

CORROSION
RESISTANT
LINING

REDUCER

CHIMNEY

CLEANOUT

Courtesy Mid-Continent Metal Products Co.

Fig. 16-10. Chimney and flue connection with horizontal draft hood.

Pitch the flue pipe with a rise toward the chimney of at least ¼ in. per ft. At approximate intervals, fasten the flue pipe securely with sheet-metal screws to prevent sagging.

Masonry is recommended for the construction of chimneys (prefab chimneys are also found suitable). Outside metal stacks are generally unsuitable for oil-fired burners.

The ordinary chimney *must* be at least 3 ft. higher than the roof or 2 ft. higher than any ridge within 10 ft. of the building in order to avoid downdrafts.

Aside from adapting the furnace or boiler for a conversion burner, changes must also be made in the passages formed by the heating surface.

Extra-large flue passages are *not* suited to the high-temperature flue gas encountered with oil. In boilers having these large flue passages, baffling must be resorted to in order to slow down the high velocities of the extra-hot gases; otherwise unburned particles of oil may lodge on the heating surface, resulting in carbon. Because carbon is an excellent insulator, the efficiency of the heating surface is lowered whenever it collects. (See the following section.)

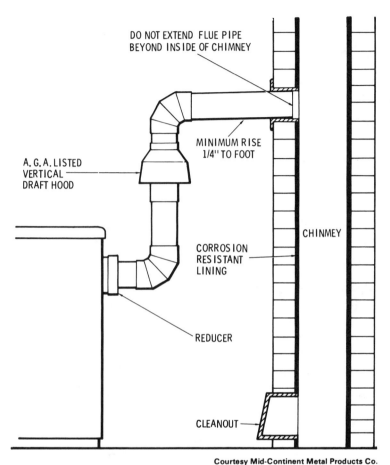

DO NOT EXTEND FLUE PIPE
BEYOND INSIDE OF CHIMNEY

MINIMUM RISE
1/4" TO FOOT

A. G. A. LISTED
VERTICAL
DRAFT HOOD

CHINMEY

CORROSION
RESISTANT
LINING

REDUCER

CLEANOUT

Courtesy Mid-Continent Metal Products Co.

Fig. 16-11. Chimney and flue connection with vertical draft hood.

BAFFLING

Baffling is a type of obstruction designed to deflect and regulate the speed of flue gases. They are especially necessary for round boilers and furnaces constructed so that the flue passes are almost direct from the firebox. Furthermore, boilers designed for burning coal are usually provided with relatively large flue passages which are not normally suited to the higher flue gas velocities encountered in oil firing.

Baffling will, in most cases, help to partly overcome the inefficient operation resulting from the usual excessively high stack-gas temperatures. In such instances it is advisable to experiment with various methods of baffling as shown in Fig. 16-12. *Note:* Any baffling of the flue passes of a nature that prevents efficient operation of the burner should not be done.

Another form of baffling is a *corbel*, or stepout, arrangement of the brickwork of the rear wall, forming a target wall which the flame strikes and is curled back to prevent short circuiting.

GAS PIPING AND PIPING CONNECTIONS

Figs. 16-13 and 16-14 illustrate the piping connections necessary for two different models of gas-fired conversion burners. *Always* refer to local installation codes for guidance because your piping *must* comply with these codes. Other helpful sources of information are the following booklets from the American Gas Association:

1. Bulletin Z21.30, *Installation of Gas Appliances and Gas Piping.*
2. Bulletin Z21.8, *Installation of Domestic Gas Conversion Burners.*

Also helpful is NFPA Bulletin No. 54, *Gas Appliances and Gas Piping*, which is available from the National Fire Protection Association.

No matter what type of gas-fired conversion burner you decide to use, it *must* be allowed to develop its rated capacity. This can be accomplished by making certain the burner is connected to a gas supply containing sufficient pressure.

In addition to providing for a gas supply with sufficient pressure, the following recommendations are also offered:

1. Provide for a separate gas-supply line of 1-in. pipe size direct from the meter to the burner (ample for runs up to 60 ft. in length).

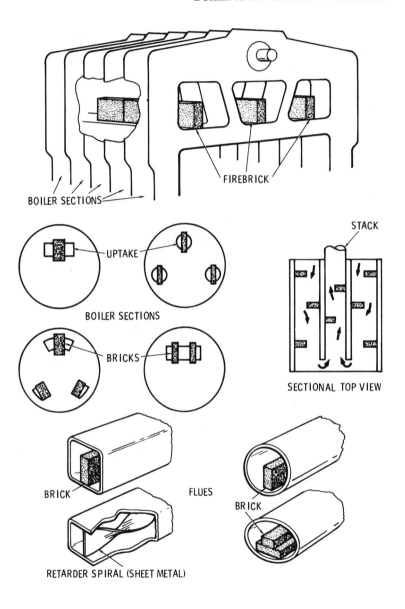

Fig. 16-12. Various methods of baffling for sectional cast-iron boilers.

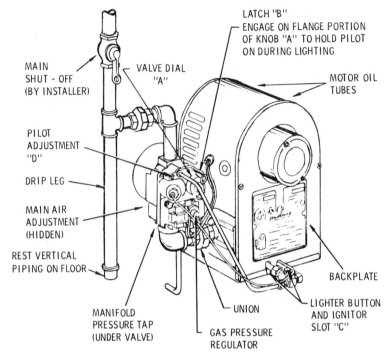

LATCH "B"
— ENGAGE ON FLANGE PORTION OF KNOB "A" TO HOLD PILOT ON DURING LIGHTING

MAIN SHUT - OFF (BY INSTALLER)

VALVE DIAL "A"

MOTOR OIL TUBES

PILOT ADJUSTMENT "D"

DRIP LEG

MAIN AIR ADJUSTMENT (HIDDEN)

REST VERTICAL PIPING ON FLOOR

BACKPLATE

MANIFOLD PRESSURE TAP (UNDER VALVE)

UNION

GAS PRESSURE REGULATOR

LIGHTER BUTTON AND IGNITOR SLOT "C"

Courtesy Mid-Continent Metal Products Co.

Fig. 16-13. Pipe connection for a conversion gas burner (with pilot light).

2. Install an intermediate regulator if the line pressure exceeds 13.5-in. W.C.

3. Connect the burner to the piping as shown in Figs. 16-13 and 16-14.

4. Install a manually operated main shutoff valve 4 to 5 ft. above the floor on the vertical pipe.

5. Make certain the pipe is clean and free of any scale.

6. Use malleable iron fittings.

7. Remove all burrs and scales from the pipe and clean it before installing it.

8. Do not tap off from the bottom of horizontal runs when branching from a pipeline.

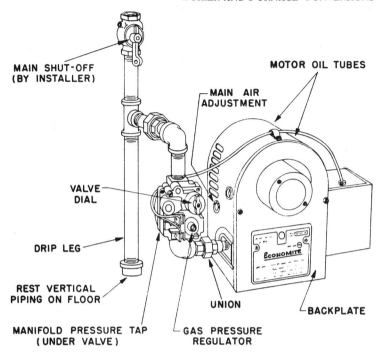

MAIN SHUT-OFF
(BY INSTALLER)

MOTOR OIL TUBES

MAIN AIR
ADJUSTMENT

VALVE
DIAL

DRIP LEG

REST VERTICAL
PIPING ON FLOOR

UNION

BACKPLATE

MANIFOLD PRESSURE TAP
(UNDER VALVE)

GAS PRESSURE
REGULATOR

ECONOMITE

Courtesy Mid-Continent Metal Products Co.

Fig. 16-14. Pipe connection for a conversion gas burner (without pilot light).

GAS INPUT SETTING

The gas-fired conversion burner illustrated in Fig. 16-13 is manufactured to operate on natural gas, but kits are available for converting to propane. A very important factor to be considered when installing these burners is a correct determination of the gas input setting, which should be based on the following two factors:

1. The heating rate of the structure.
2. The rated maximum input of the burner.

The *heating rate* of a structure is essentially the amount of

581

Table 16-3. Spud Sizing for Natural Gas

NATURAL			
Capacity Btuh		Spud Size	
Max. 3.5" W.C. Man. Pr.	Min.	Drill No.	Dia.
75,000	50,000	#20	0.161
125,000	75,000	# 4	0.209
200,000	125,000	$^{17}/_{64}$	0.266

Courtesy Mid-Continent Metal Products Co.

warm air that the heating unit must deliver in order to adequately heat it. Methods for calculating the amount of heat required (i.e., the heating rate) are described in Chapter 4 (Heating Calculations).

Another important factor to consider is the *rated maximum input* of the burner. The gas input *cannot exceed* this rating. If the burner maximum input of the burner is below the required mimimum of heat indicated by the heating rate, then the burner is inadequate for your needs.

The gas input setting for a natural gas conversion burner is regulated by the spud sizing and the pressure regulator. A burner is commonly shipped with three different spud sizes, one for each three capacity ranges (high, intermediate, and low; Table 16-3). The capacity range is changed by using a different spud sizing. After the appropriate spud has been installed, the final gas input setting is made by adjusting the pressure regulator. Be sure to carefully read the burner manufacturer's installation instructions.

A propane gas conversion burner operates directly on 11-in. W.C. gas supply pressure as determined by the propane-tank regulator. The portion of the combination control on the standard natural gas burner is blanked off with a plate provided in the propane conversion kit. Two spuds are provided: a minimum-sized main spud and a pilot spud. The main spud must be re-drilled for the required capacity (see Table 16-4).

582

Table 16-4. Spud Sizing for Propane

PROPANE		
Capacity Btuh	Spud Size	
11″ W.C. Man. Pr.	Drill No.	Dia.
50,000	#46	0.081
75,000	#40	0.098
100,000	#32	0.116
125,000	⅛″	0.125
150,000	#29	0.136
175,000	#26	0.147
200,000	#19	0.166

Courtesy Mid-Continent Metal Products Co.

INSTALLING A CONVERSION GAS BURNER

The conversion gas burners illustrated in Figs. 16-13 and 16-14 are used to convert coal-fired or oil-fired central heating plants to gas. They are adaptable for use in forced-warm-air furnaces, gravity (updraft) furnaces, or boilers (Figs. 16-15 to 16-18).

In addition to the basic combustion chamber requirements (see the appropriate section in this chapter), the following recommendations *must* be followed when installing a conversion gas burner.

1. The diameter of the burner opening into the combustion chamber must *not* be oversize. If the opening is oversized, slip the stainless steel sleeve (included with the burner) over the nozzle and fill in the remaining space with refractory cement.
2. Position the end of the burner nozzle so that it is at least 1 in. short of the inside surface of the combustion chamber. *Never* allow the nozzle to extend into the combustion chamber.
3. If the burner nozzle is too short to reach the combustion chamber, it may be lengthened with an extension tube.
4. Check the burner ports and pilot before permanently setting the burner in place, and remove any foreign matter blocking the openings.

583

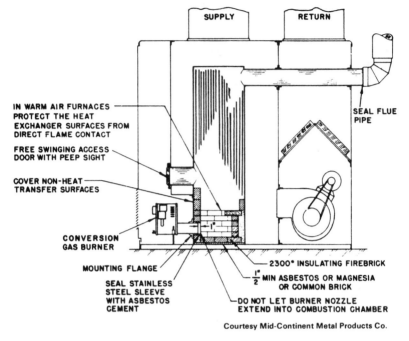

SUPPLY   RETURN

IN WARM AIR FURNACES
PROTECT THE HEAT
EXCHANGER SURFACES FROM
DIRECT FLAME CONTACT

FREE SWINGING ACCESS
DOOR WITH PEEP SIGHT

COVER NON-HEAT
TRANSFER SURFACES

CONVERSION
GAS BURNER

MOUNTING FLANGE

SEAL STAINLESS
STEEL SLEEVE
WITH ASBESTOS
CEMENT

SEAL FLUE
PIPE

2300° INSULATING FIREBRICK

$\frac{1}{2}$" MIN ASBESTOS OR MAGNESIA
OR COMMON BRICK

DO NOT LET BURNER NOZZLE
EXTEND INTO COMBUSTION CHAMBER

Courtesy Mid-Continent Metal Products Co.

Fig. 16-15. Updraft forced-warm-air furnace equipped with conversion gas
burners.

STARTING A CONVERSION GAS BURNER (WITH PILOT LIGHT)

This section covers the starting of conversion gas burners equipped with pilot lights. Before attempting to start this type of burner, the piping should be checked for leaks. The basic procedure for doing this is as follows:

1. Shut off all other gas appliances (gas water heaters, etc.) that tie into the same system.
2. Turn on the main shutoff valve (Fig. 16-13).
3. Turn valve dial A to *off* (Fig. 16-13).
4. Turn on gas pressure to the gas supply line.
5. Watch the meter test dial.

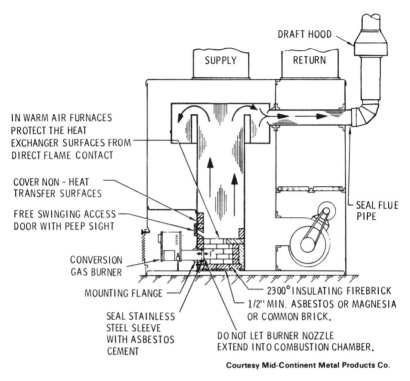

DRAFT HOOD

SUPPLY

RETURN

IN WARM AIR FURNACES
PROTECT THE HEAT
EXCHANGER SURFACES FROM
DIRECT FLAME CONTACT

COVER NON - HEAT
TRANSFER SURFACES

FREE SWINGING ACCESS
DOOR WITH PEEP SIGHT

SEAL FLUE
PIPE

CONVERSION
GAS BURNER

MOUNTING FLANGE

2300° INSULATING FIREBRICK
1/2" MIN. ASBESTOS OR MAGNESIA
OR COMMON BRICK.

SEAL STAINLESS
STEEL SLEEVE
WITH ASBESTOS
CEMENT

DO NOT LET BURNER NOZZLE
EXTEND INTO COMBUSTION CHAMBER.

Courtesy Mid-Continent Metal Products Co.

Fig. 16-16. Downdraft forced-warm-air furnace equipped with a conversion gas burner.

If the meter test dial shows no movement for *at least* 5 minutes, it is safe to say that there are no leaks in the piping. Movement of the meter test dial, however, means that you must locate the leak (or leaks) before starting the burner. The soap-suds test is a simple and effective method for doing this.

The gas line must be purged before starting the burner. To do this, the following steps are recommended:

1. Disconnect the pilot tubing.
2. Depress valve dial A as far as it will go, and turn it *counterclockwise* to pilot (Fig. 16-13). Do *not* release the dial at this point.
3. With valve dial A still fully depressed, lock the dial in posi-

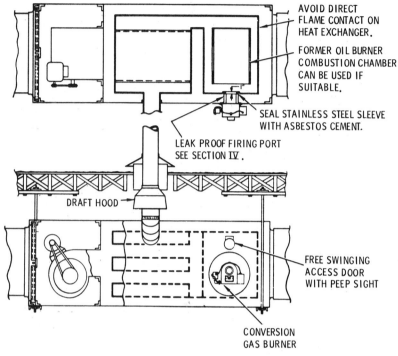

AVOID DIRECT
FLAME CONTACT ON
HEAT EXCHANGER.

FORMER OIL BURNER
COMBUSTION CHAMBER
CAN BE USED IF
SUITABLE.

SEAL STAINLESS STEEL SLEEVE
WITH ASBESTOS CEMENT.

LEAK PROOF FIRING PORT
SEE SECTION IV.

DRAFT HOOD

FREE SWINGING
ACCESS DOOR
WITH PEEP SIGHT

CONVERSION
GAS BURNER

Courtesy Mid-Continent Metal Products Co.

Fig. 16-17. Downdraft suspended heater equipped with a conversion gas burner.

tion by engaging the lighting latch B (Fig. 16-13).

4. Allow the air to escape until the gas line is completely purged (which will be indicated by gas replacing the air issuing from the pilot connection). Caution: *Never* purge the gas line into the combustion chamber.

5. Turn off valve dial A (depress and turn clockwise).

6. Reconnect the pilot tubing.

After you have checked the piping for leaks and have completely purged the gas lines, you are now ready to start the burner. First add a few drops of oil (#20 SAE oil is recommended) to the burner motor bearings. Then, check the setting of

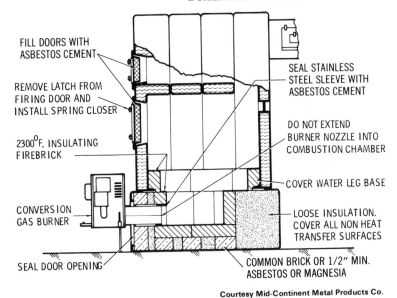

FILL DOORS WITH
ASBESTOS CEMENT

REMOVE LATCH FROM
FIRING DOOR AND
INSTALL SPRING CLOSER

2300°F. INSULATING
FIREBRICK

CONVERSION
GAS BURNER

SEAL DOOR OPENING

SEAL STAINLESS
STEEL SLEEVE WITH
ASBESTOS CEMENT

DO NOT EXTEND
BURNER NOZZLE INTO
COMBUSTION CHAMBER

COVER WATER LEG BASE

LOOSE INSULATION.
COVER ALL NON HEAT
TRANSFER SURFACES

COMMON BRICK OR 1/2″ MIN.
ASBESTOS OR MAGNESIA

Courtesy Mid-Continent Metal Products Co.

Fig. 16-18. Hot water or steam boiler equipped with a conversion gas burner.

valve dial A. At this point, it should be turned to *off* (see Step 5 in the purging procedure described above).

You are now ready for the initial startup procedure, which is accomplished as follows:

1. Set the main air shutter at approximately one-quarter open.
2. Turn on the main line switch.
3. Set the room thermostat *above* room temperature (this will cause the burner motor to start).
4. The burner motor should be allowed to run for about 5 minutes to ensure complete purging of the combustion chamber.
5. Shut off the burner motor by adjusting the thermostat setting *below* room temperature.
6. Fully depress valve dial A and turn it *counterclockwise* to *pilot*. Keep it depressed.
7. Lock valve dial A in its depressed position by engaging the lighting latch B.

8. Light a match and hold the flame under the igniter valve and at ignitor slot C while at the same time depressing button C for at least 10 seconds.
9. Release the lighter button. The pilot should now ignite. If it does not, then repeat Step 8 above.
10. After the pilot has been burning for at least 1 minute, release valve dial A and turn it *counterclockwise* to the *on* setting.

STARTING A CONVERSION GAS BURNER (PILOTLESS)

Most of the instructions given in the previous section for starting a conversion gas burner equipped with a pilot light will also apply here. The major difference is the use of direct spark ignition rather than a pilot flame.

The basic procedure for starting a pilotless conversion gas burner is as follows (Fig. 16-14):

1. Check the piping for leaks.
2. Bleed (purge) the gas line.
3. Set the air shutter about three-quarters open.
4. Depress the valve dial and turn it on the *on* position.
5. Turn on the main line switch.
6. Set the room thermostat *above* the room temperature. This should cause the burner motor to start. If it does, burner ignition should occur after 30 seconds.
7. If the burner fails to light, turn off the main line switch and then turn it on again.
8. Readjust the air shutter for a quiet, soft flame. The flame should be blue at the burner, changing to orange at the tips.
9. Start and stop the burner several times with the thermostat to check its operation.

SERVICING A CONVERSION GAS BURNER

Always be sure that the manual gas valve and burner switch are turned off before attempting to service a gas burner. *Never* attempt to remove any parts for service before taking this precaution. In direct spark ignition burners, the nozzle and electrode assembly can usually be removed as a unit. For example, the nozzle-and-electrode assembly of the Economite Model DS20A conversion gas burner can be completely removed in the following manner:

1. Remove the burner backplate.
2. Disconnect the pipe union.
3. Remove the curved orifice pipe.
4. Withdraw the nozzle assembly.
5. Disconnect the electrode leads.
6. Remove the unit.

The nozzle-and-electrode assembly can be reinstalled by reversing Steps 1 to 6. When reassembling, make certain the orifice pipe enters the nozzle casting.

With the nozzle-and-electrode assembly completely removed, you are now in a position to inspect and clean it. Check the electrode for insulator cracks and evidence of serious burning.

Removal of the burner backplate also provides access to the motor and blower, motor relay, low-voltage transformer, and terminal board. These can be removed as a unit for servicing, but you will have to disconnect the terminal board and unplug the electrical pass-through fitting to do this. The following service may be required for this unit:

1. Cleaning of the blower wheel.
2. Cleaning the motor air vents.

Does the motor run without further burner sequencing taking place? This may indicate trouble in the centrifugal interlock switch. Some burners are provided with removable motor end caps to provide ready access to the centrifugal switch. If this should be the case, remove the cap and clean the contacts by burnishing. The contacts must be open when the motor is not operating. If necessary, the burner motor can be removed and replaced as follows:

1. Remove the blower wheel.

2. Remove the retainer clips at the motor grommets.
3. Pull the motor out of the brackets.
4. Repair or substitute a new motor.
5. Place motor in brackets.
6. Make certain grommets fit well in the bracket and replace retainer clips.
7. Replace blower wheel.

The sealed motor relay and low-voltage transformer normally do not require servicing. They should be replaced when defective.

Check gas lines and valves for leaks with the soapsuds test. *Never* use a flame to locate a gas leak. If the leakage occurs around the valve, the valve seat may need cleaning. Disassemble the valve, clean the seat, and reassemble it. If the valve malfunction is due to causes other than the seat, replace the entire valve.

Replace the gas pressure regulator if it fails to maintain a constant pressure. On the conversion gas burner illustrated in Fig. 16-14, the regulator is part of a combination valve (in this case, a Robertshaw, 24-volt, combination gas valve), but is designed so that it can be replaced without replacing the entire assembly.

Pressure adjustment required for setting intermediate capacities can be accomplished as follows:

1. Remove the adjustment screw cap.
2. Take a screwdriver and turn the adjustment screw *counterclockwise* to reduce pressure.
3. Measure the pressure through the manometer connection adjacent to the outlet tapping.

Conversion gas burners equipped with a pilot light require inspection and servicing of the following:

1. The pilot orifice size will depend upon the type of gas used. For the conversion gas burner illustrated in Fig. 11-13, the pilot orifice will have a diameter of 0.018 in. for natural gas and 0.012 in. for propane.
2. Set the end of the burner nozzle at least 1 in. short of the inside of the combustion chamber (see Figs. 16-15 to 16-18). It must *not* be set flush with the chamber or be allowed to

extend into the chamber (the pilot is often snuffed out in such cases by recirculated flue products).
3. Check for a high or low gas pressure. If over 7 in. W.C., reset the pilot adjustment (see Fig 16-13) to reduce the size of the pilot flame and to increase the heat on the thermocouple.

The thermocouple and pilot-safety section of the main valve should be tested to determine if they are operating efficiently. The testing procedure is as follows:

1. Turn the burner switch to *off*.
2. Allow the combustion chamber to cool and the pilot light to operate for *at least* 5 minutes.
3. Turn off the pilot and listen for a "click" from the main valve.
4. If the "click" of the main valve is heard *more* than 30 seconds after you have shut off the pilot, then both the thermocouple and the pilot-safety section of the main valve are operating efficiently.
5. If the "click" occurs *less* than 30 seconds after you have shut off the pilot, then either the thermocouple or pilot-safety section of the main valve is faulty.
6. Test the thermocouple by disconnecting it from the main valve and checking it with a millivoltmeter. Under normal conditions (i.e., when heated by a standby pilot), it should develop at least 15 millivolts.
7. If the thermocouple develops at least 15 millivolts but the main valve "click" still occurs less than 30 seconds after you have shut off the pilot, the pilot-safety section is probably malfunctioning.
8. The thermocouple may also be tested while still connected to the valve (a closed-circuit connection) by using an adapter to connect the millivoltmeter. Under normal operating conditions, the thermocouple should develop at least 7 millivolts and the valve 4 millivolts.

The wiring diagram for a conversion gas burner equipped with a pilot light is shown in Fig. 16-19. The installation and service instructions from most manufacturers will include a wiring diagram for the burner. Study it carefully before servicing or making repairs.

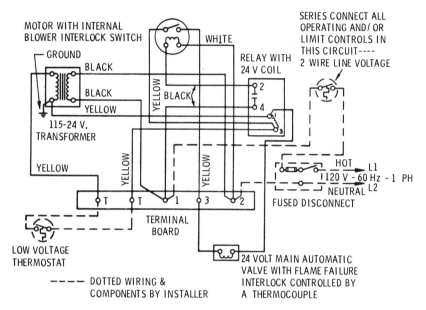

Courtesy Mid-Continent Metal Products Co.

Fig. 16-19. Wiring diagram for a conversion gas burner equipped with a pilot light.

OIL TANKS AND OIL PIPING

Local codes and the National Board of Fire Underwriters will provide information for the installation of oil tanks and piping.

In oil piping, copper tubing is commonly used for suction and return lines. A ⅜-in. OD minimum is recommended for lines under 30 ft. in length. If the suction and return lines exceed 30 ft., then ½-in. OD copper tubing is recommended.

Other suggestions for the installation of the oil storage tank and piping include:

1. Use a good-quality pipe joint compound and make your pipe thread connections as tight as possible.

2. Keep the lines between the oil tank and burner as straight as possible (avoid kinks and traps).
3. Keep the connections in the return line to a minimum.
4. Check tubing flares and connections for air leaks (an indication of a poor seal resulting from improper meeting).

The two basic types of piping arrangements are: (1) the single-line system and (2) the two-line system. These two piping arrangements are distinguished from one another primarily by the location of the oil storage tank with respect to the oil-fired conversion burner.

A *single-line piping* arrangement is used only where the oil storage tank is located *above* the level of the oil burner. No return line is necessary. The oil is taken from an outlet in the bottom of the storage tank and fed by gravity to the burner. The slope of the pipe should be gradually downward (approximately ½ in. per foot) to a point directly below the burner connection. This downward slope in the line is designed to prevent the formation of air pockets and bubbles. The installation of a shutoff valve in the line is recommended.

A *two-line piping* arrangement is required where the oil storage tank is buried at a level *below* that of the burner.

INSTALLING A CONVERSION OIL BURNER

Conversion oil burners should be installed in accordance with the regulations of the National Fire Protection Association (NFPA Bulletin No. 31) and local codes and regulations. Read these regulations and the manufacturer's installation instructions before making any attempt to install the burner.

If the furnace or boiler was originally designed to burn solid fuel and the ashpit is *not* used as a part of the combustion chamber, the ash door should be removed to prevent the accumulation of vapors in the ashpit. If removal of the ash door is not feasible, then some other means of bottom ventilation must be provided.

Before installing the burner, check the condition of the boiler or furnace. It must be in good condition or repair. The flue gas passages and combustion chamber must be tight against leaks.

Reseal or reset the sections of a cast-iron boiler or furnace. Replace any damaged parts. The boiler or furnace should also be as clean as possible before installing the burner.

A combustion chamber must be provided that is in accordance with the specifications of the burner manufacturer. Read the appropriate sections in this chapter (e.g., "Basic Combustion Chamber Requirements") for more information.

If you are satisfied that the boiler or furnace with which you are working satisfactorily meets the conditions mentioned in the previous paragraphs, then you are ready to install the conversion oil burner. The basic steps recommended for installation are as follows:

1. Position the burner on its mounting flange or pedestal so that the burner air tube is flush with the inside surface of the combustion chamber front wall. Do *not* allow the burner air tube to extend into the combustion chamber.

2. Adjust the burner tube on pedestal-mounted burners so that it pitches downward about 1½ in. Any fuel in the burner tube will then drain into the combustion chamber.

3. On flange-mounted burners, the distance from the flange to the end of the air cone will depend upon the requirements of each installation. Remove the air cone before installing the flange.

4. The distance between the center of the nozzle and the floor of the combustion chamber must be correct. If it is too close, it will result in flame impingement and carbonization. On the other hand, a nozzle placed too high will result in excessive flame noise and poor combustion.

5. The fuel pump should be adjusted for the oil pressure recommended by the manufacturer.

6. Adjust the air supply to the burner with the air inlet band. Rotate the air inlet band to the required position (the one which will deliver the smallest amount of air and still maintain clean combustion), and secure it in position.

7. Make certain the size and spray angle of the nozzle is correct for the installation (check the UL rating plate).

8. Install the nozzle, and check the electrode tip position in accordance with the manufacturer's instructions.

9. Prime the fuel pump (see the manufacturer's instructions).

STARTING A CONVERSION OIL BURNER

Before attempting to start the oil burner, you should first carefully read the manufacturer's instructions. Unfortunately, these are not always available on older heating installations. If this is the case and you do not possess the necessary expertise to operate the burner without an instruction manual, it is strongly recommended that you call a professional for services.

Always test the oil lines for possible leaks and make the necessary repairs before starting the oil burner. Consult the manufacturer's instructions for bleeding the fuel pump. The bleed valve on the fuel pump is designed to facilitate air purging, cleaning, and priming. See Chapter 1 of Volume 2 (Oil Burners) for additional information about fuel pumps.

Check the electrical wiring diagram (if one exists) to make certain that the burner is correctly wired. Problems with an oil burner can often be traced to incorrect wiring. All wiring must be in accordance with the National Electrical Code as well as local codes, standards, and regulations.

It is important to make certain that the primary and limit controls are operating properly. Test the primary control by first removing the motor lead from the burner and then energizing the ignition circuit by throwing the switch. If the primary control is operating properly, the ignition will shut off *within 2 minutes* after the switch is thrown. If the ignition fails to shut off after 2 minutes, the primary control should be replaced.

Check the fuel level in the oil storage tank. There must be enough oil to operate the burner. Open both oil valves (at the oil storage tank and at the boiler or furnace).

You *must* have an adequate air supply for efficient combustion. The basic ventilation requirements (see "Ventilation Requirements" in this chapter) must be provided for when installing the oil burner. It is also important to determine if the secondary air setting is correct. If there is smoke or soot present when the oil burner is operating, the air setting must be corrected.

If all the suggestions in the previous paragraphs have been followed and the necessary adjustments or repairs made, the oil burner can now be started. The procedure for doing this is basically as follows:

1. Make certain that all controls have been set in normal starting position.
2. Make certain that the oil valve in the oil supply line is open.
3. Set the room thermostat 10° above the room temperature.
4. Check the reset button on the primary control to make certain it has been reset.
5. Open the air control band on the burner to about half open.
6. Throw the main electrical switch to *on*. The burner should now start, ignite, and burn. If it does not, recheck Steps 1 through 5. If you still experience difficulty in starting the burner, consult "Troubleshooting Oil Burners" in Chapter 1 of Volume 2 (Oil Burners).
7. Allow the oil burner to operate for about 15 minutes; then shut it down and give it time to cool off. After it has cooled off, restart it to be sure it operates properly on a cool start.

While the oil burner is operating (Step 7), you should take the opportunity to make a number of tests. Check the fuel-pump pressure with an oil pressure gauge. These should be factory set. Adjust the air control band on the burner until 0+ smoke is obtained with a smoke tester. Adjust the draft control (if necessary) to obtain at least 0.02 in. W.C. of draft over the fire.

SERVICING A CONVERSION OIL BURNER

Protecting the oil burner motor from unnecessary wear is extremely important. *At least* once a year (but not more than three times a year), preferably at the beginning of each heating season, place about 10 drops of #20 SAE oil in each oil port of the burner motor. After you have oiled the burner motor, clean the fuel strainer and change the oil-filter cartridge. Clogged fuel strainers are a common source of trouble. In rotary- and gun-type oil burners, these components are usually located either at the point at which the oil line connects to the tank or at the oil line connection to the burner. Follow the manufacturer's instructions for removing the fuel strainer, and clean it with a stiff wire brush or a blast of compressed air.

Be sure that you are using the correct weight of oil. If the oil is too heavy, the burner motor will start but it may fail to establish a flame. Continued use of the wrong oil may cause damage to burner components.

Soot should be periodically removed from the burner and combustion chamber. Commercial soot removers are available for this purpose. When burning out a layer of soot, take care not to ignite a soot fire in the chimney. This could be very dangerous because it may lead to the igniting of combustionable materials near the chimney. The hard layer of carbon that frequently forms on the bottom of the fire pot in burners should also be removed.

Other servicing recommendations for conversion oil burners are as follows:

1. Check the furnace or boiler for possible air leaks. Any that are found should be sealed (an asbestos cement is recommended) because they interfere with combustion efficiency.

2. Check the burner nozzle in gun-type burners. If it is dirty, remove and clean it according to the manufacturer's instructions.

3. Check the position of the electrodes in rotary- and gun-type oil burners. They should be positioned near, but out of the direct spray of, the oil.

APPENDIX A

Professional and Trade Associations

Many professional and trade associations have been organized to develop and provide materials and/or services in connection with heating, ventilating, and air conditioning. These materials and services include:

1. Formulating and establishing specifications and professional standards.
2. Certifying equipment and materials.
3. Conducting product research.
4. Promoting interest in the product.

Much useful information can be obtained by contacting these associations. With that in mind. the names and addresses of the principal organizations have been included in this Appendix. They are listed in alphabetical order.

AIR-CONDITIONERS AND REFRIGERATION WHOLESALERS (ARW)
P.O. Box 640
1351 South Federal Highway
Deerfield Beach, FL 33441
Air conditioning and refrigeration equipment wholesalers.

AIR-CONDITIONING AND REFRIGERATION INSTITUTE (ARI)
1815 North Fort Meyer Drive
Arlington, VA 22209
Manufacturers of air conditioning, warm-air heating, and commercial and industrial refrigeration equipment.

AIR CONDITIONING CONTRACTORS OF AMERICA (ACCA)
1228 17th Street, NW
Washington, DC 20036
Formerly (until 1978) National Environmental Systems Contractors Association. Heating, air conditioning, and refrigeration systems contractors.

AIR DIFFUSION COUNCIL (ADC)
435 North Michigan Avenue
Chicago, IL 60611
Manufacturers of registers, grilles, diffusers, and related equipment.

AIR FILTER INSTITUTE
(*See* Air-Conditioning and Refrigeration Institute)

AIR MOVEMENT AND CONTROL ASSOCIATION (AMCA)
30 West University Drive
Arlington Heights, IL 60004
Formerly (until 1977) Air Moving and Conditioning Association. Manufacturers of air moving and conditioning equipment.

AMERICAN GAS ASSOCIATION (AGA)
1515 Wilson Boulevard
Arlington, VA 22209
Residential gas operating and performance standards from distributors and transporters of natural, manufactured, and mixed gas.

AMERICAN NATIONAL STANDARDS INSTITUTE
(ANSI)
1430 Broadway
New York, NY 10018
Clearinghouse for nationally coordinated safety, engineering, and industrial standards.

AMERICAN SOCIETY FOR TESTING AND
MATERIALS (ASTM)
1916 Race Street
Philadelphia, PA 19103
Engineering standards for materials.

AMERICAN SOCIETY OF HEATING,
REFRIGERATION AND AIR-CONDITIONING
ENGINEERS (ASHRAE)
1791 Tullie Circle
Atlanta, GA 30329
Professional association.

AMERICAN SOCIETY OF MECHANICAL
ENGINEERS (ASME)
345 East 47th Street
New York, NY 10017
Develops safety codes and standards for various types of equipment (boiler and pressure vessel codes, etc.)

AMERICAN VENTILATION ASSOCIATION (AVA)
Box 7464
Houston, TX 77008
Association of residential ventilating equipment manufacturers and dealers.

BETTER HEATING-COOLING COUNCIL
(*See* Hydronics Institute)

EDISON ELECTRIC INSTITUTE (EEI)
1111 19th Street, NW
Washington, DC 20036
Association of investor-owned electric utility companies.

ELECTRICAL ENERGY ASSOCIATION
(*See* Edison Electric Institute)

FIREPLACE INSTITUTE
Merged with Wood Energy Institute in 1980 to form Wood Heating Alliance.

HEATING AND PIPING CONTRACTORS NATIONAL
ASSOCIATION
(*See* Mechanical Contractors Association of America)

HOME VENTILATING INSTITUTE (HVI)
4300-L Lincoln Avenue
Rolling Meadows, IL 60008
Develops performance standards for residential ventilating equipment.

HYDRONICS INSTITUTE
35 Russo Place
Berkley Heights, NJ 07922
Manufacturers, installers, etc., of hot-water and steam heating and cooling equipment. Formed by a merger of the Beter Heating-Cooling Council (1956–1970) and the Institute of Boiler and Radiator Manufacturers (1915–1970).

INSTITUTE OF BOILER AND RADIATOR
MANUFACTURERS
(*See* Hydronics Institute)

INTERNATIONAL SOLAR ENERGY SOCIETY
National Science Center
P.O. box 52
Parkville, Victoria 3052
Australia
Educational and research organization

MECHANICAL CONTRACTORS ASSOCIATION
  OF AMERICA (MCAA)
5530 Wisconsin Avenue, NW, Suite 750
Washington, DC 20015
Contractors of piping and related equipment used in heating,
cooling, refrigeration, ventilating, and air conditioning.

NATIONAL ASSOCIATION OF PLUMBING-
  HEATING-COOLING CONTRACTORS
1016 20th Street, NW
Washington, DC 20005
Local plumbing, heating, and cooling contractors association.

NATIONAL BOARD OF FIRE UNDERWRITERS
  (NBFU)
85 John Street
New York, NY 10036
Safety standards.

NATIONAL BUREAU OF STANDARDS
U.S. Department of Commerce
Washington, DC 20025
Government regulatory agency.

NATIONAL ENVIRONMENTAL SYSTEMS
  CONTRACTORS ASSOCIATION
(*See* Air Conditioning Contractors of America)

NATIONAL FIRE PROTECTION ASSOCIATION (NFPA)
Batterymarch Park
Quincy, MA 02269
Fire safety standards.

NATIONAL LP-GAS ASSOCIATION (NLPGA)
1301 West 22nd Street
Oak Brook, IL 60521

NATIONAL OIL FUEL INSTITUTE
(*See* National Oil Jobbers Council)

NATIONAL OIL JOBBERS COUNCIL
1707 H Street, NW
Washington, DC 20006
Absorbed National Oil Fuel Institute in 1974. Independent wholesale petroleum marketers and retail fuel oil dealers.

NATIONAL WARM AIR HEATING AND AIR
CONDITIONING ASSOCIATION
(*See* Air Conditioning Contractors of America)

PLUMBING-HEATING-COOLING
INFORMATION BUREAU (PHCIB)
35 East Wacker Drive
Chicago, IL 60601
Association of plumbing, heating, and cooling manufacturers, contractors, wholesalers, etc.

REFRIGERATION AND AIR CONDITIONING
CONTRACTORS ASSOCIATION
(*See* Air Conditioning Contractors of America)

REFRIGERATION SERVICE ENGINEERS SOCIETY
(RSES)
960 Rand Road
Des Plaines, IL 60018
Association of refrigeration, air conditioning, and heating equipment installers, servicemen, and salesmen.

SHEET METAL AND AIR CONDITIONING
CONTRACTORS NATIONAL ASSOCIATION
(SMACNA)
8224 Old Courthouse Road
Vienna, VA 22180
Sheet-metal contractors who install ventilating, warm-air heating,
and air-handling equipment and systems.

STEAM HEATING EQUIPMENT MANUFACTURERS
ASSOCIATION (*defunct*)

STEEL BOILER INSTITUTE (*defunct*)

UNDERWRITERS' LABORATORIES
333 Pfingsten Road
Northbrook, IL 60062
Testing laboratory. Promotes safety standards for equipment.

WOOD ENERGY INSTITUTE
Merged with Fireplace Institute in 1980 to form Wood Heating
Alliance.

WOOD HEATING ALLIANCE (WHA)
1101 Connecticut Avenue, SW, Suite 700
Washington, DC 20036
Manufacturers, dealers, and suppliers of fireplaces, fireplace
components, and related equipment.

APPENDIX B

Table B-1. Equivalent Length of New Straight Pipe for Valves and Fittings for Turbulent Flow

Fittings			1/4	3/8	1/2	3/4	1	1 1/4	1 1/2	2	2 1/2	3	4	5	6	8	10	12	14	16	18	20	24
Regular 90° Ell	Screwed	Steel	2.3	3.1	3.6	4.4	5.2	6.6	7.4	8.5	9.3	11	13	—	—	—	—	—	—	—	—	—	—
	Screwed	C.I.	—	—	—	—	—	—	—	—	—	9.0	11	—	—	—	—	—	—	—	—	—	—
	Flanged	Steel	—	—	.92	1.2	1.6	2.1	2.4	3.1	3.6	4.4	5.9	7.3	8.9	12	14	17	18	21	23	25	30
	Flanged	C.I.	—	—	—	—	—	—	—	3.6	—	3.6	4.8	—	7.2	9.8	12	15	17	19	22	24	28
Long Radius 90° Ell	Screwed	Steel	1.5	2.0	2.2	2.3	2.7	3.2	3.4	3.6	3.6	4.0	4.6	—	—	—	—	—	—	—	—	—	—
	Screwed	C.I.	—	—	—	—	—	—	—	—	—	3.3	3.7	—	—	—	—	—	—	—	—	—	—
	Flanged	Steel	—	—	1.1	1.3	1.6	2.0	2.3	2.7	2.9	3.4	4.2	5.0	5.7	7.0	8.0	9.0	9.4	10	11	12	14
	Flanged	C.I.	—	—	—	—	—	—	—	—	—	2.8	3.4	—	4.7	5.7	6.8	7.8	8.6	9.6	11	11	13
Regular 45° Ell	Screwed	Steel	.34	.52	.71	.92	1.3	1.7	2.1	2.7	3.2	4.0	5.5	—	—	—	—	—	—	—	—	—	—
	Screwed	C.I.	—	—	—	—	—	—	—	—	—	3.3	4.5	—	—	—	—	—	—	—	—	—	—
	Flanged	Steel	—	—	.45	.59	.81	1.1	1.3	1.7	2.0	2.6	3.5	4.5	5.6	7.7	9.0	11	13	15	16	18	22
	Flanged	C.I.	—	—	—	—	—	—	—	—	—	2.1	2.9	—	4.5	6.3	8.1	9.7	12	13	15	17	20
Tee-Line Flow	Screwed	Steel	.79	1.2	1.7	2.4	3.2	4.6	5.6	7.7	9.3	12	17	—	—	—	—	—	—	—	—	—	—
	Screwed	C.I.	—	—	—	—	—	—	—	—	—	9.9	14	—	—	—	—	—	—	—	—	—	—
	Flanged	Steel	2.4	3.5	.69	.82	1.0	1.3	1.5	1.8	1.9	2.2	2.8	3.3	3.8	4.7	5.2	6.0	6.4	7.2	7.6	8.2	9.6
	Flanged	C.I.	—	—	—	—	—	—	—	—	—	1.9	2.2	—	3.1	3.9	4.6	5.2	5.9	6.5	7.2	7.7	8.8
Tee-Branch Flow	Screwed	Steel	2.4	3.5	4.2	5.3	6.6	8.7	9.9	12	13	17	21	—	—	—	—	—	—	—	—	—	—
	Screwed	C.I.	—	—	—	—	—	—	—	—	—	14	17	—	—	—	—	—	—	—	—	—	—
	Flanged	Steel	—	—	2.0	2.6	3.3	4.4	5.2	6.6	7.5	9.4	12	15	18	24	30	34	37	43	47	52	62
	Flanged	C.I.	—	—	—	—	—	—	—	—	—	7.7	10	—	15	20	25	30	35	39	44	49	57

Pipe Size

Table B-1. Equivalent Length of New Straight Pipe for Valves and Fittings for Turbulent Flow (Cont'd)

Fittings		1/4	3/8	1/2	3/4	1	1 1/4	1 1/2	2	2 1/2	3	4	5	6	8	10	12	14	16	18	20	24
180° Return Bend, Screwed	Steel	2.3	3.1	3.6	4.4	5.2	6.6	7.4	8.5	9.3	11	13	—	—	—	—	—	—	—	—	—	—
	C.I.	—	—	—	—	—	—	—	—	—	9.0	11	—	—	—	—	—	—	—	—	—	—
Reg. Flanged	Steel	—	—	.92	1.2	1.6	2.1	2.4	3.1	3.6	4.4	5.9	7.3	8.9	12	14	17	18	21	23	25	30
	C.I.	—	—	—	—	—	—	—	—	—	3.6	4.8	—	7.2	9.8	12	15	17	19	22	24	28
Long Rad. Flanged	Steel	—	—	—	—	—	—	—	—	—	3.4	4.2	5.0	5.7	7.0	8.0	9.0	9.4	10	11	12	14
	C.I.	—	—	—	—	—	—	—	—	—	2.8	3.4	—	4.7	5.7	6.8	7.8	8.6	9.6	11	11	13
Globe Valve, Screwed	Steel	21	22	22	24	29	37	42	54	62	79	110	—	—	—	—	—	—	—	—	—	—
	C.I.	—	—	—	—	—	—	—	—	—	65	86	—	—	—	—	—	—	—	—	—	—
Flanged	Steel	—	—	38	40	45	54	59	70	77	94	120	150	190	260	310	390	—	—	—	—	—
	C.I.	—	—	—	—	—	—	—	—	—	77	99	—	150	210	270	330	—	—	—	—	—
Gate Valve, Screwed	Steel	.32	.45	.56	.67	.84	1.1	1.2	1.5	1.7	1.9	2.5	3.1	3.2	3.2	3.2	3.2	3.2	3.2	3.2	3.2	3.2
	C.I.	—	—	—	—	—	—	—	—	—	1.6	2.0	—	—	—	—	—	—	—	—	—	—
Flanged	Steel	—	—	—	—	—	—	—	2.6	2.7	2.8	2.9	—	3.2	3.2	3.2	3.2	3.2	3.2	3.2	3.2	3.2
	C.I.	—	—	—	—	—	—	—	—	—	2.3	2.4	—	2.6	2.7	2.8	2.9	3.0	3.0	3.0	3.0	3.0
Angle Valve, Flanged	Steel	12.8	15	15	15	17	18	18	21	22	28	38	50	63	90	120	140	160	190	210	240	300
	C.I.	—	—	—	—	—	—	—	—	—	23	31	—	52	74	98	120	150	170	200	230	280
Screwed	Steel	7.2	7.3	8.0	8.8	11	13	15	19	22	27	38	—	—	—	—	—	—	—	—	—	—
	C.I.	—	—	—	—	—	—	—	—	—	22	31	—	—	—	—	—	—	—	—	—	—

Table B-1. Equivalent Length of New Straight Pipe for Valves and Fittings for Turbulent Flow (Cont'd)

Fittings		1/4	3/8	1/2	3/4	1	1 1/4	1 1/2	2	2 1/2	3	4	5	6	8	10	12	14	16	18	20	24
Swing Check Valve — Flanged	Steel	—	—	3.8	5.3	7.2	10	12	17	21	27	38	50	63	90	120	140	—	—	—	—	—
	C. I.	—	—	—	—	—	—	—	—	—	22	31	—	52	74	98	120	—	—	—	—	—
Coupling or Union — Screwed	Steel	.14	.18	.21	.24	.29	.36	.39	.45	.47	.53	.65	—	—	—	—	—	—	—	—	—	—
	C. I.	—	—	—	—	—	—	—	—	—	.44	.62	—	—	—	—	—	—	—	—	—	—
Bell Mouth Inlet	Steel	.04	.07	.10	.13	.18	.26	.31	.43	.52	.67	.95	1.3	1.6	2.3	2.9	3.5	4.0	4.7	5.3	6.1	7.6
	C. I.	—	—	—	—	—	—	—	—	—	.55	.77	—	1.3	1.9	2.4	3.0	3.6	4.3	5.0	5.7	7.0
Square Mouth Inlet	Steel	.44	.68	.96	1.3	1.8	2.6	3.1	4.3	5.2	6.7	9.5	13	16	23	29	35	40	47	53	61	76
	C. I.	—	—	—	—	—	—	—	—	—	5.5	7.7	—	13	19	24	30	36	43	50	57	70
Reentrant Pipe	Steel	.88	1.4	1.9	2.6	3.6	5.1	6.2	8.5	10	13	19	25	32	45	58	70	80	95	110	120	150
	C. I.	—	—	—	—	—	—	—	—	—	11	15	—	26	37	49	61	73	86	100	110	140
Y-Strainer		—	—	4.6	5.0	6.6	7.7	18	20	27	34	42	53	61								

Sudden Enlargement

$$h = \frac{(V_1 - V_2)^2}{2g} \text{ Feet of Liquid; IF } V_2 = 0 \quad h = \frac{V^2}{2g} \text{ Feet of Liquid}$$

Courtesy The Hydraulic Institute (reprinted from the *Standards of the Hydraulic Institute*, Eleventh Edition. Copyright 1965)

Table B-2. Schedule 80 Pipe Dimensions

Size Inches	Diameters External Inches	Diameters Internal Inches	Nominal Thickness Inches	Transverse Areas External Sq. In.	Transverse Areas Internal Sq. In.	Transverse Areas Metal Sq. In.	Length of Pipe Per Square Foot of External Surface Feet	Length of Pipe Per Square Foot of Internal Surface Feet	Cubic Feet per Foot of Pipe	Weight per Foot Pounds	Number Threads per Inch of Screw
⅛	0.405	0.215	0.095	0.129	0.036	0.093	9.431	17.750	0.00025	0.314	27
¼	0.540	0.302	0.119	0.229	0.072	0.157	7.073	12.650	0.00050	0.535	18
⅜	0.675	0.423	0.126	0.358	0.141	0.217	5.658	9.030	0.00098	0.738	18
½	0.840	0.546	0.147	0.554	0.234	0.320	4.547	7.000	0.00163	1.00	14
¾	1.050	0.742	0.154	0.866	0.433	0.433	3.637	5.15	0.00300	1.47	14
1	1.315	0.957	0.179	1.358	0.719	0.639	2.904	3.995	0.00500	2.17	14
1¼	1.660	1.278	0.191	2.164	1.283	0.881	2.301	2.990	0.00891	3.00	11½
1½	1.900	1.500	0.200	2.835	1.767	1.068	2.010	2.542	0.01227	3.65	11½
2	2.375	1.939	0.218	4.430	2.953	1.477	1.608	1.970	0.02051	5.02	11½
2½	2.875	2.323	0.276	6.492	4.238	2.254	1.328	1.645	0.02943	7.66	11½
3	3.500	2.900	0.300	9.621	6.605	3.016	1.091	1.317	0.04587	10.3	8
3½	4.000	3.364	0.318	12.56	8.888	3.678	0.954	1.135	0.06172	12.5	8
4	4.500	3.826	0.337	15.90	11.497	4.407	0.848	0.995	0.0798	14.9	8
5	5.563	4.813	0.375	24.30	18.194	6.112	0.686	0.792	0.1263	20.8	8
6	6.625	5.761	0.432	34.47	26.067	8.300	0.576	0.673	0.1810	28.6	8
8	8.625	7.625	0.500	58.42	46.663	12.76	0.442	0.501	0.3171	43.4	8
10	10.750	9.564	0.593	90.76	71.84	18.92	0.355	0.400	0.4989	64.4	
12	12.750	11.376	0.687	127.64	101.64	26.00	0.299	0.336	0.7058	88.6	
14	14.000	12.500	0.750	153.94	122.72	31.22	0.272	0.306	0.8522	107.0	
16	16.000	14.314	0.843	201.05	160.92	40.13	0.238	0.263	1.112	137.0	
18	18.000	16.126	0.937	254.85	204.24	50.61	0.212	0.237	1.418	171.0	
20	20.000	17.938	1.031	314.15	252.72	61.43	0.191	0.208	1.755	209.0	
24	24.000	21.564	1.218	452.40	365.22	87.18	0.159	0.177	2.536	297.0	

Courtesy Sarco Company, Inc.

Table B-3. Schedule 40 Pipe Dimensions

Size Inches	Diameters			Transverse Areas			Length of Pipe per Square Foot of		Cubic Feet per Foot of Pipe	Weight per Foot Pounds	Number Threads per Inch of Screw
	External Inches	Internal Inches	Nominal Thickness Inches	External Sq. In.	Internal Sq. In.	Metal Sq. In.	External Surface Feet	Internal Surface Feet			
$\frac{1}{8}$	0.405	0.269	0.068	0.129	0.057	0.072	9.431	14.199	0.00039	0.244	27
$\frac{1}{4}$	0.540	0.364	0.088	0.229	0.104	0.125	7.073	10.493	0.00072	0.424	18
$\frac{3}{8}$	0.675	0.493	0.091	0.358	0.191	0.167	5.658	7.747	0.00133	0.567	18
$\frac{1}{2}$	0.840	0.622	0.109	0.554	0.304	0.250	4.547	6.141	0.00211	0.850	14
$\frac{3}{4}$	1.050	0.824	0.113	0.866	0.533	0.333	3.637	4.635	0.00370	1.130	14
1	1.315	1.049	0.133	1.358	0.864	0.494	2.904	3.641	0.00600	1.678	$11\frac{1}{2}$
$1\frac{1}{4}$	1.660	1.380	0.140	2.164	1.495	0.669	2.301	2.767	0.01039	2.272	$11\frac{1}{2}$
$1\frac{1}{2}$	1.900	1.610	0.145	2.835	2.036	0.799	2.010	2.372	0.01414	2.717	$11\frac{1}{2}$
2	2.375	2.067	0.154	4.430	3.355	1.075	1.608	1.847	0.02330	3.652	$11\frac{1}{2}$
$2\frac{1}{2}$	2.875	2.469	0.203	6.492	4.788	1.704	1.328	1.547	0.03325	5.793	8
3	3.500	3.068	0.216	9.621	7.393	2.228	1.091	1.245	0.05134	7.575	8
$3\frac{1}{2}$	4.000	3.548	0.226	12.56	9.886	2.680	0.954	1.076	0.06866	9.109	8
4	4.500	4.026	0.237	15.90	12.73	3.174	0.848	0.948	0.08840	10.790	8
5	5.563	5.047	0.258	24.30	20.00	4.300	0.686	0.756	0.1389	14.61	8
6	6.625	6.065	0.280	34.47	28.90	5.581	0.576	0.629	0.2006	18.97	8
8	8.625	7.981	0.322	58.42	50.02	8.399	0.442	0.478	0.3552	28.55	8
10	10.750	10.020	0.365	90.76	78.85	11.90	0.355	0.381	0.5476	40.48	8
12	12.750	11.938	0.406	127.64	111.9	15.74	0.299	0.318	0.7763	53.6	
14	14.000	13.125	0.437	153.94	135.3	18.64	0.272	0.280	0.9354	63.0	
16	16.000	15.000	0.500	201.05	176.7	24.35	0.238	0.254	1.223	78.0	
18	18.000	16.874	0.563	254.85	224.0	30.85	0.212	0.226	1.555	105.0	
20	20.000	18.814	0.593	314.15	278.0	36.15	0.191	0.203	1.926	123.0	
24	24.000	22.626	0.687	452.40	402.1	50.30	0.159	0.169	2.793	171.0	

Courtesy Sarco Company, Inc.

Table B-4. Properties of Saturated Steam

Gauge Pressure psig	Temperature °F	Heat in Btu/lb. Sensible	Latent	Total	Specific Volume Cu. ft. per lb.
In Vac.					
25	134	102	1017	1119	142.0
20	162	129	1001	1130	73.9
15	179	147	990	1137	51.3
10	192	160	982	1142	39.4
5	203	171	976	1147	31.8
0	212	180	970	1150	26.8
1	215	183	968	1151	25.2
2	219	187	966	1153	23.5
3	222	190	964	1154	22.3
4	224	192	962	1154	21.4
5	227	195	960	1155	20.1
6	230	198	959	1157	19.4
7	232	200	957	1157	18.7
8	233	201	956	1157	18.4
9	237	205	954	1159	17.1
10	239	207	953	1160	16.5
12	244	212	949	1161	15.3
14	248	216	947	1163	14.3
16	252	220	944	1164	13.4
18	256	224	941	1165	12.6

Gauge Pressure psig	Temperature °F	Heat in Btu/lb. Sensible	Latent	Total	Specific Volume Cu. ft. per lb.
150	366	339	857	1196	2.74
155	368	341	885	1196	2.68
160	371	344	853	1197	2.60
165	373	346	851	1197	2.54
170	375	348	849	1197	2.47
175	377	351	847	1198	2.41
180	380	353	845	1198	2.34
185	382	355	843	1198	2.29
190	384	358	841	1199	2.24
195	386	360	839	1199	2.19
200	388	362	837	1199	2.14
205	390	364	836	1200	2.09
210	392	366	834	1200	2.05
215	394	368	832	1200	2.00
220	396	370	830	1200	1.96
225	397	372	828	1200	1.92
230	399	374	825	1201	1.89
235	401	376	823	1201	1.85
240	403	378	823	1201	1.81
245	404	380	822	1202	1.78

Courtesy Sarco Company, Inc.

Table B-4. Properties of Saturated Steam (Cont'd)

Gauge Pressure psig	Temperature °F	Heat in Btu/lb. Sensible	Heat in Btu/lb. Latent	Heat in Btu/lb. Total	Specific Volume Cu. ft per lb.	Gauge Pressure psig	Temperature °F	Heat in Btu/lb. Sensible	Heat in Btu/lb. Latent	Heat in Btu/lb. Total	Specific Volume Cu. ft. per lb.
20	259	227	939	1166	11.9	250	406	382	820	1202	1.75
22	262	230	937	1167	11.3	255	408	383	819	1202	1.72
24	265	233	934	1167	10.8	260	409	385	817	1202	1.69
26	268	236	933	1169	10.3	265	411	387	815	1202	1.66
28	271	239	930	1169	9.85	270	413	389	814	1203	1.63
30	274	243	929	1172	9.46	275	414	391	812	1203	1.60
32	277	246	927	1173	9.10	280	416	392	811	1203	1.57
34	279	248	925	1173	8.75	285	417	394	809	1203	1.55
36	282	251	923	1174	8.42	290	418	395	808	1203	1.53
38	284	253	922	1175	8.08	295	420	397	806	1203	1.49
40	286	256	920	1176	7.82	300	421	398	805	1203	1.47
42	289	258	918	1176	7.57	305	423	400	803	1203	1.45
44	291	260	917	1177	7.31	310	425	402	802	1204	1.43
46	293	262	915	1177	7.14	315	426	404	800	1204	1.41
48	295	264	914	1178	6.94	320	427	405	799	1204	1.38
50	298	267	912	1179	6.68	325	429	407	797	1204	1.36
55	300	271	909	1180	6.27	330	430	408	796	1204	1.34
60	307	277	906	1183	5.84	335	432	410	794	1204	1.33
65	312	282	901	1183	5.49	340	433	411	793	1204	1.31
70	316	286	898	1184	5.18	345	434	413	791	1204	1.29

Table B-4. Properties of Saturated Steam (Cont'd)

Gauge Pressure psig	Temperature °F	Heat in Btu/lb.			Specific Volume Cu. ft. per lb.	Gauge Pressure psig	Temperature °F	Heat in Btu/lb.			Specific Volume Cu. ft. per lb.
		Sensible	Latent	Total				Sensible	Latent	Total	
75	320	290	895	1185	4.91	350	435	414	790	1204	1.28
80	324	294	891	1185	4.67	355	437	416	789	1205	1.26
85	328	298	889	1187	4.44	360	438	417	788	1205	1.24
90	331	302	886	1188	4.24	365	440	419	786	1205	1.22
95	335	305	883	1188	4.05	370	441	420	785	1205	1.20
100	338	309	880	1189	3.89	375	442	421	784	1205	1.19
105	341	312	878	1190	3.74	380	443	422	783	1205	1.18
110	344	316	875	1191	3.59	385	445	424	781	1205	1.16
115	347	319	873	1192	3.46	390	446	425	780	1205	1.14
120	350	322	871	1193	3.34	395	447	427	778	1205	1.13
125	353	325	868	1193	3.23	400	448	428	777	1205	1.12
130	356	328	866	1194	3.12	450	460	439	766	1205	1.00
140	361	333	861	1194	2.92	500	470	453	751	1204	0.89
145	363	336	859	1195	2.84	550	479	464	740	1204	0.82
						600	489	475	728	1203	0.74

Table B-5. Friction Loss for Water in Feet per 100 Feet of Schedule 40 Steel Pipe

U.S. Gal./Min.	Vel. Ft./Sec.	hf Friction	U.S. Gal./Min.	Vel. Ft./Sec.	hf Friction
³⁄₈" PIPE			½" PIPE		
1.4	2.25	9.03	2	2.11	5.50
1.6	2.68	11.6	2.5	2.64	8.24
1.8	3.02	14.3	3	3.17	11.5
2.0	3.36	17.3	3.5	3.70	15.3
2.5	4.20	26.0	4.0	4.22	19.7
3.0	5.04	36.6	5	5.28	29.7
3.5	5.88	49.0	6	6.34	42.0
4.0	6.72	63.2	7	7.39	56.0
5.0	8.40	96.1	8	8.45	72.1
6	10.08	136	9	9.50	90.1
7	11.8	182	10	10.56	110.6
8	13.4	236	12	12.7	156
9	15.1	297	14	14.8	211
10	16.8	364	16	16.9	270
³⁄₄" PIPE			1" PIPE		
4	2.41	4.85	6	2.23	3.16
5	3.01	7.27	8	2.97	5.20
6	3.61	10.2	10	3.71	7.90
7	4.21	13.6	12	4.45	11.1
8	4.81	17.3	14	5.20	14.7
9	5.42	21.6	16	5.94	19.0
10	6.02	26.5	18	6.68	23.7
12	7.22	37.5	20	7.42	28.9
14	8.42	50.0	22	8.17	34.8
16	9.63	64.8	24	8.91	41.0
18	10.8	80.9	26	9.65	47.8
20	12.0	99.0	28	10.39	55.1
22	13.2	120	30	11.1	62.9
24	14.4	141	35	13.0	84.4
26	15.6	165	40	14.8	109
28	16.8	189	45	16.7	137
			50	18.6	168

Table B-5. Friction Loss for Water in Feet per 100 Feet of Schedule 40 Steel Pipe (Cont'd)

U.S. Gal./Min.	Vel. Ft./Sec.	hf Friction	U.S. Gal./Min.	Vel. Ft./Sec.	hf Friction
1¼″ PIPE			1½″ PIPE		
12	2.57	2.85	16	2.52	2.26
14	3.00	3.77	18	2.84	2.79
16	3.43	4.83	20	3.15	3.38
18	3.86	6.00	22	3.47	4.05
20	4.29	7.30	24	3.78	4.76
22	4.72	8.72	26	4.10	5.54
24	5.15	10.27	28	4.41	6.34
26	5.58	11.94	30	4.73	7.20
28	6.01	13.7	35	5.51	9.63
30	6.44	15.6	40	6.30	12.41
35	7.51	21.9	45	7.04	15.49
40	8.58	27.1	50	7.88	18.9
45	9.65	33.8	55	8.67	22.7
50	10.7	41.4	60	9.46	26.7
55	11.8	49.7	65	10.24	31.2
60	12.9	58.6	70	11.03	36.0
65	13.9	68.6	75	11.8	41.2
70	15.0	79.2	80	12.6	46.6
75	16.1	90.6	85	13.4	52.4
			90	14.2	58.7
			95	15.0	65.0
			100	15.8	71.6
2″ PIPE			2½″ PIPE		
25	2.39	1.48	35	2.35	1.15
30	2.87	2.10	40	2.68	1.47
35	3.35	2.79	45	3.02	1.84
40	3.82	3.57	50	3.35	2.23
45	4.30	4.40	60	4.02	3.13
50	4.78	5.37	70	4.69	4.18
60	5.74	7.58	80	5.36	5.36
70	6.69	10.2	90	6.03	6.69
80	7.65	13.1	100	6.70	8.18
90	8.60	16.3	120	8.04	11.50
100	4.34	2.72	200	5.04	12.61
120	11.5	28.5	160	10.7	20.0
140	13.4	38.2	180	12.1	25.2
160	15.3	49.5	200	13.4	30.7
			220	14.7	37.1
			240	16.1	43.8

Table B-5. Friction Loss for Water in Feet per 100 Feet of Schedule 40 Steel Pipe (Cont'd)

U.S. Gal./Min.	Vel. Ft./Sec.	hf Friction	U.S. Gal./Min.	Vel. Ft./Sec.	hf Friction
3" PIPE			4" PIPE		
50	2.17	0.762	100	2.52	0.718
60	2.60	1.06	120	3.02	1.01
70	3.04	1.40	140	3.53	1.35
80	3.47	1.81	160	4.03	1.71
90	3.91	2.26	180	4.54	2.14
100	3.34	2.75	200	5.04	2.61
120	5.21	3.88	220	5.54	3.13
140	6.08	5.19	240	6.05	3.70
160	6.94	6.68	260	6.55	4.30
180	7.81	8.38	280	7.06	4.95
200	8.68	10.2	300	7.56	5.63
220	9.55	12.3	350	8.82	7.54
240	10.4	14.5	400	10.10	9.75
260	11.3	16.9	450	11.4	12.3
280	12.2	19.5	500	12.6	14.4
300	13.0	22.1	550	13.9	18.1
350	15.2	30	600	15.1	21.4
5" PIPE			6" PIPE		
160	2.57	0.557	220	2.44	0.411
180	2.89	0.698	240	2.66	0.482
200	3.21	0.847	260	2.89	0.560
220	3.53	1.01	300	3.33	0.733
240	3.85	1.19	350	3.89	0.980
260	4.17	1.38	400	4.44	1.25
300	4.81	1.82	450	5.00	1.56
350	5.61	2.43	500	5.55	1.91
400	6.41	3.13	600	6.66	2.69
450	7.22	3.92	700	7.77	3.60
500	8.02	4.79	800	8.88	4.64
600	9.62	6.77	900	9.99	5.81
700	11.2	9.13	1000	11.1	7.10
800	12.8	11.8	1100	12.2	8.52
900	14.4	14.8	1200	13.3	10.1
1000	16.0	18.2	1300	14.4	11.7
			1400	15.5	13.6

Courtesy Sarco Company, Inc.

Table B-6. Flow of Water through Schedule 40 Steel Pipe

Pressure Drop per 1000 Feet of Schedule 40 Steel Pipe, in Pounds per Square Inch

Discharge Gal. per Min.	1" Vel. Ft. per Sec.	1" Pres. sure Drop	1¼" Vel. Ft. per Sec.	1¼" Pres. sure Drop	1½" Vel. Ft. per Sec.	1½" Pres. sure Drop	2" Vel. Ft. per Sec.	2" Pres. sure Drop	2½" Vel. Ft. per Sec.	2½" Pres. sure Drop	3" Vel. Ft. per Sec.	3" Pres. sure Drop	3½" Vel. Ft. per Sec.	3½" Pres. sure Drop	4" Vel. Ft. per Sec.	4" Pres. sure Drop	5" Vel. Ft. per Sec.	5" Pres. sure Drop	6" Vel. Ft. per Sec.	6" Pres. sure Drop	8" Vel. Ft. per Sec.	8" Pres. sure Drop
1	0.37	0.49																				
2	0.74	1.70	0.43	0.45																		
3	1.12	3.53	0.64	0.94	0.47	0.44																
4	1.49	5.94	0.86	1.55	0.63	0.74																
5	1.86	9.02	1.07	2.36	0.79	1.12																
6	2.24	12.25	1.28	3.30	0.95	1.53	0.57	0.46														
8	2.98	21.1	1.72	5.52	1.26	2.63	0.76	0.75														
10	3.72	30.8	2.14	8.34	1.57	3.86	0.96	1.14	.67	0.48												
15	5.60	64.6	3.21	17.6	2.36	8.13	1.43	2.33	1.00	0.99												
20	7.44	110.5	4.29	29.1	3.15	13.5	1.91	3.86	1.34	1.64	0.87	0.59										
25			5.36	43.7	3.94	20.2	2.39	5.81	1.68	2.48	1.08	0.67	0.81	0.42								
30			6.43	62.9	4.72	29.1	2.87	8.04	2.01	3.43	1.30	1.21	0.97	0.60								
35			7.51	82.5	5.51	38.2	3.35	10.95	2.35	4.49	1.52	1.58	1.14	0.79	0.88	0.42						
40					6.30	47.8	3.82	13.7	2.68	5.88	1.74	2.06	1.30	1.00	1.01	0.53						
45					7.08	60.6	4.30	17.4	3.00	7.14	1.95	2.51	1.46	1.21	1.13	0.67						
50					7.87	74.7	4.78	20.6	3.35	8.82	2.17	3.10	1.62	1.44	1.26	0.80						
60							5.74	29.6	4.02	12.2	2.60	4.29	1.95	2.07	1.51	1.10						
70							6.69	38.6	4.69	15.3	3.04	5.84	2.27	2.71	1.76	1.50	1.12	0.48				
80							7.65	50.3	5.37	21.7	3.48	7.62	2.59	3.53	2.01	1.87	1.28	0.63				
90							8.60	63.6	6.04	26.1	3.91	9.22	2.92	4.46	2.26	2.37	1.44	0.80				
100							9.56	75.1	6.71	32.3	4.34	11.4	3.24	5.27	2.52	2.81	1.60	0.95	1.11	0.39		
125									8.38	48.2	5.45	17.1	4.05	7.86	3.15	4.38	2.00	1.48	1.39	0.56		
150									10.06	60.4	6.51	23.3	4.86	11.3	3.78	6.02	2.41	2.04	1.67	0.78		
175									11.73	90.0	7.59	32.0	5.67	14.7	4.41	8.20	2.81	2.78	1.94	1.06		
200											8.68	39.7	6.48	19.2	5.04	10.2	3.21	3.46	2.22	1.32		
225											9.77	50.2	7.29	23.1	5.67	12.9	3.61	4.37	2.50	1.66	1.44	0.44
250											10.85	61.9	8.10	28.5	6.30	15.9	4.01	5.14	2.78	2.05	1.60	0.55
275											11.94	75.0	8.91	34.4	6.93	18.3	4.41	6.22	3.06	2.36	1.76	0.63
300											13.02	84.7	9.72	40.9	7.56	21.8	4.81	7.41	3.33	2.80	1.92	0.75
325													10.53	45.5	8.18	25.5	5.21	8.25	3.61	3.29	2.08	0.88
350													11.35	52.7	8.82	29.7	5.61	9.57	3.89	3.62	2.24	0.97
375													12.17	60.7	9.45	32.3	6.01	11.0	4.16	4.16	2.40	1.11
400													13.78	77.8	10.70	41.5	6.82	14.1	4.44	4.72	2.56	1.27
425													12.97	68.9	10.08	39.7	6.41	12.9	4.72	5.34	2.72	1.43

Table B-6. Flow of Water through Schedule 40 Steel Pipe (Cont'd)

Pressure Drop per 1000 Feet of Schedule 40 Steel Pipe, in Pounds per Square Inch

(Vel. = Velocity, Ft. per Sec.; Pres. = Pressure Drop, lb/sq in.)

Discharge Gal. per Min.	3½" Vel.	3½" Pres.	5" Vel.	5" Pres.	6" Vel.	6" Pres.	8" Vel.	8" Pres.	10" Vel.	10" Pres.	12" Vel.	12" Pres.	14" Vel.	14" Pres.	16" Vel.	16" Pres.	18" Vel.	18" Pres.	20" Vel.	20" Pres.	24" Vel.	24" Pres.
450	14.59	87.3	7.22	15.0	5.00	5.96	2.88	1.60	1.83	0.30												
475			7.62	16.7	5.27	6.66	3.04	1.69	1.93													
500			8.02	18.5	5.55	7.39	3.20	1.87	2.04	0.63												
550			8.82	22.4	6.11	8.94	3.53	2.26	2.24	0.70												
600			9.62	26.7	6.66	10.6	3.85	2.70	2.44	0.86												
650			10.42	31.3	7.21	11.8	4.17	3.16	2.65	1.01												
700			11.22	36.3	7.77	13.7	4.49	3.69	2.85	1.18	2.01	0.48										
750			12.02	41.6	8.32	15.7	4.81	4.21	3.05	1.35	2.15	0.55										
800			12.82	44.7	8.88	17.8	5.13	4.79	3.26	1.54	2.29	0.62										
850			13.62	50.5	9.44	20.2	5.45	5.11	3.46	1.74	2.44	0.70	2.02	0.43								
900			14.42	56.6	10.00	22.6	5.77	5.73	3.66	1.94	2.58	0.79	2.14	0.48								
950			15.22	63.1	10.55	23.7	6.09	6.38	3.87	2.23	2.72	0.88	2.25	0.53								
1,000			16.02	70.0	11.10	26.3	6.41	7.08	4.07	2.40	2.87	0.98	2.38	0.59								
1,100			17.63	84.6	12.22	31.8	7.05	8.56	4.48	2.74	3.16	1.18	2.61	0.68								
1,200					13.32	37.8	7.69	10.2	4.88	3.27	3.45	1.40	2.85	0.81	2.18	0.40						
1,300					14.43	44.4	8.33	11.3	5.29	3.86	3.73	1.56	3.09	0.95	2.36	0.47						
1,400					15.54	51.5	8.97	13.0	5.70	4.44	4.02	1.80	3.32	1.10	2.54	0.54						
1,500					16.65	55.5	9.62	15.0	6.10	5.11	4.30	2.07	3.55	1.19	2.73	0.62						
1,600					17.76	66.1	10.26	16.6	6.51	5.46	4.59	2.36	3.80	1.35	2.91	0.71						
1,800					19.98	79.8	11.54	21.6	7.32	6.91	5.16	2.98	4.27	1.71	3.27	0.85	2.58	0.48				
2,000					22.20	98.5	12.83	25.0	8.13	8.54	5.73	3.47	4.74	2.11	3.63	1.05	2.88	0.56				
2,500							16.03	39.0	10.18	12.5	7.17	5.41	5.92	3.09	4.54	1.63	3.59	0.88				
3,000							19.24	52.4	12.21	18.0	8.60	7.31	7.12	4.45	5.45	2.21	4.31	1.27	3.45	0.73		
3,500							22.43	71.4			10.03	9.95	8.32	6.18	6.35	3.00	5.03	1.52	4.03	0.94		
4,000							25.65	93.3	16.28	29.9	11.48	13.0	9.49	7.92	7.25	3.92	5.74	2.12	4.61	1.22	3.19	0.51
4,500									18.31	37.8	12.90	15.4	10.67	9.36	8.17	4.97	6.47	2.50	5.19	1.55	3.59	0.60
5,000									20.35	46.7	14.34	18.9	11.84	11.6	9.08	5.72	7.17	3.08	5.76	1.78	3.99	0.74
6,000									24.42	67.2	17.21	27.3	14.32	15.4	10.88	8.24	8.62	4.45	6.92	2.57	4.80	1.00
7,000									28.50	85.1	20.08	37.2	16.60	21.0	12.69	12.2	10.04	6.06	8.06	3.50	5.68	1.36
8,000											22.95	45.1	18.98	27.4	14.52	13.6	11.48	7.34	9.23	4.57	6.38	1.78
9,000											25.80	57.0	21.35	34.7	16.32	17.2	12.92	9.20	10.37	5.36	7.19	2.25
10,000											28.63	70.4	23.75	42.9	18.16	21.2	14.37	11.5	11.53	6.63	7.96	2.78
12,000											34.38	93.6	28.50	61.8	21.80	30.9	17.23	16.5	13.83	9.54	9.57	3.71
14,000													33.20	84.0	25.42	41.6	20.10	20.7	16.14	12.0	11.18	5.05
16,000															29.05	54.4	22.96	27.1	18.43	15.7	12.77	6.60

Courtesy Sarco Company, Inc.

Table B-7. Warmup Load in Pounds of Steam per 100 Feet of Steam Main (Ambient Temperature 70°F)*

Steam Pressure (psig)	MAIN SIZE														0°F Correction Factor†
	2"	2½"	3"	4"	5"	6"	8"	10"	12"	14"	16"	18"	20"	24"	
0	6.2	9.7	12.8	18.2	24.6	31.9	48	68	90	107	140	176	207	208	1.50
5	6.9	11.0	14.4	20.4	27.7	35.9	48	77	101	120	157	198	233	324	1.44
10	7.5	11.8	15.5	22.0	29.9	38.8	58	83	109	130	169	213	251	350	1.41
20	8.4	13.4	17.5	24.9	33.8	43.9	66	93	124	146	191	241	284	396	1.37
40	3.9	15.8	20.6	29.3	39.7	51.6	78	110	145	172	225	284	334	465	1.32
60	11.0	17.5	22.9	32.6	44.2	57.3	86	122	162	192	250	316	372	518	1.29
80	12.0	19.0	24.9	35.3	47.9	62.1	93	132	175	208	271	342	403	561	1.27
100	12.8	20.3	26.6	37.8	51.2	66.5	100	142	188	222	290	366	431	600	1.26
125	13.7	21.7	28.4	40.4	54.8	71.1	107	152	200	238	310	391	461	642	1.25
150	14.5	23.0	30.0	42.8	58.0	75.2	113	160	212	251	328	414	487	679	1.24
175	15.3	24.2	31.7	45.1	61.2	79.4	119	169	224	265	347	437	514	716	1.23
200	16.0	25.3	33.1	47.1	63.8	82.8	125	177	234	277	362	456	537	748	1.22
250	17.2	27.3	35.8	50.8	68.9	89.4	134	191	252	299	390	492	579	807	1.21
300	25.0	38.3	51.3	74.8	104.0	142.7	217	322	443	531	682	854	1045	1182	1.20
400	27.8	42.6	57.1	83.2	115.7	158.7	241	358	493	590	759	971	1163	1650	1.18
500	30.2	46.3	62.1	90.5	125.7	172.6	262	389	535	642	825	1033	1263	1793	1.17
600	32.7	50.1	67.1	97.9	136.0	186.6	284	421	579	694	893	1118	1367	1939	1.16

*Loads based on Schedule 40 pipe for pressures up to and including 250 psig and on Schedule 80 pipe for pressures above 250 psig.

†For outdoor temperature of 0°F, multiply load value in table for each main size by correction factor corresponding to steam pressure.

Courtesy Sarco Company, Inc.

Table B-8. Condensation Load in Pounds per Hour per 100 Feet of Insulated Steam Main (Ambient Temperature 70°F; Insulation 80% Efficient)*

Steam Pressure (psig)	MAIN SIZE														0°F Correction Factor†
	2"	2½"	3"	4"	5"	6"	8"	10"	12"	14"	16"	18"	20"	24"	
10	6	7	9	11	13	16	20	24	29	32	36	39	44	53	1.58
30	8	9	11	14	17	20	26	32	38	42	48	51	57	68	1.50
60	10	12	14	18	24	27	33	41	49	54	62	67	74	89	1.45
100	12	15	18	22	28	33	41	51	61	67	77	83	93	111	1.41
125	13	16	20	24	30	36	45	56	66	73	84	90	101	121	1.39
175	16	19	23	26	33	38	53	66	78	86	98	107	119	142	1.38
250	18	22	27	34	42	50	62	77	92	101	116	126	140	168	1.36
300	20	25	30	37	46	54	68	85	101	111	126	138	154	184	1.35
400	23	28	34	43	53	63	80	99	118	130	148	162	180	216	1.33
500	27	33	39	49	61	73	91	114	135	148	170	185	206	246	1.32
600	30	37	44	55	68	82	103	128	152	167	191	208	232	277	1.31

Courtesy Sarco Company, Inc.

*Chart loads represent losses due to radiation and convection for saturated steam.

†For outdoor temperature of 0°F, multiply load value in table for each main size by correction factor corresponding to steam pressure.

Table B-9. Flange Standards
(All dimensions are in inches)

125-lb. CAST IRON — ASA B16.1

Pipe Size	1/2	3/4	1	1 1/4	1 1/2	2	2 1/2	3	3 1/2	4	5	6	8	10	12
Diameter of Flange			4 1/4	4 5/8	5	6	7	7 1/2	8 1/2	9	10	11	13 1/2	16	19
Thickness of Flange (min)[1]			7/16	1/2	9/16	5/8	11/16	3/4	13/16	15/16	15/16	1	1 1/8	1 3/16	1 1/4
Diameter of Bolt Circle			3 1/8	3 1/2	3 7/8	4 3/4	5 1/2	6	7	7 1/2	8 1/2	9 1/2	11 3/4	14 1/4	17
Number of Bolts			4	4	4	4	4	4	8	8	8	8	8	12	12
Diameter of Bolts			1/2	1/2	1/2	5/8	5/8	5/8	5/8	5/8	3/4	3/4	3/4	7/8	7/8

[1]125-lb. flanges have plain faces.

250-lb. CAST IRON — ASA B16.2

Pipe Size	1/2	3/4	1	1 1/4	1 1/2	2	2 1/2	3	3 1/2	4	5	6	8	10	12
Diameter of Flange			4 7/8	5 1/4	6 1/8	6 1/2	7 1/2	8 1/4	9	10	11	12 1/2	15	17 1/2	20 1/2
Thickness of Flange (min)[2]			11/16	3/4	13/16	7/8	1	1 1/8	1 3/16	1 1/4	1 3/8	1 7/16	1 5/8	1 7/8	2
Diameter of Raised Face			2 11/16	3 1/16	3 9/16	4 3/16	4 15/16	5 11/16	6 5/16	6 15/16	8 5/16	9 11/16	11 15/16	14 1/16	16 7/16
Diameter of Bolt Circle			3 1/2	3 7/8	4 1/2	5	5 7/8	6 5/8	7 1/4	7 7/8	9 1/4	10 5/8	13	15 1/4	17 3/4
Number of Bolts			4	4	4	8	8	8	8	8	8	12	12	16	16
Diameter of Bolts			5/8	5/8	3/4	5/8	3/4	3/4	3/4	3/4	3/4	3/4	7/8	1	1 1/8

[2]250-lb. flanges have a 1/16" raised face, which is included in the flange thickness dimensions.

Table B-9. Flange Standards (Cont'd)
(All dimensions are in inches)

150-lb. BRONZE — ASA B16.24

Pipe Size	1/2	3/4	1	1 1/4	1 1/2	2	2 1/2	3	3 1/2	4	5	6	8	10	12
Diameter of Flange	3 1/2	3 7/8	4 1/4	4 5/8	5	6	7	7 1/2	8 1/2	9	10	11	13 1/2	16	19
Thickness of Flange (min)[3]	5/16	11/32	3/8	13/32	7/16	1/2	9/16	5/8	11/16	11/16	3/4	13/16	15/16	1	1 1/16
Diameter of Bolt Circle	2 3/8	2 3/4	3 1/8	3 1/2	3 7/8	4 3/4	5 1/2	6	7	7 1/2	8 1/2	9 1/2	11 3/4	14 1/4	17
Number of Bolts	4	4	4	4	4	4	4	4	8	8	8	8	8	12	12
Diameter of Bolts	1/2	1/2	1/2	1/2	1/2	5/8	5/8	5/8	5/8	5/8	3/4	3/4	3/4	7/8	7/8

[3]150-lb. bronze flanges have plain faces with two concentric gasket-retaining grooves between the port and the bolt holes.

300-lb. BRONZE — ASA B16.24

Pipe Size	1/2	3/4	1	1 1/4	1 1/2	2	2 1/2	3	3 1/2	4	5	6	8	10	12
Diameter of Flange	3 3/4	4 5/8	4 7/8	5 1/4	6 1/8	6 1/2	7 1/2	8 1/4	9	10	11	12 1/2	15		
Thickness of Flange (min)[4]	1/2	17/32	19/32	5/8	11/16	3/4	13/16	29/32	31/32	1 1/16	1 1/8	1 3/16	1 3/8		
Diameter of Bolt Circle	2 5/8	3 1/4	3 1/2	3 7/8	4 1/2	5	5 7/8	6 5/8	7 1/4	7 7/8	9 1/4	10 5/8	13		
Number of Bolts	4	4	4	4	4	8	8	8	8	8	8	12	12		
Diameter of Bolts	1/2	5/8	5/8	5/8	5/8	3/4	3/4	3/4	3/4	3/4	3/4	3/4	7/8		

[4]300-lb. bronze flanges have plain faces with two concentric gasket-retaining grooves between the port and the bolt holes.

Table B-9. Flange Standards (Cont'd)

(All dimensions are in inches)

150-lb. STEEL									**ASA B16.5**					
Pipe Size	$\frac{1}{2}$	$\frac{3}{4}$	1	$1\frac{1}{4}$	$1\frac{1}{2}$	2	$2\frac{1}{2}$	3	$3\frac{1}{2}$	4	5	6	8	10	12
			$4\frac{1}{4}$	$4\frac{5}{8}$	5	6	7	$7\frac{1}{2}$	$8\frac{1}{2}$	9	10	11	$13\frac{1}{2}$	16	19
Thickness of Flange (min)[5]			$\frac{7}{16}$	$\frac{1}{2}$	$\frac{9}{16}$	$\frac{5}{8}$	$\frac{13}{16}$	$\frac{3}{4}$	$\frac{13}{16}$	$\frac{15}{16}$	$\frac{15}{16}$	1	$1\frac{1}{8}$	$1\frac{3}{16}$	$1\frac{1}{4}$
Diameter of Raised Face			2	$2\frac{1}{2}$	$2\frac{7}{8}$	$3\frac{5}{8}$	$4\frac{1}{8}$	5	$5\frac{1}{2}$	$6\frac{3}{16}$	$7\frac{5}{16}$	$8\frac{1}{2}$	$10\frac{5}{8}$	$12\frac{3}{4}$	15
Diameter of Bolt Circle			$3\frac{1}{8}$	$3\frac{1}{2}$	$3\frac{7}{8}$	$4\frac{3}{4}$	$5\frac{1}{2}$	6	7	$7\frac{1}{2}$	$8\frac{1}{2}$	$9\frac{1}{2}$	$11\frac{3}{4}$	$14\frac{1}{4}$	17
Number of Bolts			4	4	4	4	4	4	8	8	8	8	8	12	12
Diameter of Bolts			$\frac{1}{2}$	$\frac{1}{2}$	$\frac{1}{2}$	$\frac{5}{8}$	$\frac{5}{8}$	$\frac{5}{8}$	$\frac{5}{8}$	$\frac{5}{8}$	$\frac{3}{4}$	$\frac{3}{4}$	$\frac{3}{4}$	$\frac{7}{8}$	$\frac{7}{8}$

[5] 150-lb. steel flanges have a $\frac{1}{16}$" raised face, which is included in the flange thickness dimensions.

300-lb. STEEL									**ASA B16.5**					
Pipe Size	$\frac{1}{2}$	$\frac{3}{4}$	1	$1\frac{1}{4}$	$1\frac{1}{2}$	2	$2\frac{1}{2}$	3	$3\frac{1}{2}$	4	5	6	8	10	12
Diameter of Flange			$4\frac{7}{8}$	$5\frac{1}{4}$	$6\frac{1}{8}$	$6\frac{1}{2}$	$7\frac{1}{2}$	$8\frac{1}{4}$	9	10	11	$12\frac{1}{2}$	15	$17\frac{1}{2}$	$20\frac{1}{2}$
Thickness of Flange (min)[6]			$\frac{11}{16}$	$\frac{3}{4}$	$\frac{13}{16}$	$\frac{7}{8}$	1	$1\frac{1}{8}$	$1\frac{3}{16}$	$1\frac{1}{4}$	$1\frac{3}{8}$	$1\frac{7}{16}$	$1\frac{5}{8}$	$1\frac{7}{8}$	2
Diameter of Raised Face			2	$2\frac{1}{2}$	$2\frac{7}{8}$	$3\frac{5}{8}$	$4\frac{1}{8}$	5	$5\frac{1}{2}$	$6\frac{3}{16}$	$7\frac{5}{16}$	$8\frac{1}{2}$	$10\frac{5}{8}$	$12\frac{3}{4}$	15
Diameter of Bolt Circle			$3\frac{1}{2}$	$3\frac{7}{8}$	$4\frac{1}{2}$	5	$5\frac{7}{8}$	$6\frac{5}{8}$	$7\frac{1}{4}$	$7\frac{7}{8}$	$9\frac{1}{4}$	$10\frac{5}{8}$	13	$15\frac{1}{4}$	$17\frac{3}{4}$
Number of Bolts			4	4	4	8	8	8	8	8	8	12	12	16	16
Diameter of Bolts			$\frac{5}{8}$	$\frac{5}{8}$	$\frac{3}{4}$	$\frac{5}{8}$	$\frac{3}{4}$	$\frac{3}{4}$	$\frac{3}{4}$	$\frac{3}{4}$	$\frac{3}{4}$	$\frac{3}{4}$	$\frac{7}{8}$	1	$1\frac{1}{8}$

[6] 300-lb. steel flanges have a $\frac{1}{16}$" raised face, which is included in the flange thickness dimensions.

Table B-9. Flange Standards (Cont'd)
(All dimensions are in inches)

400-lb. STEEL

ASA B16.5

Pipe Size	1/2	3/4	1	1 1/4	1 1/2	2	2 1/2	3	3 1/2	4	5	6	8	10	12
Diameter of Flange	3 3/4	4 5/8	4 7/8	5 1/4	6 1/8	6 1/2	7 1/2	8 1/4	9	10	11	12 1/2	15	17 1/2	20 1/2
Thickness of Flange (min)[7]	9/16	5/8	11/16	13/16	7/8	1	1 1/8	1 1/4	1 3/8	1 3/8	1 1/2	1 5/8	1 7/8	2 1/8	2 1/4
Diameter of Raised Face	1 3/8	1 11/16	2	2 1/2	2 7/8	3 5/8	4 1/8	5	5 1/2	6 3/16	7 5/16	8 1/2	10 5/8	12 3/4	15
Diameter of Bolt Circle	2 5/8	3 1/4	3 1/2	3 7/8	4 1/2	5	5 7/8	6 5/8	7 1/4	7 7/8	9 1/4	10 5/8	13	15 1/4	17 3/4
Number of Bolts	4	4	4	4	4	8	8	8	8	8	8	12	12	16	16
Diameter of Bolts	1/2	5/8	5/8	5/8	3/4	5/8	3/4	3/4	7/8	7/8	7/8	7/8	1	1 1/8	1 1/4

[7]400-lb. steel flanges have a 1/4" raised face, which is NOT included in the flange dimensions.

600-lb. STEEL

ASA B16.5

Pipe Size	1/2	3/4	1	1 1/4	1 1/2	2	2 1/2	3	3 1/2	4	5	6	8	10	12
Diameter of Flange	3 3/4	4 5/8	4 7/8	5 1/4	6 1/8	6 1/2	7 1/2	8 1/4	9	10 3/4	13	14	16 1/2	20	22
Thickness of Flange (min)[8]	5/8	5/8	11/16	13/16	7/8	1	1 1/8	1 1/4	1 3/8	1 1/2	1 3/4	1 7/8	2 3/16	2 1/2	2 5/8
Diameter of Raised Face	1 3/8	1 11/16	2	2 1/2	2 7/8	3 5/8	4 1/8	5	5 1/2	6 3/16	7 5/16	8 1/2	10 5/8	12 3/4	15
Diameter of Bolt Circle	2 5/8	3 1/4	3 1/2	3 7/8	4 1/2	5	5 7/8	6 5/8	7 1/4	8 1/2	10 1/2	11 1/2	13 3/4	17	19 1/2
Number of Bolts	4	4	4	4	4	8	8	8	8	8	8	12	12	16	20
Diameter of Bolts	1/2	5/8	5/8	5/8	3/4	5/8	3/4	3/4	7/8	7/8	1	1	1 1/8	1 1/4	1 1/4

[8]600-lb. steel flanges have a 1/4" raised face, which is NOT included in the flange dimensions.

Table B-10. Pressure Drop in Schedule 40 Pipe

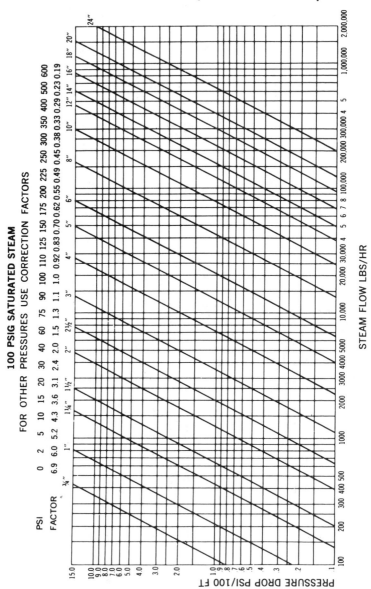

Courtesy Sarco Company, Inc.

626

Table B-11. Steam Velocity Chart

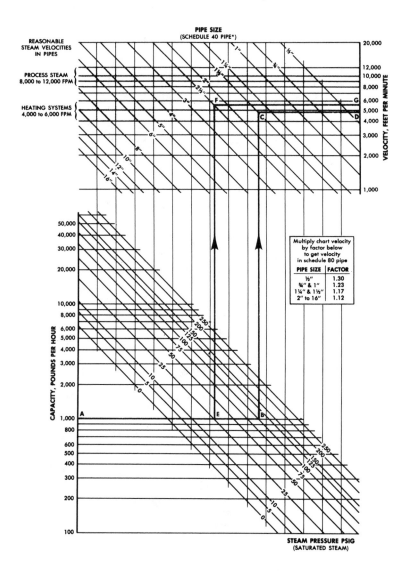

Courtesy Sarco Company, Inc.

Table B-12. Pressure Drop in Schedule 80 Pipe

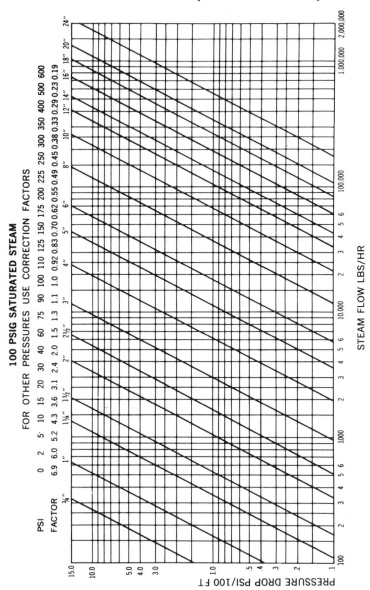

Courtesy Sarco Company, Inc.

628

Table B-13. Chimney Connector and Vent Connector Clearance from Combustible Materials

Description of Appliance	Minimum Clearance, Inches*
RESIDENTIAL APPLIANCES	
Single-Wall, Metal Pipe Connectors	
Electric, Gas, and Oil Incinerators	18
Oil and Solid-Fuel Appliances	18
Oil Appliances Listed as Suitable for Use with Type L	
Venting Systems, but only when connected to chimneys	9
Type L Venting-System Piping Connectors	
Electric, Gas, and Oil Incinerators	9
Oil and Solid-Fuel Appliances	9
Oil Appliances Listed as Suitable for Use with	
Type L Venting Systems	†
COMMERCIAL AND INDUSTRIAL APPLIANCES	
Low-Heat Appliances	
Single-Wall, Metal Pipe Connectors	
Gas, Oil, and Solid-Fuel Boilers, Furnaces,	
and Water Heaters	18
Ranges, Restaurant Type	18
Oil Unit Heaters	18
Other Low-Heat Industrial Appliances	18
Medium-Heat Appliances	
Single-Wall, Metal Pipe Connectors	
All Gas, Oil, and Solid-Fuel Appliances	36

*These clearances apply except if the listing of an appliance specifies different clearance, in which case the listed clearance takes precedence.

†If listed Type L venting-system piping is used, the clearance may be in accordance with the venting-system listing.

If listed Type B or Type L venting-system piping is used, the clearance may be in accordance with the venting-system listing.

The clearances from connectors to combustible materials may be reduced if the combustible material is protected in accordance with Table 1C.

Courtesy National Oil Fuel Institute

Table B-14. Standard Clearances for Heat-Producing Appliances in Residential Installations

Residential Type Appliances for Installation in Rooms that are Large*	APPLIANCE				
	Above Top of Casing or Appliance	From Top and Sides of Warm-Air Bonnet or Plenum	From Front†	From Back	From Sides
Boilers and Water Heaters Steam Boilers — 15 psi; Water Boilers — 250°F; Water Heaters — 200°F; All Water Walled or Jacketed Automatic Oil or Comb. Gas-Oil	6	—	24	6	6
Furnaces — Central Gravity, Upflow, Downflow, Horizontal and Duct. Warm-Air — 250°F Max. Automatic Oil or Comb. Gas-Oil	6‡	6‡	24	6	6
Furnaces — Floor For Mounting in Combustible Floors Automatic Oil or Comb. Gas-Oil	36	—	12	12	12

* Rooms that are large in comparison to the size of the appliance are those having a volume equal to at least 12 times the total volume of a furnace and at least 16 times the total volume of a boiler. If the actual ceiling height of a room is greater than 8 ft., the volume of a room shall be figured on the basis of a ceiling height of 8 ft.

† The minimum dimension should be that necessary for servicing the appliance including access for cleaning and normal care, tube removal, etc.

‡ For a listed oil, combination gas-oil, gas, or electric furnace, this dimension may be 2 in. if the furnace limit control cannot be set higher than 250°F or 1 in. if the limit control cannot be set higher than 200°F.

Table B-15. Standard Clearances for Heat-Producing Appliances in Commercial and Industrial Installations

Commercial-Industrial Type Low Heat Appliances (Any and All Physical Sizes Except As Noted)		APPLIANCE				
		Above Top of Casing or Appliance*	From Top and Sides of Warm-Air Bonnet or Plenum	From Front	From Back*	From Sides*
Boiler and Water Heaters						
100 cu. ft. or less, Any psi, Steam	All Fuels	18	—	48	18	18
50 psi or less, Any Size	All Fuels	18	—	48	18	18
Unit Heaters						
Floor Mounted or Suspended — Any Size	Steam or Hot Water	1	—	—	1	1
Suspended — 100 cu. ft. or less	Oil or Comb. Gas-Oil	6	—	24	18	18
Suspended — Over 100 cu. ft.	All Fuels	18	—	48	18	18
Floor Mounted Any Size	All Fuels	18	—	48	18	18

*If the appliance is encased in brick, the 18 in. clearance above and at sides and rear may be reduced to not less than 12 in.

Courtesy National Oil Fuel Institute

Table B-16. Clearance (in Inches) with Specified Forms of Protection

Input Heat Units	Eff. %	Usable Btuh	gph 100°F Rise	Tank Size, Gal.	AVAILABLE HOT-WATER STORAGE PLUS RECOVERY 100°F RISE				Continuous Draw, gph
					15 Min.	30 Min.	45 Min.	60 Min.	
ELECTRICITY, kWh									
1.5	92.5	4,750	5.7*	20	21.4	22.8	24.3	25.7	5.7
2.5	92.5	7,900	9.5*	20	32.4	34.8	37.1	39.5	9.5
4.5	92.5	14,200	17.1*	30	44.3	48.6	52.9	57.1	17.1
4.5	92.5	14,200	17.1*	50	54.3	58.6	62.9	67.1	17.1
6.0	92.5	19,000	22.8*	66	71.6	77.2	82.8	88.8	22.8
7.0	92.5	22,100	26.5	80	86.6	93.2	99.8	106.5	26.5
GAS, btuh									
34,000	75	25,500	30.6	30	37.7	45.3	53.0	55.6†	25.6†
42,000	75	31,600	38.0	30	39.5	49.0	58.8	61.7†	31.7†
50,000	75	37,400	45.0	40	51.3	62.6	73.9	77.6†	37.6†
60,000	75	45,000	54.0	50	63.5	77.0	90.5	95.0†	45.0†
OIL, gph									
0.50	75	52,500	63.0	30	45.8	61.6	77.4	82.5†	52.5†
0.75	75	78,700	94.6	30	53.6	77.2	100.8	109.0†	79.0†
0.85	75	89,100	107.0	30	57.7	83.4	110.1	119.1†	89.0†
1.00	75	105,000	126.0	50	81.5	113.0	144.5	155.0†	105.0†
1.20	75	126,000	151.5	50	87.9	125.8	163.7	176.0†	126.0†
1.35	75	145,000	174.0	50	93.5	137.0	180.5	195.0†	145.0†
1.50	75	157,000	188.5	85	132.1	179.2	226.3	242.0†	157.0†
1.65	75	174,000	204.5	85	136.1	187.2	238.4	259.0†	174.0†

*Assumes simultaneous operation of upper and lower elements.
†Based on 50 minute-per-hour operation.

Courtesy National Oil Fuel Institute

Table B-17. Performance of Storage Water Heaters

Where the required clearance with no protection is*

TYPE OF PROTECTION Applied to the combustible material unless otherwise specified and covering all surfaces within the distance specified as the required clearance with no protection (see Fig. 1A). Thicknesses are minimum.	36 Inches			18 Inches			12 Inches		9 Inches	6 Inches		
	Above	Sides & Rear	Chimney or Vent Connector	Above	Sides & Rear	Chimney or Vent Connector	Above	Sides & Rear	Chimney or Vent Connector	Above	Sides & Rear	Chimney or Vent Connector
(a) ¼ in. asbestos millboard spaced out 1 in.†	30	18	30	15	9	12	9	6	6	3	2	3
(b) 28-gauge sheet metal on ¼ in. asbestos millboard	24	18	24	12	9	12	9	6	4	3	2	2
(c) 28-gauge sheet metal spaced out 1 in.†	18	12	18	9	6	9	6	4	4	2	2	2
(d) 28-gauge sheet metal on ⅛ in. asbestos millboard spaced out 1 in.†	18	12	18	9	6	9	6	4	4	2	2	2
(e) 1½ in. asbestos cement covering on heating appliance	18	12	36	9	6	18	6	4	9	2	1	6
(f) ¼ in. asbestos millboard on 1 in. mineral fiber bats reinforced with wire mesh or equivalent	18	12	18	6	6	6	4	4	4	2	2	2
(g) 22-gauge sheet metal on 1 in. mineral fiber bats reinforced with wire or equivalent	18	12	12	4	3	3	2	2	2	2	2	2
(h) ¼ in. asbestos millboard	36	36	36	18	18	18	12	12	9	4	4	4
(i) ¼ in. cellular asbestos	36	36	36	18	18	18	12	12	9	3	3	3

*Except for the protection described in (e), all clearances should be measured from the outer surface of the appliance to the combustible material, disregarding any intervening protection applied to the combustible material.

†Spacers shall be of noncombustible material.

NOTE: Asbestos millboard referred to above is a different material from asbestos cement board. It is not intended that asbestos cement board be used in complying with these requirements when asbestos millboard is specified.

Courtesy National Oil Fuel Institute

Table B-18. Conversion Tables

TO CONVERT	INTO	MULTIPLY BY
	A	
Acres	sq. feet	43,560.0
Atmospheres	cms. of mercury	76.0
Atmospheres	ft. of water (at 4°C)	33.90
Atmospheres	in. of mercury (at 0° C)	29.92
Atmospheres	kgs./sq. cm.	1.0333
Atmospheres	pounds/sq. in.	14.70
	B	
Barrels (U.S., liquid)	gallons	31.5
Barrels (oil)	gallons (oil)	42.0
Btu	foot-lb.	778.3
Btu	gram-calories	252.0
Btu	horsepower-hr.	3.931×10^{-4}
Btu	kilowatt-hr.	2.928×10^{-4}
Btuh	horsepower	3.931×10^{-4}
Btuh	watts	0.2931
	C	
Calories, gram (mean)	Btu (mean)	3.9685×10^{-3}
Centigrade	Fahrenheit	(C° + 40)9/5-40
Centimeters	feet	3.281×10^{-2}
Centimeters	inches	0.3937
Centimeters	mils	393.7
Centimeters of mercury	atmospheres	0.01316
Centimeters of mercury	feet of water	0.4461
Centimeters of mercury	pounds/sq. in.	0.1934
Circumference	radians	6.283
Cubic centimeters	cu. feet	3.531×10^{-5}
Cubic centimeters	cu. inches	0.06102
Cubic centimeters	gallons (U.S. liq.)	2.642×10^{-4}
Cubic feet	cu. cm.	28,320.0
Cubic feet	cu. inches	1,728.0
Cubic feet	gallons (U.S. liq.)	7.481
Cubic feet	liters	28.32
Cubic feet	quarts (U.S. liq.)	29.92
Cubic feet/min.	gallons/sec.	0.1247
Cubic feet/min.	pounds of water/min.	62.43
Cubic inches	cu. cm.	16.39
Cubic inches	gallons	4.329×10^{-3}
Cubic inches	quarts (U.S. liq.)	0.01732
Cubic meters	cu. feet	35.31
Cubic meters	gallons (U.S. liq.)	264.2
Cubic yards	cu. feet	27.0

Table B-18. Conversion Tables (Cont'd)

TO CONVERT	INTO	MULTIPLY BY
Cubic yards	cu. meters	0.7646
Cubic yards	gallons (U.S.liq.)	202.0

D

Degrees (angle)	radians	0.01745
Drams (apothecaries' or troy)	ounces (avoirdupois)	0.13714
Drams (apothecaries' or troy)	ounces (troy)	0.125
Drams (U.S., fluid or apoth.)	cubic cm.	3.6967
Drams	grams	1.772
Drams	grains	27.3437
Drams	ounces	0.0625

F

Fahrenheit	centigrade	(F+40)5/9-40
Feet	centimeters	30.48
Feet	kilometers	3.048×10^{-4}
Feet	meters	0.3048
Feet	miles (naut.)	1.645×10^{-4}
Feet	miles (stat.)	1.894×10^{-4}
Feet of water	atmospheres	0.02950
Feet of water	in. of mercury	0.8826
Feet of water	kg./sq. cm.	0.03045
Feet of water	kg./sq. meter	304.8
Feet of water	pounds/sq. ft.	62.43
Feet of water	pounds/sq. in.	0.4335
Foot-pounds	Btu	1.286×10^{-3}
Foot-pounds	gram-calories	0.3238
Foot-pounds	hp-hr.	5.050×10^{-7}
Foot-pounds	kilowatt-hr.	3.766×10^{-7}
Foot-pounds/min.	Btu/min.	1.286×10^{-3}
Foot-pounds/min.	horsepower	3.030×10^{-5}
Foot-pounds/sec.	Btu-hr.	4.6263
Furlongs	miles (U.S.)	0.125
Furlongs	feet	660.0

G

Gallons	cu. cm.	3,785.0
Gallons	cu. feet	0.1337
Gallons	cu. inches	231.0
Gallons	cu. meters	3.785×10^{-3}
Gallons	cu. yards	4.951×10^{-3}
Gallons	liters	3.785

Table B-18. Conversion Tables (Cont'd)

TO CONVERT	INTO	MULTIPLY BY
Gallons (liq. Br. Imp.)	gallons (U.S. liq.)	1.20095
Gallons (U.S.)	gallons (Imp.)	0.83267
Gallons of water	pounds of water	8.3453
Gallons/min.	cu. ft./sec.	2.228×10^{-3}
Gallons/min.	liters/sec.	0.06308
Gallons/min.	cu. ft./hr.	8.0208
Grains (troy)	grains (avdp.)	1.0
Grains (troy)	grams	0.06480
Grains (troy)	ounces (avdp.)	2.286×10^{-3}
Grains (troy)	pennyweight (troy)	0.04167
Grains/U.S. gal.	parts/million	17.118
Grains/U.S. gal.	pounds/million gal.	142.86
Grains/Imp. gal.	parts/million	14.286
Grams	grains	15.43
Grams	ounces (avdp.)	0.03527
Grams	ounces (troy)	0.03215
Grams	poundals	0.07093
Grams	pounds	2.205×10^{-3}
Grams/liter	parts/million	1,000.0
Gram-calories	Btu	3.9683×10^{-3}
Gram-calories	foot-pounds	3.0880
Gram-calories	kilowatt-hr.	1.1630×10^{-6}
Gram-calories	watt-hr.	1.1630×10^{-3}

H

Horsepower	Btu/min.	42.40
Horsepower	foot-lb./min.	33,000.0
Horsepower	foot-lb./sec.	550.0
Horsepower (metric) (542.5 ft.-lb./sec.)	horsepower (550 ft.-lb./sec.)	0.9863
Horsepower (550 ft.-lb./sec.)	horsepower (metric) (542.5 ft.-lb./sec.)	1.014
Horsepower	kilowatts	0.7457
Horsepower	watts	745.7
Horsepower (boiler)	Btu/hr.	33.520
Horsepower (boiler)	kilowatts	9.803
Horsepower-hr.	Btu	2,547
Horsepower-hr.	foot-lb.	1.98×10^{6}
Horsepower-hr.	kilowatt-hr.	0.7457

I

Inches	centimeters	2.540
Inches	meters	25.40
Inches	millimeters	2.540×10^{-2}
Inches	yards	22.778×10^{-2}
Inches of mercury	atmospheres	0.03342

Table B-18. Conversion Tables (Cont'd)

TO CONVERT	INTO	MULTIPLY BY
Inches of mercury	feet of water	1.133
Inches of mercury	kg./sq. cm.	0.03453
Inches of mercury	kg./sq. meter	345.3
Inches of mercury	pounds/sq. ft.	70.73
Inches of mercury	pounds/sq. in.	0.4912
Inches of water (at 4°C)	atmospheres	2.458×10^{-3}
Inches of water (at 4°C)	inches of mercury	0.07355
Inches of water (at 4°C)	kg./sq. cm.	2.538×10^{-3}
Inches of water (at 4°C)	ounces/sq. in.	0.5781
Inches of water (at 4°C)	pounds/sq. ft.	5.204
Inches of water (at 4°C)	pounds/sq. in.	0.03613

J

Joules	Btu	9.480×10^{-4}

K

Kilograms	grams	1,000.0
Kilograms	pounds	2.205
Kilograms/cu. meter	pounds/cu. ft.	0.06243
Kilograms/cu. meter	pounds/cu. in.	3.613×10^{-5}
Kilograms/sq. cm.	atmospheres	0.9678
Kilograms/sq. cm.	feet of water	32.84
Kilograms/sq. cm.	inches of mercury	28.96
Kilograms/sq. cm.	pounds/sq. ft.	2,048.0
Kilograms/sq. cm.	pounds/sq. in.	14.22
Kilograms/sq. meter	atmospheres	9.678×10^{-5}
Kilograms/sq. meter	feet of water	3.281×10^{-3}
Kilograms/sq. meter	inches of mercury	2.896×10^{-3}
Kilograms/sq. meter	pounds/sq. ft.	0.2048
Kilograms/sq. meter	pounds/sq. in.	1.422×10^{-3}
Kilograms/sq. mm.	kg./sq. meter	10^{6}
Kilogram-calories	Btu	3.968
Kilogram-calories	foot-pounds	3,088.0
Kilogram-calories	hp-hr.	1.560×10^{-3}
Kilogram-calories	kilowatt-hr.	1.163×10^{-3}
Kilogram meters	Btu	9.294×10^{-3}
Kilometers	centimeters	10^{5}
Kilometers	feet	3,281.0
Kilometers	miles	0.6214
Kilowatts	Btu/min.	56.87
Kilowatts	foot-lb./min.	4.426×10^{4}
Kilowatts	foot-lb./sec.	737.6
Kilowatts	horsepower	1,341.0
Kilowatts	watts	1,000.0
Kilowatt-hr.	Btu	3,413.0
Kilowatt-hr.	foot-lb.	2.655×10^{6}

Table B-18. Conversion Tables (Cont'd)

TO CONVERT	INTO	MULTIPLY BY
Kilowatt-hr.	horsepower-hr.	1.341
Knots	statute miles/hr.	1.151

L

Liters	cu. cm.	1,000.0
Liters	cu. feet	0.03531
Liters	cu. inches	61.02
Liters	gallons (U.S. liq.)	0.2642

M

Meters	centimeters	100.0
Meters	feet	3.281
Meters	inches	39.37
Meters	millimeters	1,000.0
Meters	yards	1.094
Microns	inches	39.37×10^{-6}
Microns	meters	1×10^{-6}
Miles (statute)	feet	5,280.0
Miles (statute)	kilometers	1.609
Miles/hr.	cm./sec.	44.70
Miles/hr.	feet/min.	88.0
Mils	inches	0.001
Mils	yards	2.778×10^{-5}

N

Nepers	decibels	8.686

O

Ohms	megohms	10^{-6}
Ohms	microhms	10^{6}
Ounces (avoirdupois)	drams	16.0
Ounces (avoirdupois)	grains	437.5
Ounces (avoirdupois)	grams	28.35
Ounces (avoirdupois)	pounds	0.0625
Ounces (avoirdupois)	ounces (troy)	0.9115
Ounces (troy)	grains	480.0
Ounces (troy)	grams	31.10
Ounces (troy)	ounces (avdp.)	1.09714
Ounces (troy)	pounds (troy)	0.08333

P

Parts/million	grains/U.S. gal.	0.0584
Parts/million	grains/Imp. gal.	0.07016

Table B-18. Conversion Tables (Cont'd)

TO CONVERT	INTO	MULTIPLY BY
Parts/million	pounds/million gal.	8.33
Pounds (avoirdupois)	ounces (troy)	14.58
Pounds (avoirdupois)	drams	256.0
Pounds (avoirdupois)	grains	7,000.0
Pounds (avoirdupois)	grams	28.35
Pounds (avoirdupois)	kilograms	0.02835
Pounds (avoirdupois)	ounces	16.0
Pounds (avoirdupois)	tons (short)	0.0005
Pounds (troy)	ounces (avdp.)	13.1657
Pounds of water	cu. feet	0.01602
Pounds of water	cu. inches	27.68
Pounds of water	gallons	0.1198
Pounds of water/min.	cu. ft./sec.	2.670×10^{-4}
Pounds/cu. ft.	kg./cu. meter	0.01602
Pounds/cu. ft.	grams/cu.cm.	16.02
Pounds/cu. ft.	pounds/cu. in.	5.787×10^{-4}
Pounds/cu. in.	pounds/cu. ft.	1,728.0
Pounds/sq. ft.	atmospheres	4.725×10^{-4}
Pounds/sq. ft.	feet of water	0.01602
pounds/sq. ft.	inches of mercury	0.01414
Pounds/sq. in.	atmospheres	0.06804
pounds/sq. in.	feet of water	2.307
Pounds/sq. in.	inches of mercury	2.036
Pounds/sq. in.	kg./sq. meter	703.1
Pounds/sq. in.	pounds/sq. ft.	144.0

R

Radians	degrees	57.30
Revolutions/min.	degrees/sec.	6.0
Revolutions/min.	radians/sec.	0.1047
Revolutions/min.	rev./sec.	0.01667

S

Square centimeters	sq. feet	1.076×10^{-3}
Square centimeters	sq. inches	0.1550
Square centimeters	sq. meters	0.0001
Square centimeters	sq. millimeters	100.0
Square feet	acres	2.296×10^{-5}
Square feet	sq. cm.	929.0
Square feet	sq. inches	144.0
Square feet	sq. miles	3.587×10^{-8}
Square inches	sq. cm.	6.452
Square inches	sq. feet	6.944×10^{-3}
Square inches	sq. yards	7.716×10^{-4}
Square meters	sq. feet	10.76
Square meters	sq. inches	1,550.0

Table B-18. Conversion Tables (Cont'd)

TO CONVERT	INTO	MULTIPLY BY
Square meters	sq. millimeters	10^6
Square meters	sq. yards	1.196
Square millimeters	sq. inches	1.550×10^{-3}
Square yards	sq. feet	9.0
Square yards	sq. inches	1,296.0
Square yards	sq. meters	0.8361

T

Temperature (°C) + 273	absolute temperature (°C)	1.0
Temperature (°C) + 17.78	temperature (°F)	1.8
Temperature (°F) + 460	absolute temperature (°F)	1.0
Temperature (°F) − 32	temperature (°C)	$^5/_9$
Tons (long)	kilograms	1,016.0
Tons (long)	pounds	2,240.0
Tons (long)	tons (short)	1.120
Tons (metric)	kilograms	1,000.0
Tons (metric)	pounds	2,205.0
Tons (short)	kilograms	907.2
Tons (short)	pounds	2,000.0
Tons (short)	tons (long)	0.89287
Tons of water/24 hr.	pounds of water/hr.	83.333
Tons of water/24 hr.	gallons/min.	0.16643
Tons of water/24 hr.	cu. ft./hr.	1.3349

W

Watts	Btu/hr.	3.4129
Watts	Btu/min.	0.05688
Watts	horsepower	1.341×10^{-3}
Watts	horsepower (metric)	1.360×10^{-3}
Watts	kilowatts	0.001
Watts (abs.)	Btu (mean)/min.	0.056884
Watt-hours	Btu	3.413
Watt-hours	horsepower-hr.	1.341×10^{-3}

Y

Yards	centimeters	91.44
Yards	kilometers	9.144×10^{-4}
Yards	meters	0.9144

Index

**Over a Century of Excellence
for the Professional
and
Vocational Trades and the Crafts**

Order now from your local bookstore
or use the convenient order form
at the back of this book.

AUDEL

These fully illustrated, up-to-date guides and manuals mean a better job done for mechanics, engineers, electricians, plumbers, carpenters, and all skilled workers.

CONTENTS

ELECTRICAL

HOUSE WIRING (Sixth Edition)
ROLAND E. PALMQUIST
5 1/2 x 8 1/4   Hardcover   256 pp.   150 Illus.
ISBN: 0-672-23404-1 $14.95
The rules and regulations of the National Electrical Code as they apply to residential wiring fully detailed with examples and illustrations.

PRACTICAL ELECTRICITY
(Fifth Edition)
ROBERT G. MIDDLETON;
revised by L. DONALD MEYERS
5 1/2 x 8 1/4   Hardcover   512 pp.   335 Illus.
ISBN: 0-02-584561-6 $19.95
The fundamentals of electricity for electrical workers, apprentices, and others requiring concise information about electric principles and their practical applications.

GUIDE TO THE 1987 NATIONAL ELECTRICAL CODE
ROLAND E. PALMQUIST
5 1/2 x 8 1/4   Hardcover   664 pp.   225 Illus.
ISBN: 0-02-594560-2 $22.50
The most authoritative guide available to interpreting the National Electrical Code for electricians, contractors, electrical inspectors, and homeowners. Examples and illustrations.

MATHEMATICS FOR ELECTRICIANS AND ELECTRONICS TECHNICIANS
REX MILLER
5 1/2 x 8 1/4   Hardcover   312 pp.   115 Illus.
ISBN: 0-8161-1700-4 $14.95
Mathematical concepts, formulas, and problem-solving techniques utilized on-the-job by electricians and those in electronics and related fields.

FRACTIONAL-HORSEPOWER ELECTRIC MOTORS
REX MILLER and
MARK RICHARD MILLER
5 1/2 x 8 1/4   Hardcover   436 pp.   285 Illus.
ISBN: 0-672-23410-6 $15.95
The installation, operation, maintenance, repair, and replacement of the small-to-moderate-size electric motors that power home appliances and industrial equipment.

ELECTRIC MOTORS (Fourth Edition)
EDWIN P. ANDERSON;
revised by REX MILLER
5 1/2 x 8 1/4   Hardcover   656 pp.   405 Illus.
ISBN: 0-672-23376-2 $14.95
Installation, maintenance, and repair of all types of electric motors.

HOME APPLIANCE SERVICING (Fourth Edition)
EDWIN P. ANDERSON;
revised by REX MILLER
5 1/2 x 8 1/4   Hardcover   640 pp.   345 Illus.
ISBN: 0-672-23379-7 $22.50
The essentials of testing, maintaining, and repairing all types of home appliances.

TELEVISION SERVICE MANUAL (Fifth Edition)
ROBERT G. MIDDLETON;
revised by JOSEPH G. BARRILE
5 1/2 x 8 1/4   Hardcover   512 pp.   395 Illus.
ISBN: 0-672-23395-9 $16.95
A guide to all aspects of television transmission and reception, including the operating principles of black and white and color receivers. Step-by-step maintenance and repair procedures.

ELECTRICAL COURSE FOR APPRENTICES AND JOURNEYMEN (Third Edition)
ROLAND E. PALMQUIST
5 1/2 x 8 1/4   Hardcover   478 pp.   290 Illus.
ISBN: 0-02-594550-5 $19.95
This practical course in electricity for those in formal training programs or learning on their own provides a thorough understanding of operational theory and its applications on the job.

QUESTIONS AND ANSWERS FOR ELECTRICIANS EXAMINATIONS (Ninth Edition)
ROLAND E. PALMQUIST
5 1/2 x 8 1/4   Hardcover   316 pp.   110 Illus.
ISBN: 0-02-594691-9 $18.95
Based on the 1987 National Electrical Code, this book reviews the subjects included in the various electricians examinations—apprentice, journeyman, and master. Question and Answer format.

MACHINE SHOP AND MECHANICAL TRADES

MACHINISTS LIBRARY
(Fourth Edition, 3 Vols.)
REX MILLER
5 1/2 x 8 1/4   Hardcover   1352 pp.   1120 Illus.
ISBN: 0-672-23380-0 $52.95
An indispensable three-volume reference set for machinists, tool and die makers, machine operators, metal workers, and those with home workshops. The principles and methods of the entire field are covered in an up-to-date text, photographs, diagrams, and tables.

Volume I: Basic Machine Shop
REX MILLER
5 1/2 x 8 1/4   Hardcover   392 pp.   375 Illus.
ISBN: 0-672-23381-9 $17.95
Volume II: Machine Shop
REX MILLER
5 1/2 x 8 1/4   Hardcover   528 pp.   445 Illus.
ISBN: 0-672-23382-7 $19.95
Volume III: Toolmakers Handy Book
REX MILLER
5 1/2 x 8 1/4   Hardcover   432 pp.   300 Illus.
ISBN: 0-672-23383-5 $14.95

MATHEMATICS FOR MECHANICAL TECHNICIANS AND TECHNOLOGISTS
JOHN D. BIES
5 1/2 x 8 1/4   Hardcover   342 pp.   190 Illus.
ISBN: 0-02-510620-1 $17.95
The mathematical concepts, formulas, and problem-solving techniques utilized on the job by engineers, technicians, and other workers in industrial and mechanical technology and related fields.

MILLWRIGHTS AND MECHANICS GUIDE
(Fourth Edition)
CARL A. NELSON
5 1/2 x 8 1/4   Hardcover   1,040 pp.   880 Illus.
ISBN: 0-02-588591-x $29.95
The most comprehensive and authoritative guide available for millwrights, mechanics, maintenance workers, riggers, shop workers, foremen, inspectors, and superintendents on plant installation, operation, and maintenance.

WELDERS GUIDE (Third Edition)
JAMES E. BRUMBAUGH
5 1/2 x 8 1/4   Hardcover   960 pp.   615 Illus.
ISBN: 0-672-23374-6 $23.95
The theory, operation, and maintenance of all welding machines. Covers gas welding equipment, supplies, and process; arc welding equipment, supplies, and process; TIG and MIG welding; and much more.

WELDERS/FITTERS GUIDE
JOHN P. STEWART
8 1/2 x 11   Paperback   160 pp.   195 Illus.
ISBN: 0-672-23325-8 $7.95
Step-by-step instruction for those training to become welders/fitters who have some knowledge of welding and the ability to read blueprints.

SHEET METAL WORK

JOHN D. BIES

5 1/2 x 8 1/4   Hardcover   456 pp.   215 Illus.
ISBN: 0-8161-1706-3 $19.95

An on-the-job guide for workers in the manufacturing and construction industries and for those with home workshops. All facets of sheet metal work detailed and illustrated by drawings, photographs, and tables.

POWER PLANT ENGINEERS GUIDE (Third Edition)

FRANK D. GRAHAM;
revised by CHARLIE BUFFINGTON

5 1/2 x 8 1/4   Hardcover   960 pp.   530 Illus.
ISBN: 0-672-23329-0 $27.50

This all-inclusive, one-volume guide is perfect for engineers, firemen, water tenders, oilers, operators of steam and diesel-power engines, and those applying for engineer's and firemen's licenses.

MECHANICAL TRADES POCKET MANUAL
(Second Edition)

CARL A. NELSON

4 x 6   Paperback   364 pp.   255 Illus.
ISBN: 0-672-23378-9   10.95

A handbook for workers in the industrial and mechanical trades on methods, tools, equipment, and procedures. Pocket-sized for easy reference and fully illustrated.

PLUMBING

PLUMBERS AND PIPE FITTERS LIBRARY
(Fourth Edition, 3 Vols.)

CHARLES N. McCONNELL

5 1/2 x 8 1/4   Hardcover   952 pp.   560 Illus.
ISBN: 0-02-582914-9 $68.45

This comprehensive three-volume set contains the most up-to-date information available for master plumbers, journeymen, apprentices, engineers, and those in the building trades. A detailed text and clear diagrams, photographs, and charts and tables treat all aspects of the plumbing, heating, and air conditioning trades.

Volume I: Materials, Tools, Roughing-In

CHARLES N. McCONNELL;
revised by TOM PHILBIN

5 1/2 x 8 1/4   Hardcover   304 pp.   240 Illus.
ISBN: 0-02-582911-4 $20.95

Volume II: Welding, Heating, Air Conditioning

CHARLES N. McCONNELL;
revised by TOM PHILBIN

5 1/2 x 8 1/4   Hardcover   384 pp.   220 Illus.
ISBN: 0-02-582912-2 $22.95

Volume III: Water Supply, Drainage, Calculations

CHARLES N. McCONNELL;
revised by TOM PHILBIN

5 1/2 x 8 1/4   Hardcover   264 pp.   100 Illus.
ISBN: 0-02-582913-0 $20.95

HOME PLUMBING HANDBOOK
(Third Edition)

CHARLES N. McCONNELL

8 1/2 x 11   Paperback   200 pp.   100 Illus.
ISBN: 0-672-23413-0 $13.95

An up-to-date guide to home plumbing installation and repair.

THE PLUMBERS HANDBOOK
(Seventh Edition)

JOSEPH P. ALMOND, SR.

4 x 6   Paperback   352 pp.   170 Illus.
ISBN: 0-672-23419-x $11.95

A handy sourcebook for plumbers, pipe fitters, and apprentices in both trades. It has a rugged binding suited for use on the job, and fits in the tool box or conveniently in the pocket.

QUESTIONS AND ANSWERS FOR PLUMBERS EXAMINATIONS (Second Edition)

JULES ORAVITZ

5 1/2 x 8 1/4   Paperback   256 pp.   145 Illus.
ISBN: 0-8161-1703-9 $9.95

A study guide for those preparing to take a licensing examination for apprentice, journeyman, or master plumber. Question and answer format.

HVAC

AIR CONDITIONING: HOME AND COMMERCIAL
(Second Edition)

EDWIN P. ANDERSON;
revised by REX MILLER

5 1/2 x 8 1/4   Hardcover   528 pp.   180 Illus.
ISBN: 0-672-23397-5 $15.95

A guide to the construction, installation, operation, maintenance, and repair of home, commercial, and industrial air conditioning systems.

HEATING, VENTILATING, AND AIR CONDITIONING LIBRARY
(Second Edition, 3 Vols.)
JAMES E. BRUMBAUGH
5 1/2 x 8 1/4 Hardcover 1,840 pp. 1,275 Illus.
ISBN: 0-672-23388-6 $53.95
An authoritative three-volume reference library for those who install, operate, maintain, and repair HVAC equipment commercially, industrially, or at home.
Volume I: Heating Fundamentals, Furnaces, Boilers, Boiler Conversions
JAMES E. BRUMBAUGH
5 1/2 x 8 1/4 Hardcover 656 pp. 405 Illus.
ISBN: 0-672-23389-4 $17.95
Volume II: Oil, Gas and Coal Burners, Controls, Ducts, Piping, Valves
JAMES E. BRUMBAUGH
5 1/2 x 8 1/4 Hardcover 592 pp. 455 Illus.
ISBN: 0-672-23390-8 $17.95
Volume III: Radiant Heating, Water Heaters, Ventilation, Air Conditioning, Heat Pumps, Air Cleaners
JAMES E. BRUMBAUGH
5 1/2 x 8 1/4 Hardcover 592 pp. 415 Illus.
ISBN: 0-672-23391-6 $17.95

OIL BURNERS (Fourth Edition)
EDWIN M. FIELD
5 1/2 x 8 1/4 Hardcover 360 pp. 170 Illus.
ISBN: 0-672-23394-0 $16.95
An up-to-date sourcebook on the construction, installation, operation, testing, servicing, and repair of all types of oil burners, both industrial and domestic.

REFRIGERATION: HOME AND COMMERCIAL (Second Edition)
EDWIN P. ANDERSON;
revised by REX MILLER
5 1/2 x 8 1/4 Hardcover 768 pp. 285 Illus.
ISBN: 0-672-23396-7 $19.95
A reference for technicians, plant engineers, and the home owner on the installation, operation, servicing, and repair of everything from single refrigeration units to commercial and industrial systems.

PNEUMATICS AND HYDRAULICS

HYDRAULICS FOR OFF-THE-ROAD EQUIPMENT (Second Edition)
HARRY L. STEWART;
revised by TOM PHILBIN
5 1/2 x 8 1/4 Hardcover 256 pp. 175 Illus.
ISBN: 0-8161-1701-2 $13.95

This complete reference manual on heavy equipment covers hydraulic pumps, accumulators, and motors; force components; hydraulic control components; filters and filtration, lines and fittings, and fluids; hydrostatic transmissions; maintenance; and troubleshooting.

PNEUMATICS AND HYDRAULICS (Fourth Edition)
HARRY L. STEWART;
revised by TOM STEWART
5 1/2 x 8 1/4 Hardcover 512 pp. 315 Illus.
ISBN: 0-672-23412-2 $19.95
The principles and applications of fluid power. Covers pressure, work, and power; general features of machines; hydraulic and pneumatic symbols; pressure boosters; air compressors and accessories; and much more.

PUMPS (Fourth Edition)
HARRY STEWART;
revised by TOM PHILBIN
5 1/2 x 8 1/4 Hardcover 508 pp. 360 Illus.
ISBN: 0-672-23400-9 $17.95
The principles and day-to-day operation of pumps, pump controls, and hydraulics are thoroughly detailed and illustrated.

CARPENTRY AND CONSTRUCTION

CARPENTERS AND BUILDERS LIBRARY (Fifth Edition, 4 Vols.)
JOHN E. BALL; revised by TOM PHILBIN
5 1/2 x 8 1/4 Hardcover 1,224 pp. 1,010 Illus.
ISBN: 0-672-23369-x $43.95

Also available as a boxed set at no extra cost:
ISBN: 0-02-506450-9 $43.95

This comprehensive four-volume library has set the professional standard for decades for carpenters, joiners, and woodworkers.

Volume I: Tools, Steel Square, Joinery
JOHN E. BALL; revised by TOM PHILBIN
5 1/2 x 8 1/4 Hardcover 384 pp. 345 Illus.
ISBN: 0-672-23365-7 $10.95

Volume II: Builders Math, Plans, Specifications
JOHN E. BALL; revised by TOM PHILBIN
5 1/2 x 8 1/4 Hardcover 304 pp. 205 Illus.
ISBN: 0-672-23366-5 $10.95

Volume III: Layouts, Foundations, Framing
JOHN E. BALL; revised by TOM PHILBIN
5 1/2 x 8 1/4 Hardcover 272 pp. 215 Illus.
ISBN: 0-672-23367-3 $10.95

Volume IV: Millwork, Power Tools, Painting

JOHN E. BALL; revised by TOM PHILBIN

5 1/2 x 8 1/4   Hardcover   344 pp.   245 Illus.
ISBN: 0-672-23368-1 $10.95

COMPLETE BUILDING CONSTRUCTION (Second Edition)

JOHN PHELPS; revised by TOM PHILBIN

5 1/2 x 8 1/4   Hardcover   744 pp.   645 Illus.
ISBN: 0-672-23377-0 $22.50

Constructing a frame or brick building from the footings to the ridge. Whether the building project is a tool shed, garage, or a complete home, this single fully illustrated volume provides all the necessary information.

COMPLETE ROOFING HANDBOOK

JAMES E. BRUMBAUGH

5 1/2 x 8 1/4   Hardcover   536 pp.   510 Illus.
ISBN: 0-02-517850-4 $29.95

Covers types of roofs; roofing and reroofing; roof and attic insulation and ventilation; skylights and roof openings; dormer construction; roof flashing details; and much more.

COMPLETE SIDING HANDBOOK

JAMES E. BRUMBAUGH

5 1/2 x 8 1/4   Hardcover   512 pp.   450 Illus.
ISBN: 0-02-517880-6 $24.95

This companion volume to the *Complete Roofing Handbook* includes comprehensive step-by-step instructions and accompanying line drawings on every aspect of siding a building.

MASONS AND BUILDERS LIBRARY (Second Edition, 2 Vols.)

LOUIS M. DEZETTEL;
revised by TOM PHILBIN

5 1/2 x 8 1/4   Hardcover   688 pp.   500 Illus.
ISBN: 0-672-23401-7 $27.95

This two-volume set provides practical instruction in bricklaying and masonry. Covers brick; mortar; tools; bonding; corners, openings, and arches; chimneys and fireplaces; structural clay tile and glass block; brick walls; and much more.

Volume I: Concrete, Block, Tile, Terrazzo

LOUIS M. DEZETTEL;
revised by TOM PHILBIN

5 1/2 x 8 1/4   Hardcover   304 pp.   190 Illus.
ISBN: 0-672-23402-5 $13.95

Volume 2: Bricklaying, Plastering, Rock Masonry, Clay Tile

LOUIS M. DEZETTEL;
revised by TOM PHILBIN

5 1/2 x 8 1/4   Hardcover   384 pp.   310 Illus.
ISBN: 0-672-23403-3 $13.95

WOODWORKING

WOOD FURNITURE: FINISHING, REFINISHING, REPAIRING (Second Edition)

JAMES E. BRUMBAUGH

5 1/2 x 8 1/4   Hardcover   352 pp.   185 Illus.
ISBN: 0-672-23409-2 $12.95

A fully illustrated guide to repairing furniture and finishing and refinishing wood surfaces. Covers tools and supplies; types of wood; veneering; inlaying; repairing, restoring, and stripping; wood preparation; and much more.

WOODWORKING AND CABINETMAKING

F. RICHARD BOLLER

5 1/2 x 8 1/4   Hardcover   360 pp.   455 Illus.
ISBN: 0-02-512800-0 $18.95

Essential information on all aspects of working with wood. Step-by-step procedures for woodworking projects are accompanied by detailed drawings and photographs.

MAINTENANCE AND REPAIR

BUILDING MAINTENANCE (Second Edition)

JULES ORAVETZ

5 1/2 x 8 1/4   Hardcover   384 pp.   210 Illus.
ISBN: 0-672-23278-2 $11.95

Professional maintenance procedures used in office, educational, and commercial buildings. Covers painting and decorating; plumbing and pipe fitting; concrete and masonry; and much more.

GARDENING, LANDSCAPING AND GROUNDS MAINTENANCE (Third Edition)

JULES ORAVETZ

5 1/2 x 8 1/4   Hardcover   424 pp.   340 Illus.
ISBN: 0-672-23417-3 $15.95

Maintaining lawns and gardens as well as industrial, municipal, and estate grounds.

HOME MAINTENANCE AND REPAIR: WALLS, CEILINGS AND FLOORS

GARY D. BRANSON

8 1/2 x 11   Paperback   80 pp.   80 Illus.
ISBN: 0-672-23281-2 $6.95

The do-it-yourselfer's guide to interior remodeling with professional results.

PAINTING AND DECORATING

REX MILLER and GLEN E. BAKER

5 1/2 x 8 1/4   Hardcover   464 pp.   325 Illus.
ISBN: 0-672-23405-x $18.95

A practical guide for painters, decorators, and homeowners to the most up-to-date materials and techniques in the field.

TREE CARE (Second Edition)

JOHN M. HALLER

8 1/2 x 11   Paperback   224 pp.   305 Illus.
ISBN: 0-02-062870-6 $9.95

The standard in the field. A comprehensive guide for growers, nursery owners, foresters, landscapers, and homeowners to planting, nurturing and protecting trees.

UPHOLSTERING (Updated)

JAMES E. BRUMBAUGH

5 1/2 x 8 1/4   Hardcover   400 pp.   380 Illus.
ISBN: 0-672-23372-x $15.95

The essentials of upholstering fully explained and illustrated for the professional, the apprentice, and the hobbyist.

AUTOMOTIVE AND ENGINES

DIESEL ENGINE MANUAL
(Fourth Edition)

PERRY O. BLACK;
revised by WILLIAM E. SCAHILL

5 1/2 x 8 1/4   Hardcover   512 pp.   255 Illus.
ISBN: 0-672-23371-1 $15.95

The principles, design, operation, and maintenance of today's diesel engines. All aspects of typical two- and four-cycle engines are thoroughly explained and illustrated by photographs, line drawings, and charts and tables.

GAS ENGINE MANUAL
(Third Edition)

EDWIN P. ANDERSON;
revised by CHARLES G. FACKLAM

5 1/2 x 8 1/4   Hardcover   424 pp.   225 Illus.
ISBN: 0-8161-1707-1 $12.95

How to operate, maintain, and repair gas engines of all types and sizes. All engine parts and step-by-step procedures are illustrated by photographs, diagrams, and troubleshooting charts.

SMALL GASOLINE ENGINES

REX MILLER and MARK RICHARD MILLER

5 1/2 x 8 1/4   Hardcover   640 pp.   525 Illus.
ISBN: 0-672-23414-9 $16.95

Practical information for those who repair, maintain, and overhaul two- and four-cycle engines—including lawn mowers, edgers, grass sweepers, snowblowers, emergency electrical generators, outboard motors, and other equipment with engines of up to ten horsepower.

TRUCK GUIDE LIBRARY (3 Vols.)

JAMES E. BRUMBAUGH

5 1/2 x 8 1/4   2,144 pp.   1,715 Illus.
ISBN: 0-672-23392-4 $45.95

This three-volume set provides the most comprehensive, profusely illustrated collection of information available on truck operation and maintenance.

Volume 1: Engines

JAMES E. BRUMBAUGH

5 1/2 x 8 1/4   Hardcover   416 pp.   290 Illus.
ISBN: 0-672-23356-8 $16.95

Volume 2: Engine Auxiliary Systems

JAMES E. BRUMBAUGH

5 1/2 x 8 1/4   Hardcover   704 pp.   520 Illus.
ISBN: 0-672-23357-6 $16.95

Volume 3: Transmissions, Steering, and Brakes

JAMES E. BRUMBAUGH

5 1/2 x 8 1/4   Hardcover   1,024 pp.   905 Illus.
ISBN: 0-672-23406-8 $16.95

DRAFTING

INDUSTRIAL DRAFTING

JOHN D. BIES

5 1/2 x 8 1/4   Hardcover   544 pp.   Illus.
ISBN: 0-02-510610-4 $24.95

Professional-level introductory guide for practicing drafters, engineers, managers, and technical workers in all industries who use or prepare working drawings.

ANSWERS ON BLUEPRINT
READING (Fourth Edition)

ROLAND PALMQUIST;
revised by THOMAS J. MORRISEY

5 1/2 x 8 1/4   Hardcover   320 pp.   275 Illus.
ISBN: 0-8161-1704-7 $12.95

Understanding blueprints of machines and tools, electrical systems, and architecture. Question and answer format.

HOBBIES

COMPLETE COURSE IN STAINED GLASS

PEPE MENDEZ

8 1/2 x 11   Paperback   80 pp.   50 Illus.
ISBN: 0-672-23287-1 $8.95

The tools, materials, and techniques of the art of working with stained glass.